Techniques in
Mineral Exploration

Techniques in
Mineral Exploration

J. H. REEDMAN
B.Sc., M.Phil., M.I.M.M.
Noranda Exploration Company Ltd, Winnipeg, Canada

APPLIED SCIENCE PUBLISHERS LTD
LONDON

APPLIED SCIENCE PUBLISHERS LTD
RIPPLE ROAD, BARKING, ESSEX, ENGLAND

British Library Cataloguing in Publication Data

Reedman, J H
   Techniques in mineral exploration.
   1. Prospecting
   I. Title
   622'.1'024553   TN270

   ISBN 0-85334-817-0

WITH 56 TABLES AND 213 ILLUSTRATIONS

© APPLIED SCIENCE PUBLISHERS LTD 1979

Printed in Great Britain by Galliard (Printers) Ltd, Great Yarmouth

Preface

For some years I have felt there was a need for a single, comprehensive, reference book on exploration geology. Numerous textbooks are available on subjects such as geophysical prospecting, exploration geochemistry, mining geology, photogeology and general economic geology, but, for the geologist working in mineral exploration, who does not require a specialist's knowledge, a general book on exploration techniques is needed. Many undergraduate university courses tend to neglect economic geology and few deal with the more practical aspects in any detail. Graduate geologists embarking on a career in economic geology or mineral exploration are therefore often poorly equipped and have to learn a considerable amount 'on the job'. By providing a book that includes material which can be found in some of the standard texts together with a number of practical aspects not to be found elsewhere, I hope that both recent graduates and more experienced exploration geologists will find it a useful reference work and manual. In addition, students of economic geology and personnel working in related fields in the mining and mineral extraction industries will find it informative.

J. H. REEDMAN

Acknowledgements

The author would like to thank Dr K. Fletcher, geochemist with the Department of Geology, University of British Columbia, and Kari Savario, geophysicist with Finnish Technical Aid to Zambia, for reading the original drafts and offering constructive criticism and advice on the chapters on geochemical and geophysical prospecting respectively. The following organizations are thanked for allowing figures and drawings, which have appeared in their various publications, to be copied: The Institution of Mining and Metallurgy, The Society of Exploration Geophysicists, The European Association of Exploration Geophysicists, The Society of Mining Engineers of A.I.M.E., and the Canadian Institute of Mining and Metallurgy. Hunting Surveys Ltd kindly supplied an illustration of a seismic survey in the North Sea, Hunting Geology and Geophysics Ltd provided an excellent example of an SLAR image, and Spectral Africa Pty. Ltd, South Africa, permitted the copying of an example of thermal imagery that appeared in one of their publications. The author is indebted to the Geological Survey and Mines Department of Uganda, Noranda Exploration (Ireland) Ltd and Mindeco Noranda Ltd, Lusaka, Zambia, for unpublished material and data used in the book. Aerial photographs used as illustrations all came from the Survey Department, Republic of Zambia. Finally, the author is particularly grateful to Miss Esther Zulu, who typed a large part of the manuscript, and to Ladreck Lungu, who draughted many of the illustrations.

Contents

CHAPTER 1

Introduction: Mineral Resources and Exploration

Mining is one of man's oldest activities and may even be considered older than agriculture, since Stone Age hunters, who worked flints into arrowheads and other tools and weapons, often obtained suitable material from carefully selected sites which were worked by pits and small underground workings. With the advent of the ancient civilizations in Egypt, Mesopotamia, China and elsewhere metals were widely worked to produce tools, weapons and ornaments. The earliest metals used by man were gold, copper, bronze (copper + tin), silver and lead. The use of iron was discovered later and from it far superior tools and weapons could be manufactured. During the time of the Roman Empire, mining for gold, silver, copper, tin, lead and iron was well established in various parts of Europe and many of these ancient mining centres in Israel, Cyprus, Spain and Britain are still producing today. After the fall of the Roman Empire, mining in Europe went into decline, but by the Middle Ages it was flourishing again in areas such as the Harz Mountains and the Erzgebirge in Germany and Cornwall in England. Agricola's famous treatise *De Re Metallica* is a fascinating account, beautifully illustrated with woodcuts, of mining practices in the 1500's. In Book One he gives an eloquent defence of mining as an honourable and essential occupation, carefully and systematically refuting the various arguments of critics.

1.1 GROWTH IN MINERAL PRODUCTION

Although iron and the common base metals, together with the monetary metals, gold and silver, were in common usage in the civilized world for hundreds of years before the Industrial Revolution, consumption was very low and it was only with the advent of industrialization that demand increased significantly. Even so, consumption of base metals during most of the 1800's was extremely low by present-day standards and it was not until the twentieth century

1

that mining really expanded. For example, total world copper production in the early 1800's has been estimated at about 10 000 tonnes per year, equivalent to only 11 hours' world production in 1974. Even by the 1850's total world copper production was still only about 50 000 tonnes per year. With the demands brought about by development of the electrical industry and the discovery of large copper deposits in Chile and the United States, copper production rose rapidly to over 300 000 tonnes per year by the end of the nineteenth century with the other old base metals showing a similar increase in production. The late eighteenth and early nineteenth centuries were times of great scientific discovery compared to earlier centuries and a number of new metals were discovered and separated from their ores (Table 1.1). Most of these were to remain scientific curiosities until the early twentieth century, but experimental work had shown that some of the new metals such as nickel, molybdenum and chromium could be alloyed with steel to produce metals with properties superior to those of ordinary steel.

TABLE 1.1

METALS LISTED ACCORDING TO THE YEARS OF THEIR DISCOVERY OR ISOLATION

Year	Metal	Discoverer	Year	Metal	Discoverer
1735	Cobalt	Brandt	1801	Vanadium	Roscoe
1751	Nickel	Cronstedt	1801	Niobium	Hatchett
1774	Manganese	Scheele	1802	Tantalum	Ekeberg
1778	Molybdenum	Scheele	1804	Osmium Iridium	Tennant
1783	Tungsten	d'Ehujar brothers	1808	Magnesium	Davy
1789	Uranium Zirconium	Klaproth	1817	Selenium	Berzelius
1791	Titanium	Gregor	1817	Cadmium	Stromeyer
1797	Beryllium	Vanquelin	1827	Aluminium	Wöhler
1798	Chromium	Vanquelin	1886	Germanium	Winkler

Source: Van Nostrand's Scientific Encyclopedia

After the turn of the present century, industrialization spread rapidly in North America and Europe and created an increasing demand for the traditional base metals, copper, tin, lead and zinc, which was met by a steady increase in mine production (Figs. 1.1 and 1.2). There was also increased use of the new base metals, nickel and aluminium, which resulted in a truly spectacular increase in their production (Figs. 1.1 and 1.2). In the case of aluminium a break-

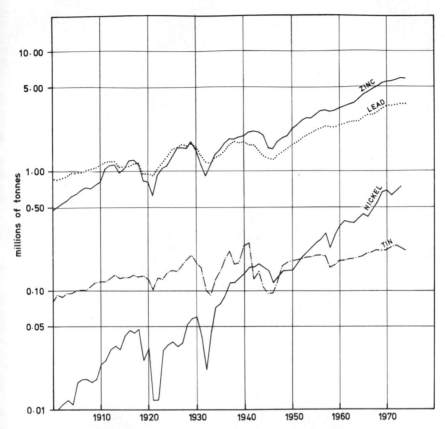

FIG. 1.1. Growth in primary production of lead, zinc, nickel and tin since 1900. (Source: Metallgesellschaft AG, 1975.)

through in extraction technology made simultaneously in the late 1880's by the American, Hall, and the Frenchman, Heroult, meant that aluminium could be produced at a much lower cost than earlier techniques of extraction which had kept aluminium as a scientific curiosity and jewellers' metal. Aluminium found increased application as a major structural metal, replacing steel in many of its traditional roles: in the electrical industry, aircraft industry, domestic appliance manufacture and packaging it has become pre-eminent so that production has risen from a few thousand tonnes per year in 1900 to over 13 million tonnes per year in the mid-1970's, making aluminium the second most widely used metal after iron and steel. In the case of nickel, growth of consumption and production has not been quite as

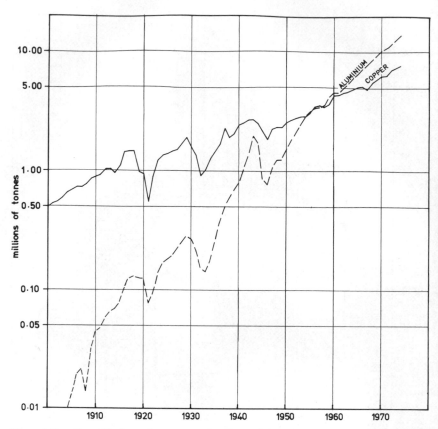

FIG. 1.2. Growth in primary production of copper and aluminium since 1900. (Source: Metallgesellschaft AG, 1975.)

spectacular as that of aluminium, but its use as an essential ingredient for stainless and tough strong steels has meant that there has been a big increase in production over the past forty years (Fig. 1.1). From the early 1900's to the present day, growth in production of the major non-ferrous metals has been over 1000-fold in the case of aluminium, 240-fold in the case of nickel, 12-fold in the case of copper, 10-fold in the case of zinc, 4-fold in the case of lead and just over 2-fold in the case of tin. The increased demand for both non-ferrous metals and iron and steel (Table 1.2) has been met by an equally impressive increase in the consumption of energy (Table 1.3), in particular petroleum which has become a dominant factor in world economic order in the 1970's. In addition the growth of consumption and

TABLE 1.2
WORLD IRON ORE AND STEEL PRODUCTION IN
MILLIONS OF TONNES SINCE 1950

	1950	1960	1970	1976
Iron ore	246	468	750	890
Steel	189	356	585	680

Source: Mining Annual Review 1952, 1962, 1971, 1977

TABLE 1.3
WORLD COAL AND OIL PRODUCTION IN MILLIONS
OF TONNES SINCE 1950

	1950	1960	1970	1975
Coal	1 550	2 161	2 964	3 267
Oil	594	1 085	2 500	2 765

Source: Mining Annual Review 1952, 1962, 1972, 1977

production of industrial minerals has been equally impressive (Table 1.4), but they have not received the same publicity as they lack the 'glamour' of metals or petroleum.

To meet the increased demand for various minerals enormous increases in mine production were required. These were met both through advances in extraction technology which permitted the beneficiation of lower grade ores and through major discoveries made by concerted exploration campaigns in different parts of the world. In the case of copper, grades as high as 10–20% were required in the late

TABLE 1.4
WORLD ASBESTOS, PHOSPHATE AND POTASH PRODUCTION IN THOU-
SANDS OF TONNES SINCE 1950

	1950	1960	1970	1975
Asbestos	1 300*	2 098	4 780	5 712
Phosphates	25 000*	41 200	80 088	106 563
Potash (contained K_2O)	5 700*	9 100	17 727	24 453

*Estimate
Source: Mining Annual Review 1952, 1962, 1972, 1977

1800's, but by the 1930's ores grading 3% or less were commonly worked, and by the 1950's large low-grade deposits of 0·7% Cu and less were being worked on an increasingly large scale. Examples of some of the major discoveries which made it possible for production to meet increased demand are the large copper deposits of Chile and the southwest United States developed between 1850 and 1900, the fabulous goldfields of the Witwatersrand which have dominated world gold production since their discovery at the end of the nineteenth century, the Broken Hill lead–zinc–silver deposit in New South Wales discovered in 1883, the copper–nickel deposits of Sudbury, Ontario discovered in the 1880's (the world's major source of nickel since the early 1900's) and the copper belt of Zambia and Zaire which was discovered and developed between 1900 and 1940, becoming the greatest single copper-producing area in the world. More recent major discoveries include the Jamaican bauxite deposits (1940's), the enormous bauxite deposits of northern Queensland (1950's), the Pilbara iron deposits of Western Australia (1950's), and the uranium deposits of the Northern Territory, Australia (1970's).

After the Second World War there were great advances in technology and a consequent demand for new 'space age' metals such as beryllium, niobium and titanium. Developments in electronics also created an increased demand for metals such as selenium, indium, silver, tantalum, gallium, rhenium, germanium and the rare earths. The post-war period saw the burgeoning of the atomic energy industry and its need for uranium which had remained very much a scientific curiosity since its discovery in 1789. This element has now become a much sought-after commodity with the so-called energy crisis of recent years and major exploration efforts are being mounted to locate supplies to meet projected future demand.

1.2 METAL PRICES

Like any other commodity, metals and ores are subject to the law of supply and demand and the prices they can command on the world markets may vary considerably (Fig. 1.3). There are three basic components to metal prices: (i) a long-term increase to keep pace with general rising costs or inflation, (ii) short-term, cyclical changes due to fluctuating economic controls, and (iii) relative price changes vis-à-vis one metal and another. The overall increases are seen on the price graphs as a general upward trend and can be determined by averaging prices out over a number of years. For example, the average price of copper in the 1960's was 40¢/lb compared with only 13¢/lb in the1920's. The cyclical changes are seen as sharp and sudden upward and

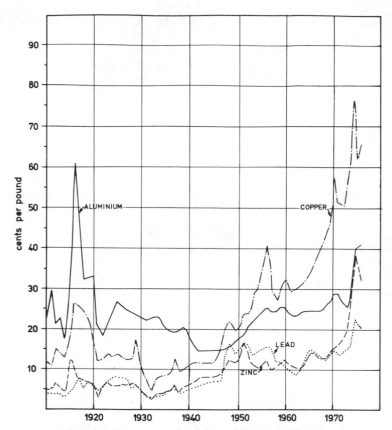

FIG. 1.3. Average yearly metal prices of copper, lead, zinc and aluminium since 1910. New York quotations. (Source: *Engineering and Mining Journal*.)

downward movements on the graphs and they can have serious repercussions for mining and mineral exploration. The recession of the early 1920's and the Great Depression of the early 1930's are clearly seen on both the production and price graphs as production outstripped demand. The marked fall in mine production of base metals immediately after the Second World War is not accompanied by a fall in price, since consumption did not outstrip production at that time and, in fact, shortage of supply kept prices steadily rising. The mid-1970's were once again a time of falling prices and output since production facilities, greatly expanded in the late 1960's, resulted in world production outstripping consumption in the industrial nations whose economies have fallen into recession due to the huge price increases in petroleum imposed by the OPEC cartel. The relative price change in metals is exemplified by aluminium. A major reduction in the cost of producing aluminium from its ores in the late nineteenth century was responsible for creating demand. Production costs fell steadily from 1925 until 1945 when it became cheaper than copper (Fig. 1.3) and in the 1950's it finally overtook copper as the world's major non-ferrous metal. For interest, the average prices pertaining in 1975 for a number of metals are given in Table 1.5.

TABLE 1.5
AVERAGE PRICES FOR SOME METALS IN 1975 IN TERMS OF DOLLARS
PER lb

Metal	Price	Metal	Price	Metal	Price
Pig iron	0·07	Nickel	2·00	Bismuth	7·72
Lead	0·21	Mercury	2·08	Niobium	21·50
Zinc	0·39	Antimony	2·09	Tantalum	39·95
Aluminium	0·39	Vanadium	2·18	Beryllium	61·00
Copper	0·58	Tin	3·12	Silver	65·11
Manganese	0·59	Cadmium	3·35	Palladium	1244·00
Magnesium	0·82	Cobalt	3·85	Platinum	2297·00
Titanium	1·17	Molybdenum	5·25	Gold	2355·00
Chromium	1·90	Tungsten	7·13	Osmium	2730·00
				Iridium	7366·00

Source: Engineering and Mining Journal

1.3 PATTERNS OF PRODUCTION AND CONSUMPTION

Mineral resources tend to be very unevenly distributed over the world with the result that some countries are richly endowed while others are almost entirely lacking in mineral wealth. No country is self-sufficient in mineral resources. Even the United States, which is a major producer of many metals, has hardly any tin, chromium, nickel or manganese deposits and has to rely heavily on imports. Table 1.6 lists the major producers of a number of metals and minerals and it is interesting to see how a few countries, or even one country in some instances, dominate production of certain commodities. Single countries that produce over 50% of the world supply of a particular metal include Brazil (niobium 83%), South Africa (gold 65%), USA (molybdenum 64%), Zaire (cobalt 59%) and USSR (platinum 55%). Patterns of production change considerably over the years with the depletion of deposits and the discovery and development of new ore

TABLE 1.6
MAJOR PRODUCERS OF A NUMBER OF DIFFERENT METALS AND
MINERALS IN 1973 UNLESS OTHERWISE STATED. (FIGURES ARE IN
THOUSANDS OF TONNES UNLESS OTHERWISE STATED)

Asbestos (1976 figures)		Antimony		Chromite (1976 figures)	
USSR	2528	South Africa	14·8	South Africa	2177
Canada	1694	China	14·0	USSR	1542
South Africa	370	Bolivia	13·1	Turkey	637
Rhodesia	350	USSR	7·5	Rhodesia	550
China	145	Mexico	3·0	Philippines	287
USA	135	Yugoslavia	2·0	Others	?
Others	400	Others	15·4		5193 +
	5622		69·8		

Cobalt		Copper (1974 figures)		Diamonds (1976 figures) (millions of carats)	
Zaire	15·1	USA	1446	Zaire	17·0
Zambia	2·0	USSR	1200	USSR	12·0
Canada	1·8	Chile	902	South Africa	7·6
USSR	1·7	Canada	826	Botswana	2·4
Others	5·0	Zambia	698	Ghana.	2·2
	25·6	Zaire	544	Namibia	1·7
		Australia	256	Sierra Leone	1·1
		Peru	213	Venezuela	0·6
		Philippines	209	Tanzania	0·5
		South Africa	205	Liberia	0·5
		Others	1387	Others	1·6
			7886		47·2

Gold (tonnes)		Lead (1974 figures)		Manganese (ore)	
South Africa	909·6	USA	602	USSR	8000
USSR	214·6	USSR	590	South Africa	4176
Canada	64.7	Australia	377	Gabon	2244
USA	45·1	Canada	296	India	1535
Australia	29·9	Mexico	218	China	998
Ghana	22·6	Peru	193	Mexico	364
Philippines	18·1	China	130	Zaire	334
Rhodesia	15·5	Yugoslavia	120	Others	4385
Others	70·6	Others	1048		22036
	1390·7		3574		

TABLE 1.6 *(Contd.)*

Mercury		Molybdenum (Mo in ore)		Nickel (1976 figures)	
Spain	2·09	USA	52·6	Canada	249·0
USSR	1·90	Canada	12·5	USSR	110·0
Italy	1·15	USSR	8·5	New Caledonia	119·0
China	1·00	Chile	5·9	Australia	37·0
Yugoslavia	0·54	China	1·5	Dominican	
Others	2·24	Others	1·2	Republic	27·0
	8·92		82·2	Cuba	26·0
				Indonesia	15·8
				South Africa	19·4
				Others	73·3
					676·5

Niobium (tonnes concentrate)		Platinum-group metals (tonnes)	
Brazil	7893	USSR	73·1
Canada	913	South Africa	45·1
Africa (mainly		Canada	12·6
Nigeria)	572	Others	1·5
Others	101		132·3
	9479		

Phosphate (1976 figures) (millions of tonnes)		Silver		Tantalum (tonnes concentrate)	
USA	44·15	Canada	1·48	Africa (mainly	
USSR	24·23	USSR	1·45	Nigeria)	167
Morocco	15·29	Peru	1·31	Canada	43
China	3·90	Mexico	1·21	Brazil	39
Tunisia	3·29	USA	1·16	Zaire	13
Togo	2·07	Japan	0·27	Malaysia	8
Jordan	1·70	Others	2·78	Others	79
South Africa	1·70		9·66		349
Senegal	1·58				
Vietnam	1·50				
Others	7·13				
	106·54				

TABLE 1.6 (*Contd.*)

Tin		Tungsten		Vanadium (*V in ore*)	
Malaysia	72·3	China	8·0	South Africa	8·9
Bolivia	35·4	USSR	7·4	USA	4·0
Indonesia	22·5	Asia (mainly		USSR	3·4
China	22·0	Korea)	5·3	Finland	1·3
Thailand	15·5	USA	3·4	Chile	1·0
USSR	12·0	Bolivia	2·2	Others	3·8
Australia	10·8	Canada	2·1		22·4
Zaire	5·4	Others	10·3		
Brazil	4·0		38·7		
Nigeria	3·8				
UK	3·8				
South Africa	3·3				
Others	15·5				
	226·3				

Zinc (1974 *figures*)

Canada	1122
USSR	950
Australia	449
USA	449
Peru	387
Mexico	263
Japan	241
Poland	200
China	130
Others	1691
	5882

Sources: 1973 *and* 1974: *Metallgesellschaft AG*, 1975
1976: *Mining Annual Review*, 1977

bodies; former major producers may be relegated to the position of minor producers or even non-producers and vice versa when previously non-producing countries become major producers. An example of this is the production of bauxite of which the two present leading producers, Australia and Jamaica, were virtual non-producers in 1950. Table 1.7 compares the major iron ore and oil producers in 1960 and 1976 and it is interesting to see how the ranking of producers has changed. In the case of iron ore both Brazil and Australia, who produced very little in 1960, have become the second and third ranked producers respectively. On the other hand, production in the case of

TABLE 1.7

LEADING IRON ORE AND OIL PRODUCERS IN 1960 AND 1976.
FIGURES IN MILLIONS OF TONNES

1960		1976	
Iron ore			
USSR	107	USSR	234
USA	87	Brazil	170
France	66	Australia	92
Canada	22	USA	78
Sweden	21	China	65
West Germany	19	Canada	56
UK	17	France	43
Others	76	India	40
	415	Sweden	29
		Liberia	23
		Others	60
			890
Oil			
USA	383	USSR	513
Venezuela	151	USA	487
USSR	147	Saudi Arabia	429
Kuwait	82	Iran	296
Saudi Arabia	61	Iraq	109
Iran	52	Kuwait	107
Iraq	47	Nigeria	103
Canada	26	UAE	97
Indonesia	20	Libya	96
Mexico	15	China	83
Others	101	Others	634
	1085		2954

Source: Mining Annual Review, 1961, 1977

West Germany and the UK, who both appear in the 1960 column of
Table 1.7, has decreased due to depletion of ores. In the case of oil
the second-ranked country in 1960, Venezuela, does not even appear
in the top ten in 1976; Nigeria, a virtual non-producer in 1960, is
ranked seventh in 1976 with a production very close to that of Iraq
and Kuwait in fifth and sixth places respectively.

If we consider consumption of mineral wealth there is an even
bigger disparity between the nations of the world. Industrialization
and development are largely concentrated in North America, Europe,
Japan, Australasia and the USSR who between them consume 95% of

the world's nickel, 92% of its aluminium and copper, 87% of its lead and zinc and 85% of its tin. If consumption of partly developed nations such as Brazil, Argentina, Korea, Taiwan, South Africa, China and Mexico, who all have significant industrial capacity, is taken into account, there is very little left for the vast numbers of people in the developing nations of Asia, Africa and South America, the so-called Third World. In addition, many of the industrialized nations, particularly Japan and those of Western Europe, have to rely heavily on imports of raw materials and unworked metals to meet their domestic consumption. Many of the world's major metal and ore producers are in the Third World and consume a negligible amount of

TABLE 1.8

COPPER PRODUCTION AND CONSUMPTION OF MAJOR INDUSTRIAL NATIONS IN 1974 (IN MILLIONS OF TONNES)

	Mine production	Smelter production	Primary consumption	Use of scrap
EEC	0·016	0·194	1·817	1·074
USSR	1·200	1·200	1·025	?
Japan	0·082	0·900	0·716	0·445
USA	1·446	1·424	1·519	1·299
Totals	2·731	3·718	5·077	2·818+

Source: Metallgesellschaft AG, 1975

TABLE 1.9

COPPER PRODUCTION AND CONSUMPTION OF MAJOR COPPER EXPORTERS IN 1974 (IN MILLIONS OF TONNES)

	Mine production	Smelter production	Primary consumption	Use of scrap
Chile	0·902	0·724	0·029	—
Canada	0·826	0·537	0·270	0·099
Zambia	0·698	0·709	0·004	—
Zaire	0·544	0·470	0·002	—
Australia	0·256	0·196	0·088	0·034
Peru	0·213	0·180	0·013	—
Totals	3·439	2·816	0·406	0·133

Source: Metallgesellschaft AG, 1975

their own products. This can be seen in Tables 1.8 and 1.9, which show the copper production and consumption of the major industrial nations and copper exporters respectively. Only the USSR produces sufficient copper for its own needs, the USA is very nearly self sufficient, but Japan and the EEC countries produce only 11% and 1% of their copper needs respectively. In addition Japan has an enormous smelting capacity in excess of her needs which has to be supplied with large quantities of copper concentrates, much of the contained copper of which is finally exported as refined metal. Amongst the major copper exporters only Canada and Australia are highly developed nations. A further example of the disparity between producers and consumers is given by bauxite and aluminium production shown in Fig. 1.4. The interesting facts to emerge are that the Third World bauxite producers consume negligible amounts of aluminium, few of them even have any capacity to produce aluminium metal and there are a number of developed nations, notably Canada and Norway, which produce no bauxite, but are major exporters of primary aluminium.

If the consumption of various metals is put in *per capita* terms, the disparity between the developed and the Third World countries is illustrated very forcibly. Table 1.10 shows the *per capita* consumption for aluminium, copper, lead and zinc for a number of nations and it is clear that the majority of people in the world consume very little of the world's mineral wealth. Similar tables for energy consumption would show an even bigger disparity. If this consumption of metals is used as a measure of wealth, it means that there will have to be an enormous increase in mineral production if peoples of the Third World are to move even slightly towards the living standards of the industrialized and developed nations. This is possible, but it is essential that the Third World uses its mineral wealth wisely by investing in agriculture and other basic development projects money earned from mining. Unfortunately, many Third World producers of raw materials are not doing this and are spending their mineral export earnings on current living expenses which can only lead to tragedy. Many people advocate that nations should not hurry to exploit mineral wealth, but should rather conserve it for the future when they may be in a better position to use the minerals for their own development rather than export them to more developed countries. This approach has little to commend it in most cases and any Third World country fortunate enough to find a large mineral deposit should consider it as an important capital asset, which can be used to assist in the development of the country, even if most of its mineral production has to be exported in a raw state, at least in the early years.

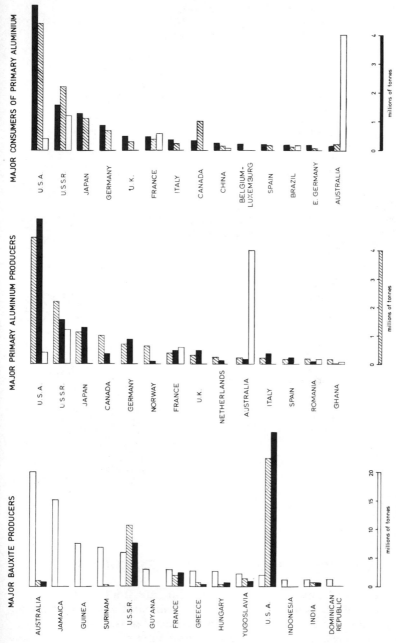

FIG. 1.4.   The world's leading bauxite producers, primary aluminium producers and aluminium consumers in 1974. (Source: Metallgesellschaft AG, 1975.)

TABLE 1.10

PER CAPITA CONSUMPTION OF ALUMINIUM, COPPER, LEAD AND ZINC
FOR A NUMBER OF SELECTED COUNTRIES IN 1974

Per capita consumption in kg	Aluminium	Copper	Lead	Zinc
>10	USA Canada Sweden EEC Japan Australia	Sweden Canada	—	—
5–10	USSR	Australia USA EEC Japan	Australia	Canada Japan USA EEC Australia
2–5	Poland Argentina South Africa Brazil	USSR Poland South Africa Mexico	USA Sweden EEC Canada Poland USSR Japan	Poland Sweden USSR South Africa
1–2	Mexico Turkey	Argentina Brazil	Mexico Argentina South Africa	Argentina Mexico Brazil
0·1–1·0	China India	China Turkey	Brazil Turkey China	Turkey China India
<0·1	Third World countries	India Third World countries	India Third World countries	Third World countries

Source: Metallgesellschaft AG, 1975 (consumption)
        UNESCO (population)

1.4 THE NATURE OF MINERAL EXPLORATION

The reserves of any mine, however great, are finite and will eventually be depleted. Thus, any country or company involved in mining should undertake a continual search for new deposits and extensions to existing ones. This can be considered akin to the research and development work undertaken by manufacturing companies which is necessary to improve existing products or develop new ones to maintain an edge over competitors. Mining is a risky business compared to other industrial activities, however, and mineral exploration carries an even higher element of risk. It is for this reason that accountants and business men, who often control or exercise a major influence on the affairs of a mining company, may cut off all exploration in times of economic recession as they see it as a totally unproductive venture with no return on capital expended. This is an extremely short-sighted policy since mineral exploration is essential to the future of any mining company.

Although mining carries an element of risk which can mean heavy financial loss to the owners of a mine, it also holds the possibility of a big return on capital invested and a much higher profitability than can be obtained from most other industrial or agricultural ventures. This point was put across very well by Agricola in the 16th century:

It is not my intention to detract anything from the dignity of agriculture, and that the profits of mining are less stable I will always and readily admit, for the veins do in time cease to yield metals, whereas the fields bring forth fruits every year. But though the business of mining may be less reliable it is more productive, so that in reckoning up, what is wanting in stability is found to be made up by productiveness. Indeed the yearly profit of a lead mine in comparison with the fruitfulness of the best fields, is three times or at least twice as great.

This potential for a high rate of return from a mining venture is the incentive which encourages individuals and companies to engage in mineral exploration. Destroy this incentive and investment in mineral exploration will decline and future mineral development will suffer. There is a growing trend in the world today to penalize mining through heavy taxation and other government controls for what are seen as 'excessive profits'. It is certainly true that some companies have made enormous profits out of mining, but what is often conveniently forgotten are the companies and individuals who invested in mineral exploration ventures which failed to discover viable deposits and were thus total losses to the people concerned. Also forgotten are the times of recession when falling metal prices turned profitable mines into loss-making operations for their owners. It cannot be denied that mineral wealth is a national resource and as many people in the country as possible should benefit from its exploitation, but at the same time it is important to achieve a satisfactory balance so that the incentives for mineral exploration and development are not removed completely. It is often thought that the socialist countries with centrally controlled economies do not suffer from these problems and that a mineral deposit will be developed for the national good regardless of short term economic considerations. This is not necessarily any more satisfactory, however, than systems pertaining in countries with free enterprise, since mineral exploration and development in the socialist countries fall directly under the umbrella of government bodies and organizations, which have a tendency in any nation to become bureaucratic and highly inefficient.

In Third World countries, which lack technological expertise and financial resources, it is necessary to attract outside interests to develop mineral wealth. Unless sufficient incentives can be offered, however, foreign capital and know-how, whether it be from capitalist or socialist countries, will be withheld. The really big problem is to create the necessary climate of confidence so that a Third World country can attract investment and still ensure that it receives a fair and equitable share of its own resources. Too often instability or fear of instability in Third World countries causes a foreign investor to expect a higher rate of return on an investment than he might expect in his own country. On the other hand, excessive profits made by a foreign investor will result in a nation feeling that it is being robbed of its national wealth and heritage.

Table 1.11 lists the sequence of operations involved in starting from a reconnaissance exploration survey and concluding with the successful development of a mine, indicating the risk element at the various stages. Hypothetical costs are given to illustrate the escalating nature of costs with each advancing stage. When risk is highest, expenditure is lowest, but even at an advanced stage, when significant sums of money may have been spent, the risk is still high. Using the hypothetical costs, a project terminated at the end of the reconnaissance phase will have incurred a loss of $0·5 million. Few projects are cut off at this stage purely for geological reasons since there is usually sufficient encouragement to proceed with a certain amount of initial follow-up work. In many cases expenditure at this stage may be no more, and may even be less, than the reconnaissance stage, but, if results of preliminary work are sufficiently encouraging, expenditure may be considerable. The risk is still very high at this stage and termination of a project due to discouraging results will involve a loss of $1·5 million using our hypothetical figures. Encouraging indications of mineralization may justify detailed follow-up work which can be very expensive as it is likely to be based heavily on drilling. The risk factor is still high and, if indicated grades and tonnage are considered insufficient, termination of the project will result in a total loss of $5·5 million. If the detailed follow-up work indicates a viable deposit, a feasibility study to determine the mining methods to be employed, the rate of production, treatment of ore,

TABLE 1.11
SEQUENCE OF OPERATIONS IN EXPLORATION AND DEVELOPMENT OF
AN ORE DEPOSIT

Stage	Type of work	Possible methods employed	Hypothetical costs (millions of $)	Risk
Exploration	reconnaissance	geological mapping, prospecting, geochemistry, geophysics, airborne surveys	0·5	extremely high
	initial follow-up	geological mapping, geochemistry, geophysics, limited drilling	1·0	very high
	detailed follow-up	drilling, limited metallurgical testing	4·0	high
Development	feasibility study	drilling, metallurgical testing, mine design, trial mining	10·0	moderate
	construction and mine development	site construction, drilling, underground and/or surface mining	100·0	low
Mining	extraction and beneficiation of ore	various mining and concentrating methods depending on deposit	operating costs	low

disposal of the product, etc. will be necessary. In the case of remote deposits, this may involve planning an entire township with all the necessary amenities. By this stage the risk factor is considerably reduced, but it may still be necessary to terminate the project if results of the feasibility study indicate serious problems. Termination at this stage involves a total loss of $15.5 million using our hypothetical figures. When the decision to mine the deposit is finally taken, the risk factor should be very low, but there is no certainty in mining and a fall in metal prices or unforeseen difficulties in transporting the product can have a disastrous effect, especially if they occur in the first few years of the new mine's life. Total exploration and development costs now stand at $115·5 million for our hypothetical example. Even if the mine does prove to be below expectations, a considerable recovery of the total investment can still be made, but profits may be poor or non-existent. A good deposit, of course, will result in rapid recovery of exploration and development investments plus profits for the shareholders. For every successful discovery, however, there will be countless projects that will have to be terminated at earlier stages and, as stated earlier, it is for this reason that returns on mining have to be better than other industrial ventures which carry a lower risk factor.

It is clear from the great disparity between consumption of mineral wealth in the developed and Third World countries that an enormous increase in mineral production will be needed if Third World countries are ever to attain the living standards enjoyed by the affluent nations. The pessimists would say that this is impossible, since the known reserves of almost every mineral are insufficient to allow expansion of production on the scale that would be required. This may be true, but to be over-pessimistic is to ignore a number of factors. Firstly, it must be remembered that the world has passed several 'deadlines' at which pundits in the past had predicted exhaustion of the reserves of various natural resources. Discoveries of new deposits have always managed to keep reserves in step with increasing consumption. Secondly, advances in technology have made it possible to work lower and lower grades so that material formerly considered waste becomes the ore reserves of tomorrow. Mining techniques have also improved so that underground mines have gone deeper than ever before and opencast mines have become bigger and are worked on a scale that would hardly have been imagined 50 years ago. Thirdly, and very importantly, patterns of consumption change. Just because the average person in Sweden accounts for 12 kg of primary copper production *per annum*, for example, does not mean that the impoverished inhabitant of a central African state will have to consume the same amount of copper to enjoy a good standard of living. Aluminium was only used for a few specialized jobs 50 years ago, but today it is the world's most important non-ferrous metal. Steel has been considered the bell-wether of an economy, but production and consumption have fallen well below forecasts made 30 or 40 years ago since a number of traditional uses of steel have been replaced by new materials such as plastics, aluminium and prestressed concrete which were not envisaged at the time the forecasts were made. In the case of tin, consumption today is only just over twice what it was in 1900 as a result of substitutes that were not foreseen 75 years ago. Many of the metal containers that are used to preserve beverages and foodstuffs are now made of aluminium instead of tinplate. Also, as a result of improved techniques, the tin coating used on tinplate today is much thinner than it was in the past so that a given amount of tin goes further. Lead is another metal whose traditional uses, such as plumbing and roofing, have been replaced by substitutes so that production today is only four times what it was in 1900. If lead accumulators used in vast numbers by the motor industry are ever replaced, the demand for lead would drop drastically. These are just some of the examples of changes in the patterns of consumption that have occurred and will continue to occur.

Thus, the potential for mineral exploration is very good as new discoveries will have to be made to keep pace with the increasing demands of the developing and developed nations. Nevertheless, short-term cyclical changes can have drastic effects on mineral exploration and big slumps will occur in times of economic recession due to shortages of funds. At times like these it is easy to be pessimistic and feel that mineral exploration has a poor future, but such times of depression are always short lived. Nevertheless, the extremely sensitive nature of mineral exploration to economic change means that exploration geology can be a rather precarious profession. 'Boom' times may also occur such as the famous Australian nickel boom of the late 1960's, when an increase in the free market price for nickel from £986 per ton in 1968 to £7000 per ton in 1969 resulted in frantic speculation and much unsound exploration activity. Such 'booms' are not really in the long-term interests of the industry, however lucrative they may seem to geologists at the time. Ideally, a more balanced approach is required, but it is difficult to see how it can be achieved. Systems pertaining in the socialist countries with centrally controlled economies move some way towards rationalizing mineral exploration, but there are other disadvantages which may outweigh any gains.

Although it is certain that production of almost all minerals can be increased significantly and exploration will result in future major discoveries, it remains a fact that mineral resources are finite. There is thus a pressing need to use the world's resources wisely and conserve minerals for future generations. One way of doing this is to recycle materials and it is interesting to note that this is already being done on a large scale (Table 1.12).

In comparison to minerals, the future of energy supplies causes much more concern. At the rate oil and natural gas are being consumed, reserves may well be depleted in the not too distant future. There is a very real need for the major energy consumers to conserve energy and use it more wisely than is being done at present. Nevertheless, further oilfields will be found by future petroleum exploration and there are a number of alternative sources of energy which can be

TABLE 1.12
USE OF SCRAP METALS BY WESTERN COUNTRIES

	Scrap as % of production	Scrap as % of consumption
Tin	21	19
Zinc	16	16
Copper	43	49
Lead	37	31
Aluminium	20	17

Source: Metallgesellschaft AG, 1975

exploited on a much larger scale. These include hydro-electricity, solar energy, wind energy, geothermal energy and, most important of all for the immediate future, atomic energy. Considerable amounts of uranium will be required to meet the world's growing consumption of atomic energy and the search for uranium deposits will account for a large proportion of expenditure on mineral exploration. The biggest problems with the widespread use of atomic energy are the threat of the proliferation of nuclear weapons and the disposal of radioactive wastes. It will be necessary to overcome these problems, however, if the world is to cope with the so-called energy crisis.

Finally, it is worth considering another source of mineral wealth which has not been exploited to any great extent—the sea. Plans are well advanced to mine deep sea manganese nodules which contain significant amounts of Ni, Cu, Zn and Co in addition to Mn. Metalliferous muds or brines, similar to those of the Red Sea which contain large tonnages at 1% Cu and 5% Zn (Bignell, 1975), might be located by prospecting and worked in the future. In shallower offshore waters a number of minerals are already being worked. These include sands and gravels, calcium carbonate in the form of shells, shell sands, aragonite muds and coral, iron sands (magnetite), cassiterite and diamonds. There is considerable scope for expanding production and working new minerals which will require concerted exploration efforts (Siegel, 1971).

1.5 MINING AND THE ENVIRONMENT

There can be no doubt that the mining industry has earned itself a bad reputation as a despoiler of the countryside. In many parts of the world scarred hillsides, derelict buildings, slag heaps and sterile ground are ugly reminders of past mining activities. Vast areas have been devastated by uncontrolled or poorly controlled strip mining and pollution and contamination from smelters and tailings dams have rendered good farmland unproductive. Not all mining activities need be like this and there are numerous examples of mines all over the world that are well planned and worked so as to make as little impact on the countryside as possible. Nevertheless, it has become fashionable in many quarters to bandy about the word 'ecology' and condemn mining in general and mining companies in particular as destroyers of our natural heritage. Taken to the extreme this outlook is quite absurd since our whole way of life in the modern world is ultimately dependent on mining. Agricola even made this point to critics in the sixteenth century:

If we remove metals from the service of man, all methods of protecting and sustaining health and more carefully preserving the course of life are done away with.

Much of the thinking of the anti-mining lobby was as illogical and irrational then as it is now and some of the most outspoken critics are people who more than most enjoy the benefits of our technological age. Due to the depth of feeling that the subject of mining arouses, explorationists often find themselves at the centre of controversy and much bitter argument. To resolve many of the differences a more balanced approach is needed. It can be conceded that certain areas of natural beauty should be preserved from development of any kind, but at the same time it must be recognized that metals and minerals are a vital necessity and that mining of them can be controlled so that relatively little damage is done to the environment. Even in the area of a national park a mining operation can be conducted in such a manner that its presence may be hardly noticed.

REFERENCES AND BIBLIOGRAPHY

Agricola (1550). *De Re Metallica* (translated by H. C. Hoover and L. H. Hoover), Dover Publications, New York, 1950.

Archer, A. A. (1974). Progress and prospects of marine mining, *Min. Mag.*, **130**(3), 150–163.

Bignell, R. D. (1975). Timing, distribution and origin of submarine mineralization in the Red Sea, *Trans. Instn. Min. Metall.*, Lond., **84**, B1–6.

Bullard, E. (1974). Minerals from the deep sea, *Endeavour*, 33(119), 80–85.

Jones, M. J. (ed.) (1975). *Minerals and the Environment*, Instn. Min. Metall., Lond., 803 pp.

Jones, W. R. (1955). *Minerals in Industry* (3rd edn), Penguin Books Ltd, England, 238 pp.

24 TECHNIQUES IN MINERAL EXPLORATION

Mero, J. (1965). *The Mineral Resources of the Sea*, Elsevier, Amsterdam, 312 pp.
Metallgesellschaft AG (1975). *Metal Statistics 1964–1974*, Frankfurt am Main, 348 pp.
Rickard, T. A. (1932). *Man and Metal. A History of Mining in Relation to the Development of Civilization* (2 vols.), McGraw–Hill Book Co., New York, 1068 pp.
Rickard, T. A. (1944). *The Romance of Mining*, The Macmillan Company of Canada, Toronto, 450 pp.
Scientific American (1965). *Technology and Economic Development*, Penguin Books Ltd, England, 237 pp.
Siegel, F. R. (1971). Marine geochemical prospecting—present and future, *Geochem. Explor.* C.I.M. Spec. 11, 251–257.
Skinner, B. J. (1976). *Earth Resources* (2nd edn), Prentice-Hall Inc., New Jersey, 151 pp.
Street, A. and Alexander, W. (1976). *Metals in the Service of Man* (6th edn), Penguin Books Ltd, England, 346 pp.
Tooms, J. S. (1972). Potentially exploitable marine minerals, *Endeavour*, 31(114), 113–117.
US Bureau of Mines (1975). *Mineral Facts and Problems*, US Dept. Interior, US Government Printing Office, Washington DC, 1259 pp.
Warren, K. (1973). *Mineral Resources*, Penguin Books Ltd, England, 272 pp.

CHAPTER 2

Geological Mapping and Prospecting

2.1 THE IMPORTANCE OF GEOLOGICAL MAPPING AND PROSPECTING

With the rapid growth and development in the fields of applied geochemistry and geophysics there is a tendency to underestimate the importance of geological mapping and prospecting in mineral exploration. It is not uncommon to find examples of exploration projects where extensive areas have been covered by soil sampling and have hardly even been visited by a geologist. Interesting and important targets may be missed by the neglect of a fundamental geological approach to exploration, for there are numerous examples of outcropping mineral occurrences which do not show up on reconnaissance geochemical surveys. In addition to the possibility of finding areas of mineralization, geological mapping provides the necessary framework within which an exploration programme can be conducted. Different types of mineralization tend to occur in definite geological environments, and features such as faults, folds, zones of alteration or particular lithologies may have important bearings on mineralization and the recognition of such features in the field at the reconnaissance stage may assist in defining target areas for detailed investigation by more costly methods. In short, basic geological mapping and prospecting must be regarded as the foundation of any thorough exploration venture.

In the 1920's and 1930's a massive prospecting campaign was carried out over a large part of what was then the British Protectorate of Northern Rhodesia and today is the independent nation of Zambia. The discoveries of rich copper deposits in what is now the Copperbelt of Zaire and Zambia were a strong stimulus to mineral exploration in this large and at that time virtually unmapped territory. The whole operation was conducted on the lines of a military campaign with the object of systematically mapping and prospecting all potentially interesting areas. Field parties consisting of two or more geologists together with large numbers of assistants and labourers were assigned to various parts of the country. After cutting two parallel base lines 5 miles (8 km) apart, traverses were run along cut lines at quarter-mile (400 m) or half-mile (800 m) intervals between base lines. Geologists working in pairs traversed alternate lines as a control to the mapping, and flank men walking 100 m or more on either side of the traverse lines searched for outcrops so that each traverse effectively covered a swath of country at least 200 m wide. Samples were taken from all mineralized outcrops and float and sent to headquarters for analysis, panned concentrates were collected from streams, pits were dug in areas of poor exposure and all gossans and many quartz veins were sampled. Although the field parties worked in the bush, often in remote areas for many months at a time, contact was maintained with headquarters at the end of each month by a messenger service carrying reports and samples. All information from the field was collated at the main office where the final geological and mineral occurrence maps were prepared. Although no producing mines were discovered off the Copperbelt as an immediate result of this work, the campaign produced the first detailed geological, albeit essentially lithological, map of a large part of the country which has been improved only in recent years. In addition to the major deposits discovered on the Copperbelt itself, the work laid down the basic geological framework which made subsequent major discoveries possible.

Although no one would advocate strictly following this line of approach today, since modern methods of geochemistry, geophysics, remote sensing and greatly improved aerial photography have put mineral exploration on a much sounder scientific basis than in the 1930's, it must always be remembered that discoveries are made on the ground and not in offices or field camps. Field assistants and samplers who are engaged in routine sampling should have impressed upon them that the job entails far more than collecting a certain number of samples per day. They should note all outcrops on their traverses as a matter of course and should be taught the recognition of common ore minerals as well as the importance of noting anything unusual such as conspicuous iron staining. It is often advantageous to pay a bonus for all finds of mineral showings and even finds of outcrop in areas of very poor exposure. Any outcrops or features of possible interest found by field assistants or prospectors while laying out a soil sampling grid, for example, can be visited by a geologist at a later date and examined in detail if necessary. Even in areas of poor exposure, careful and detailed float mapping may be instrumental in finding an ore body. In areas of Scandinavia covered by glacial till, the tracing of the provenance of mineralized float in boulder trains by thorough and systematic prospecting and mapping has contributed to the discovery of a number of deposits (Grip, 1953; Hyvarinen *et al*, 1973).

A good example of a recent successful intensive prospecting programme is the discovery of the Windarra nickel deposits in Western Australia (Robinson *et al*, 1973). Using available aerial magnetic data and the government geological maps all magnetic anomalies and ultrabasic rocks were systematically examined on the ground by Ken Shirley, a local prospector. As it is very difficult to distinguish between laterite, weathered iron formation and gossan in the area, Shirley sampled all interesting ferruginous rocks. All localities were numbered and pegged and carefully described in a field notebook. The samples were analyzed for copper and nickel and claims were staked over all ultrabasics with anomalous values. Shirley recognized the Windarra claims as the most interesting and a nickel ore body was confirmed after further sampling, geophysical surveys and drilling carried out by an exploration company.

The importance of geological mapping and conventional prospecting in mineral exploration is pointed out in an interesting paper by Derry (1968). Of 76 discoveries made in Canada between 1951 and 1966, during which time both geochemistry and geophysics were being widely employed, 27 discoveries were directly attributed to prospecting, 23 to geology, 25 to geophysics and only one to geochemistry; that is, two thirds of all discoveries were due to geology and prospecting. In other parts of the world a similar situation exists, though the importance of the roles of geophysics and geochemistry is often reversed, as for example in central Africa where geophysics has met with little success and geochemistry has proved extremely useful. In southern Africa of the 20 base metal mine discoveries made between 1954 and 1974 only two, Selebi-Pikwe (Cu–Ni) in Botswana and Shangani (Ni) in Rhodesia, had no surface expression (Buhlmann *et al*, 1974).

No particular technique is required for mapping in economic geology except to be extra alert to the recognition of minerals of possible economic interest. For reference some standard geological symbols are given in Figs. 2.1 and 2.2. A number of simple, reliable, chemical field tests for the identification of various elements and minerals are given in the following section, the only equipment required being several test tubes (2×20 cm is a useful size), a spirit lamp and a number of readily available reagents.

Some useful field tests

Beryl
A useful test to distinguish beryl from quartz or feldspar is to place the grains to be tested in a test tube, add a solution of sodium

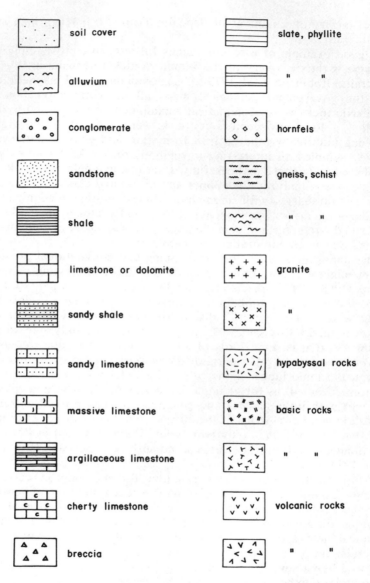

FIG. 2.1.   Some standard lithological symbols.

hydroxide and boil for several minutes. Drain off the liquid, add an aqueous solution of quinalizarin and boil for several minutes. Any grains of beryl present will be stained an intense blue. Feldspar may take on a violet stain, but it is not permanent and will disappear with continued boiling.

Cassiterite
Place the grains to be tested in a test tube together with some powdered zinc. Add dilute hydrochloric acid and leave to react for several minutes. Any grains that become a dull silvery grey (due to a coating of metallic tin) are cassiterite. If the cassiterite grains are coated with iron oxides, positive results may not be obtained. This can be overcome, however, by heating the grains in concentrated hydrochloric acid before carrying out the test. Instead of using powdered zinc it is convenient to place the grains to be tested in a zinc cup to which the dilute hydrochloric acid can be added. Heavy zinc cups can be used for many tests before they are finally etched through by the acid. This test is diagnostic for cassiterite.

Copper
Place fragments to be tested in a test tube. Add 2–3 ml concentrated nitric acid and boil gently for a few minutes. Dilute to double volume with water and slowly add 1–2 ml concentrated ammonia solution. A deep blue solution indicates copper.

Manganese oxides
Place a few drops of 3% hydrogen peroxide solution on the grains to be tested. Effervescence without any noticeable attack of the grains confirms the presence of one of the manganese oxides: hausmanite, manganite, psilomelane or pyrolusite.

Nickel
Place the grains to be tested in a test tube and add 1 ml concentrated nitric acid and 3 ml concentrated hydrochloric acid. Boil gently for several minutes and dilute to double volume with water. Add 1 ml of 1% dimethyl-glyoxime (DMG) solution in 50:50 ammonia and alcohol. A scarlet precipitate indicates nickel. Iron, copper and cobalt interfere to some extent with this test, but cobalt and copper give a brown colour and ferrous iron a red-violet colour. Concentrated ammonia can be added instead of the DMG solution. This gives a blue-green solution if nickel is present.

Niobium
Carry out the test as for tungsten. A blue solution which disappears on dilution indicates the presence of niobium.

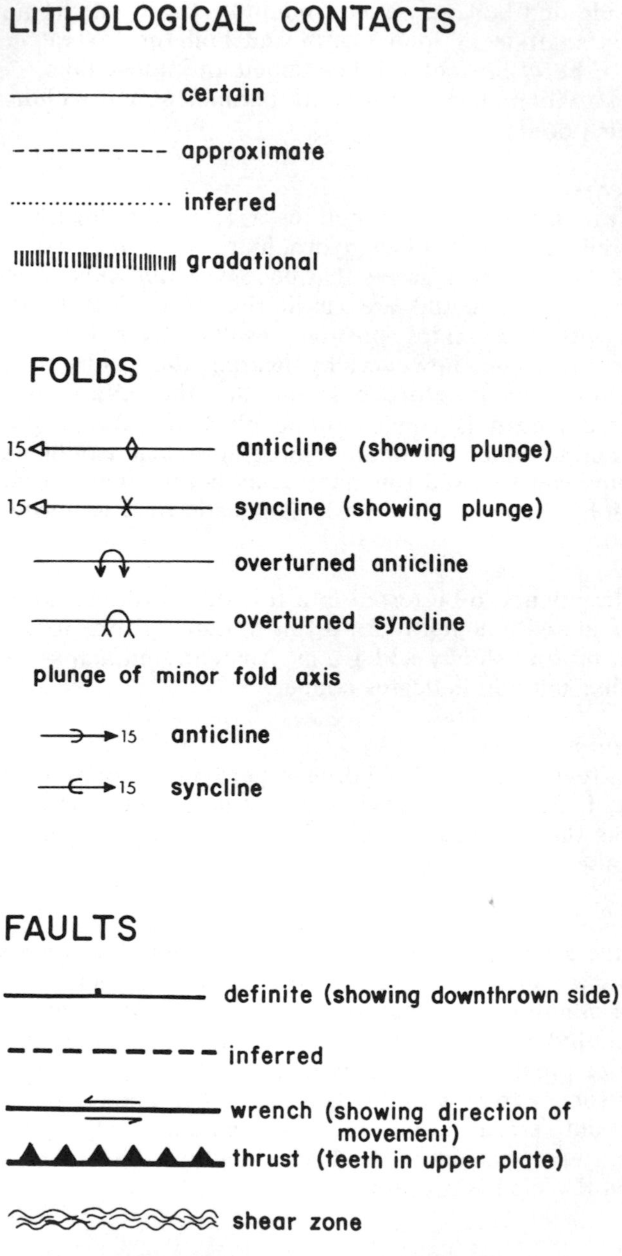

LITHOLOGICAL CONTACTS

———————— certain

– – – – – – – – approximate

······················ inferred

|||||||||||||||||||||||||||| gradational

FOLDS

15◄———◊——— anticline (showing plunge)

15◄———✕——— syncline (showing plunge)

———⋔↓——— overturned anticline

———⋏λ——— overturned syncline

plunge of minor fold axis

——Ɔ→15 anticline

——Ϲ→15 syncline

FAULTS

———————•———————— definite (showing downthrown side)

– – – – – – – – inferred

———⇇———— wrench (showing direction of movement)

▲▲▲▲▲▲▲ thrust (teeth in upper plate)

≈≈≈≈≈≈≈ shear zone

FIG. 2.2.   Standard geological symbols.

STRIKE AND DIP SYMBOLS

	Bedding	Foliation	Joints or veins	Cleavage or schistosity
strike and dip				
horizontal				
vertical				
overturned				

OTHER SYMBOLS

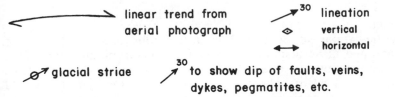

linear trend from aerial photograph

glacial striae

³⁰ lineation

vertical

horizontal

³⁰ to show dip of faults, veins, dykes, pegmatites, etc.

Letters may be used with the symbols for further clarification.

For example: ²⁵ₛ strain–slip cleavage

h ²⁰ mineral lineation (hornblende)

Some of the symbols may be combined.

For example: lineation trend in plane of foliation ₁₅

FIG. 2.2.—contd.

Phosphate
Place the grains to be tested in a test tube and add 2–3 ml concentrated nitric acid. Boil gently for a few minutes and dilute to double volume. Add 1–2 ml 6% ammonium molybdate solution. A bright yellow, cloudy precipitate indicates phosphate. On first adding the ammonium molybdate reagent, it may appear as though a yellow solution has formed, but the fine precipitate will soon be seen as a cloudy suspension and as a coating on the test tube wall. The presence of arsenates will also result in the formation of a yellow precipitate.

Titanium

Carry out the test as for tungsten. A violet solution indicates the presence of titanium.

Tungsten

Crush the grains to be tested, mix with an equal amount of sodium carbonate and place in a test tube. Fuse strongly for several minutes. Allow to cool and add 6 ml water and 3 ml concentrated hydrochloric acid. Drop two or three small granules of tin into the test tube and boil strongly for about five minutes. An inky blue, cloudy solution, which retains its colour after dilution with an equal volume of water, indicates the presence of tungsten. The 'solution' is in fact a fine precipitate which becomes apparent upon dilution and will settle out if left for several hours.

Uranium

Place the grains to be tested in a test tube and add 2–3 ml of concentrated nitric acid. Boil gently for several minutes and dilute with water to two to three times original volume. Allow to settle, decant several ml of solution to a clean test tube and slowly add concentrated ammonia solution. A pale yellow, flocculent precipitate indicates uranium.

Vanadium

Carry out the test as for tungsten, but add powdered zinc instead of tin. A blue solution which turns to green and then to blue-violet indicates the presence of vanadium.

For more thorough and detailed testing an excellent scheme for the identification of mineral grains is given by Jones and Fleming (1965) and is based on the systematic grouping of specific gravity, refractive index, hardness, colour of powder, magnetic permeability and chemical reaction with various concentrated acids; spot tests to confirm the presence of a number of elements are also described. In addition to these various testing procedures, grains of transparent minerals can be readily identified by crushing them on a glass slide, adding a drop of clove oil (refractive index similar to Canada balsam) and applying routine optical mineralogical techniques with a petrographic microscope. The phenomenon of fluorescence exhibited by some minerals when irradiated by ultra-violet rays can also be of use in mineral identification. Scheelite, hydrozincite and secondary uranium minerals almost always fluoresce and a wide range of minerals exhibit the property, but often only in rare instances and to varying degrees depending upon the locality and impurities. If it is known that a mineral from a particular locality fluoresces, it can be a valuable aid in prospecting. Table 2.1 lists a number of minerals and their colours of fluorescence.

In addition to the recognition of mineralization particular attention should be paid to certain significant features depending upon the area and environment. For example, it may be known that lead–zinc mineralization in a certain district is associated with a shale/dolomite contact, or that copper mineralization occurs in a definite sedimentary unit. Another example is the recognition of potassic metasomatic alteration which is often associated with porphyry copper deposits. There are numerous other examples and in all instances the geologist should pay particular heed to the features considered to be of most importance in the area concerned. For reference a list of elements and the types of deposits in which they occur are given in Table 2.2.

TABLE 2.1

SOME FLUORESCENT MINERALS UNDER SHORT WAVELENGTH (2 500Å) ULTRA-VIOLET LIGHT (FROM SMITH, 1953).

Mineral	Colour	Remarks concerning fluorescence
Autunite*	yellow-green	very common
Calcite	red, pink, yellow, rarely blue	not common
Fluorite	blue or cream yellow	common (better under long wavelength—3 000Å)
Gummite*	violet	very common
Gypsum	green	rare
Halite	red	rare
Hydrozincite	blue	very common
Monazite	emerald green	rare
Powellite	yellow to greenish-yellow	very common
Scapolite (Wernerite)	bright yellow, sometimes red	common
Scheelite	bluish-white	very common—becomes yellowish with impurities (usually Mo)
Sphalerite	bright orange	rare
Spodumene	red	common
Strontianite	bluish-green	not common
Torbernite*	yellow-green	very common
Uranophane*	yellow-green	very common
Willemite	green	common
Zippeite*	yellowish	very common
Zircon	orange	common

*Secondary uranium minerals

TABLE 2.2

LIST OF ELEMENTS AND THE TYPES OF ORE DEPOSITS IN WHICH THEY OCCUR. THE HEADINGS SHOULD NOT BE REGARDED AS FIRM GENETIC CLASSIFICATIONS, BUT RATHER AS A LISTING OF GEOLOGICAL ENVIRONMENTS, AS THERE IS A GREAT DEAL OF CONTROVERSY IN THE GENETIC CLASSIFICATION OF MANY DEPOSITS. MINIMUM ECONOMIC GRADES ARE ALSO INDICATED IN MOST CASES, BUT THESE FIGURES SHOULD ONLY BE REGARDED AS A ROUGH GUIDE AS THE ECONOMIC VIABILITY OF ANY PARTICULAR DEPOSIT DEPENDS UPON MANY LOCAL FACTORS. TYPES OF DEPOSITS OF PARTICULAR IMPORTANCE ARE INDICATED BY THE USE OF CAPITAL LETTERS.

ALUMINIUM

1. *Weathering*
BAUXITE* FORMED BY WEATHERING OF FELDSPATHIC OR NEPHELINE-BEARING ROCKS, ALSO OFTEN RESIDUAL DEPOSITS OVER LIMESTONE AS IN JAMAICA.
(45–50% Al_2O_3, <20% Fe_2O_3, 3–5% SiO_2)

3. *Hydrothermal*
The mineral, alunite $(KAl_3(SO_4)_2(OH)_6)$, which forms as a hydrothermal alteration product of trachytes and rhyolites in some volcanic regions, contains up to 37% Al_2O_3 and could be used as a source of aluminium.

2. *Magmatic*
Alumina-rich rocks such as anorthosites and nepheline syenites which contain more than 20% Al_2O_3 have been used as a source of aluminium in the USSR.

4. *Metamorphic*
The aluminium silicates andalusite, sillimanite and kyanite (Al_2SiO_5) are all potential sources of aluminium. They are not economic at present, but processes to extract alumina have been worked out.

Bauxite is an earthy grey, brown, yellow or red-brown mixture of clays, free silica, iron hydroxides and the aluminium hydroxides gibbsite, boehemite and diaspore.

ANTIMONY

1. *Hydrothermal*
STIBNITE OCCURS IN QUARTZ–STIBNITE VEINS. Stibnite and other antimony minerals occur in association with lead, lead–zinc, silver and copper (tetrahedrite) ores (ores generally >4% Sb).

TABLE 2.2 (*Contd.*)

BARIUM

1. *Hydrothermal*
BARITE and witherite IN VEIN DEPOSITS ASSOCIATED WITH LEAD–ZINC OR FLUORITE. ($>93\%$ $BaSO_4$, SG > 4.5)

3. *Weathering*
Since barite is a resistant mineral, important residual deposits may result from the weathering of sedimentary or vein deposits.

2. *Sedimentary*
Barite often occurs in limestone or shale beds where it may form important deposits.

4. *Magmatic*
Barite and witherite occur in carbonatites, often in significant amounts, but it is generally late stage and much of it may be considered to be hydrothermal.

BERYLLIUM

1. *Pegmatites*
OCCURS IN GRANITIC PEGMATITES AS BERYL ($Be_3Al_2Si_6O_{18}$)—also as other berylosilicates which may be of importance. Beryllium minerals also occur in alkaline pegmatites, but none of the associations with alkaline rocks has proved economic. (Saleable ore $> 10\%$ BeO, grades worked $> 0.1\%$ BeO)

3. *Metamorphic*
Beryl occurs as possible 'skarn-type' deposits in limestones and schists. IMPORTANT FOR EMERALDS where traces of Cr give the beryl its green colour.

2. *Pneumatolytic*
Beryl and other beryllium minerals occur in quartz–muscovite and quartz–topaz greisens associated with cassiterite and wolfram. Not an economic source of beryllium at present.

BISMUTH

1. *Hydrothermal*
Generally occurs in veins associated with minerals of Sn, Cu, Ag, W, Au. Most commercial Bi is obtained as a by-product of lead refining and copper, gold, tin and tungsten mining. With the exception of a few very small deposits, there are no mines worked for Bi alone. The main minerals are native bismuth, bismuthinite (Bi_2S_3), bismutite ($Bi_2CO_5.H_2O$).

2. *Pegmatites*
Native bismuth, bismuthinite and bismutite occur in some pegmatites. Small amounts are occasionally won by small-workers from these pegmatite sources.

TABLE 2.2 (*Contd.*)

BORON

1. *Sedimentary*
BORATES SUCH AS BORAX
($Na_2B_4O_7.10H_2O$), KERNITE
($Na_2B_4O_7.4H_2O$) AND
COLEMANITE ($Ca_2B_6O_{11}.5H_2O$)
OCCUR IN PLAYA-DEPOSITS
DERIVED FROM OLD SALT
LAKES ($>45\%$ B_2O_5 required). Also
occurs as borates such as ulexite and
boracite in lenses within evaporite
deposits.

2. *Volcanogenic*
Boric acid (sassoline) (H_3BO_3) also
occurs around fumaroles and in
waters of hot springs of volcanic
areas. Not an economic source at
present.

CADMIUM

1. *Hydrothermal*
Mainly associated with lead–zinc
deposits, rarely as greenockite (CdS)
in vein deposits ($>0.1\%$ Cd required
to be worthwhile as a by-product).

2. *Sedimentary*
Associated with lead–zinc
mineralization as for hydrothermal
deposits.

CARBON (*DIAMONDS AND GRAPHITE*)

1. *Magmatic*
Diamonds occur in kimberlites
(economic grades depend on quality
and size of stones, but usually >0.4–
1.8 carats per tonne; 5 carats = 1 g)
Graphite sometimes occurs in
igneous rocks.

2. *Metamorphic*
Graphite is a common constituent of
many schists, limestones and
gneisses. Sometimes the graphite
content can be quite high (grades
usually $>25\%$ C to be economic;
pencil graphite $>83\%$ C; reactor
graphite 97–99% C).

3. *Weathering*
Important alluvial diamond deposits
occur in many parts of the world.

4. *Hydrothermal*
Graphite may occur in vein deposits
some of which are economic sources
of good quality graphite.

5. *Pegmatites*
Small amounts of graphite sometimes
occur in pegmatites.

CHROMIUM

1. *Magmatic*
CHROMITE OCCURS IN
LAYERED ULTRABASIC
INTRUSIVES AND IN ALPINE
ULTRABASIC BODIES. PURE

1. *Magmatic (Contd.)*
CHROMITE SEAMS OR PODS
ARE GENERALLY REQUIRED
TO BE ECONOMIC, BUT
DISSEMINATED CHROMITE

TABLE 2.2 (Contd.)

CHROMIUM (Contd.)

1. Magmatic (Contd.)
ORES ARE WORKED IN
FINLAND. (Metallurgical grade
> 42% Cr_2O_3 and $Cr:Fe > 2\cdot5:1$,
high grade $Cr:Fe > 3:1$ and
$Cr_2O_3 > 50\%$; refractory grade
$Cr:Fe < 2\cdot5:1$.)

2. Placers
Being a resistant mineral chromite
has been concentrated in detrital
sands at a few localities, but none of
these deposits is of real economic
importance.

COBALT

1. Hydrothermal
Cobalt occurs in vein deposits as the
arsenide smaltite ($CoAs_2$) and the
sulpharsenide cobaltite (CoAsS)
associated with Ag, Ni and Cu.

3. Magmatic
Cobaltiferous pyrrhotite may occur
in nickel sulphide deposits such as at
Sudbury, Ontario.

4. Volcanogenic
Cobaltiferous pyrite sometimes
occurs in massive sulphide deposits
in sedimentary–volcanic associations.

2. Sedimentary
Minor amounts of cobalt often occur
in stratiform sulphide deposits which
may be syngenetic in origin.
Cobaltiferous pyrite and sulphides
such as linnaeite (Co_3S_4) and
carrolite ($CuCo_2S_4$) occur with
copper sulphides. (> 0·2% Co
generally required for recovery as a
by-product.) Manganese nodules
from the seabed often contain
significant amounts of cobalt and
may be an important future source.

5. Weathering
Lateritic cobalt deposits analogous to
lateritic nickel deposits sometimes
form over cobalt-rich basic and
ultrabasic rocks. Asbolite is a cobalt-
rich manganese wad found in such
deposits.

COPPER

1. Sedimentary
THE LARGE STRATIFORM
COPPER DEPOSITS IN SHALES
AND SANDSTONES ARE A
VERY IMPORTANT SOURCE OF
COPPER. (Grades typically vary
from 0·5 to 5·0%, but at least 1% is
usually required to be economic.)

2. Hydrothermal
THE LARGE LOW-GRADE
PORPHYRY COPPER DEPOSITS

2. Hydrothermal (Contd.)
GENERALLY ASSOCIATED
WITH GRANODIORITE AND
QUARTZ MONZONITE
INTRUSIONS ARE EXTREMELY
IMPORTANT SOURCES OF
COPPER. (Grades typically vary
from 0·2 to 1·0%, but at least 0·5% is
usually required to be economic.)
Copper minerals occur in a wide
range of vein deposits often

TABLE 2.2 (Contd.)

COPPER (Contd.)

2. Hydrothermal (Contd.)
associated with zinc, lead, gold and
silver. SOME VEIN DEPOSITS
ARE IMPORTANT COPPER
ORES.

3. Volcanogenic
Copper occurs in massive sulphide
deposits associated with volcanic
rocks such as the cupriferous pyrite
deposits of Cyprus and the massive
sulphide deposits of the Canadian
Shield. Sulphides include pyrite,
pyrrhotite, sphalerite, chalcopyrite
and galena. (Copper contents
typically range from 0·5 to 2·0%.)

4. Magmatic
Chalcopyrite occurs with pyrrhotite
and pentlandite in magmatic
segregation deposits such as at
Sudbury, Ontario, and at many of the
smaller Archaean nickel deposits.

5. Metamorphic
Sulphide minerals occur in many
skarn deposits around igneous
intrusions. Most of these deposits
are small and irregular, but some are
important sources of copper.
Sulphide minerals include pyrite,
sphalerite, galena, chalcopyrite and
bornite.

FLUORINE

1. Hydrothermal
VEIN DEPOSITS OFTEN
ASSOCIATED WITH LEAD–ZINC
AND BARIUM. (25–40% CaF_2
required, > 5% CaF_2 as a by-product;
acid grade 97–98% CaF_2;
metallurgical grade > 60% CaF_2.)
Also late-stage hydrothermal
deposits associated with some
carbonatites.

3. Pneumatolytic
Fluorite often occurs in gneisens
with cassiterite, topaz and
tourmaline. Not an important source.

2. Pegmatites
Fluorite occurs in many pegmatites,
but it is not an important source.

4. Magmatic
Small amounts of fluorite occur in
some granites. No economic deposits
known.

5. Sedimentary
Fluorite cements are known in some
sandstones. No economic deposits
known.

GERMANIUM

1. Hydrothermal
Ge occurs in some vein-type deposits
as the minerals germanite—Cu_3(Fe,
Ge, Zn, Ga) (S, As)$_4$—and renierite—
(Cu, Fe)$_3$(Fe, Ge, Zn, Sn) (S, As)$_4$.

2. Sedimentary
Ge occurs in quite high
concentrations in certain coal ashes.

GOLD

1. Sedimentary
GOLD OCCURS IN ANCIENT
CONGLOMERATES WHICH CAN

1. Sedimentary (Contd.)
BE CLASSED AS FOSSIL
PLACERS. Very finely disseminated

TABLE 2.2 (Contd.)

GOLD (Contd.)

1. Sedimentary (Contd.)
gold may occur in argillaceous limestones as at Carlin, Nevada. (> 10 g/t, lower grades sometimes worked.)

3. Placer
Alluvial gold deposits are quite common and SOME ARE IMPORTANT SOURCES.

4. Volcanogenic
Traces of gold sometimes occur in massive sulphide deposits and can be recovered as a by-product during smelting and refining.

2. Hydrothermal
GOLD OCCURS IN QUARTZ VEINS, often with sulphides, sometimes with copper, silver, bismuth and tungsten minerals. Some vein deposits contain gold tellurides. GOLD OFTEN OCCURS IN PORPHYRY COPPER DEPOSITS (generally <0·5 g/t) WHICH CONSTITUTE AN IMPORTANT SOURCE—RECOVERED DURING SMELTING AND REFINING.

IRON

1. Sedimentary
PRECAMBRIAN IRON FORMATIONS (taconite, itabirite, jaspilite) ARE EXTREMELY IMPORTANT SOURCES OF IRON ORE. IRON OCCURS AS HAEMATITE AND MAGNETITE, also as siderite, limonite, pyrite and iron silicates. (20–65% Fe, ores worked at 30%, but >60% required for export.)
Phanerozoic deposits include the oolitic iron ores containing chamosite, goethite, haematite, siderite and the black-band iron mudstones containing siderite (<50% Fe). IMPORTANT SOURCES IN SOME INDUSTRIAL COUNTRIES (ores worked at 30% Fe).
Bog iron deposits formed by precipitation of ferric oxides and hydroxides in bogs and lakes. Generally small and low grade—only of minor economic significance.

2. Magmatic
Large magmatic segregations of magnetite may form important iron ore deposits AS AT KIRUNA, SWEDEN. Associated with basic and syenitic intrusions (grades up to 70% Fe, often contaminated with Ti and P).

3. Metamorphic
Skarn-type deposits of magnetite often associated with basic and syenitic intrusions (grades up to 70%, often deleterious contaminants such as P and S).

4. Weathering
Tropical weathering of iron-rich rocks may result in formation of laterite iron ores. CAN BE AN IMPORTANT SOURCE.

5. Hydrothermal
Haematite of probable hydrothermal origin occurs as replacements in sediments such as limestone, also occurs in veins. Often an economic source.

TABLE 2.2 (Contd.)

LEAD

1. Sedimentary
IMPORTANT STRATIFORM LEAD–ZINC ORES OCCUR IN CARBONATE SEDIMENTS. Galena sometimes occurs disseminated in sandstones. (Grades usually > 5% Pb to be economic.)

3. Volcanogenic
Small amounts of galena occur in some massive sulphide deposits in volcano–sedimentary sequences.

2. Hydrothermal
VEIN DEPOSITS IN WHICH GALENA AND OTHER LEAD MINERALS MAY OCCUR WITH SPHALERITE, CHALCOPYRITE, TETRAHEDRITE, SILVER MINERALS, FLUORITE, BARITE ARE IMPORTANT SOURCES OF LEAD. EXTENSIVE REPLACEMENT DEPOSITS MAY OCCUR IN COUNTRY ROCKS, PARTICULARLY CARBONATES.

4. Metamorphic
Galena may occur in contact metamorphic deposits with other sulphides such as chalcopyrite, pyrite, sphalerite and bornite.

LITHIUM

1. Pegmatites
Occurs as a number of lithium-bearing minerals in many granitic pegmatites. Minerals include spodumene ($LiAlSi_2O_6$), amblygonite ($Li(F,OH)AlPO_4$), petalite ($LiAlSi_4O_{10}$) and lepidolite (lithium mica). (Saleable ore: >4% LiO_2.)

MAGNESIUM

1. Sedimentary
THE MAIN SOURCE OF MAGNESIUM METAL IS SEA WATER WHICH CONTAINS ABOUT 0·5% $MgCl_2$. Magnesium salts used in industry are obtained from evaporite deposits which may contain minerals such as epsomite ($MgSO_4.7H_2O$), kieserite ($MgSO_4.H_2O$) and carnallite ($MgCl_2.KCl.6H_2O$).

3. Metamorphic
Periclase (MgO) and brucite found in some contact metamorphosed carbonate rocks.

2. Hydrothermal
Magnesite ($MgCO_3$) and brucite ($Mg(OH)_2$) formed as a result of hydrothermal alteration of rocks such as serpentinites which are rich in magnesium silicates. They may also be formed as a result of the action of hydrothermal fluids on limestones and dolomites. Used as sources of metal, but main uses as industrial minerals. (Grade required: >40% MgO.)

TABLE 2.2 (*Contd.*)

MANGANESE

1. *Sedimentary*
MANGANESE OXIDE MINERALS— PYROLUSITE, PSILOMELANE, BRAUNITE, MANGANITE, WAD*—OCCUR IN LARGE IMPORTANT SEDIMENTARY DEPOSITS SOMETIMES ASSOCIATED WITH PYROCLASTIC SEDIMENTS AND SOMETIMES WITH SEDIMENTARY IRON DEPOSITS (ores worked at 20% Mn, saleable ore: >45% Mn).
Smaller bog-manganese deposits may form by the precipitation of manganese oxides in lakes and swamps.

2. *Weathering*
IMPORTANT ORES ARE FORMED BY THE WEATHERING AND SECONDARY ENRICHMENT OF LOWER GRADE PRIMARY DEPOSITS. THE MINERALS IN THESE LATERITIC DEPOSITS INCLUDE WAD*, BRAUNITE, PSILOMELANE, HAUSMANITE, RHODOCHROSITE.

3. *Metamorphic*
The metamorphism of low-grade sedimentary or residual deposits may result in workable deposits of braunite, hausmanite and other manganese minerals.
Some manganese minerals such as rhodonite, rhodochrosite and hausmanite are formed by metasomatism or contact metamorphism, but are of little importance as ores.

4. *Hydrothermal*
Some manganese minerals such as rhodochrosite, rhodonite and hausmanite may occur in vein deposits. Rarely of any importance. Many vein-like deposits are of doubtful hydrothermal origin.

*WAD—a mixture of oxides and hydrous oxides of manganese analogous to limonite.

MERCURY

1. *Hydrothermal*
Cinnabar occurs in epithermal deposits as disseminations and small veins often in association with pyrite, native mercury, stibnite and realgar. (Ores generally 0·6–2·0% Hg, very rarely up to 6%.)
Small amounts of mercury may also occur in 'grey copper' ores where it may be produced as a by-product, though it is generally considered a deleterious contaminant.

TABLE 2.2 (*Contd.*)

MOLYBDENUM

1. *Hydrothermal*
MOLYBDENITE OCCURS IN
DEPOSITS IMTIMATELY
ASSOCIATED WITH ACID
IGNEOUS INTRUSIONS SUCH
AS AT CLIMAX, COLORADO,
WHERE MOLYBDENITE
OCCURS IN QUARTZ VEINLETS
IN GRANITE. MANY PORPHYRY
COPPER DEPOSITS ALSO
CONTAIN MOLYBDENUM
WHICH IS A VALUABLE BY-
PRODUCT. (Economic grades
usually $> 0.06\%$ MoS$_2$.)
Molybdenite often occurs in deep-
seated veins associated with
scheelite, wolframite, topaz and fluorite.

2. *Magmatic*
Molybdenite often occurs
disseminated in acid igneous
intrusives.

3. *Pegmatites*
Molybdenite may occur in
pegmatites which are often rich and
may contain large, coarse
molybdenite crystals. Due to the
small size of such deposits they are
not an important source and are
generally sub-economic.

4. *Metamorphic*
Molybdenite occurs in some contact
metamorphic deposits together with
other sulphides and sometimes
scheelite. These occurrences are not
of economic importance.

NICKEL

1. *Magmatic*
MASSIVE SULPHIDE DEPOSITS
CONTAINING PYRRHOTITE,
PENTLANDITE AND OTHER
MINOR SULPHIDES MAY BE
ASSOCIATED WITH NORITIC
INTRUSIVES AS AT SUDBURY,
ONTARIO, OR WITH SMALLER
ULTRABASIC BODIES IN
ARCHAEAN BLOCKS AS IN
WESTERN AUSTRALIA OR IN
'MOBILE BELTS' ADJACENT TO
ARCHAEAN BELTS AS IN
MANITOBA. Small minor deposits
have been found associated with
younger basic/ultrabasic intrusives.
(Grades vary typically from 0·5–5%
Ni, usually 1·5% required to be
economic.)

2. *Weathering*
Tropical weathering of nickel-rich
ultrabasic rocks may result in the
formation of important lateritic
nickel ores with silicate minerals
such as garnierite and genthite as in
Cuba and New Caledonia. (Grades
generally vary from 0·3% to 3%, but
may reach 10%.)

3. *Hydrothermal*
Minor amounts of nickel sulphides
and arsenides occur in a number of
vein deposits associated with cobalt,
silver and copper. Only of minor
importance.

TABLE 2.2 (*Contd.*)

NIOBIUM

1. *Magmatic*
THE NIOBATE MINERAL
PYROCHLORE OCCURS IN
CARBONATITES AND OTHER
ALKALINE ROCKS AND IS NOW
THE MAIN SOURCE OF
NIOBIUM. CARBONATITES
REPRESENT IMMENSE
POTENTIAL RESERVES.
Columbite also occurs in some
granites and alkaline rocks.
(Economic grade at least 0·2%
Nb_2O_5.)

2. *Weathering*
LARGE RESERVES OF
PYROCHLORE OCCUR IN
ELUVIAL DEPOSITS OF
RESIDUAL SOIL FORMED OVER
CARBONATES IN TROPICAL
REGIONS. (Grades enriched with
respect to parent carbonatite.)
Columbite also occurs in some
eluvial deposits formed from
pegmatites.

3. *Pegmatites*
Columbite—$(Fe, Mn)Nb_2O_6$—occurs
in granitic pegmatites often in
association with cassiterite and beryl.
Once an important source. Often a
by-product of tin mining.

PHOSPHORUS

1. *Sedimentary*
THE MOST IMPORTANT
SOURCES OF PHOSPHATE ARE
THE EXTENSIVE BEDDED
ROCK PHOSPHATE OR
PHOSPHORITE DEPOSITS SUCH
AS THOSE IN NORTH AFRICA
OR THE UNITED STATES.
THESE DEPOSITS ARE OF
MARINE ORIGIN AND
PHANEROZOIC IN AGE. Late
Precambrian and freshwater deposits
do occur. Some rock phosphate
deposits are due to diagenetic
replacement of existing sediments.
($> 30\%$ P_2O_5 required.)

2. *Magmatic*
Apatite is a common accessory of
many igneous rocks. Phosphorus is
enriched in alkaline rocks in
particular and apatite-rich rocks and
even rocks formed wholly of apatite

2. *Magmatic* (*Contd.*)
occur in association with alkaline
complexes and carbonatites. AN
IMPORTANT SOURCE IN SOME
PARTS OF THE WORLD. (Rocks
at 5% P_2O_5 can be worked under
favourable circumstances.)

3. *Weathering*
Weathering of apatite-bearing rocks
may result in residual accumulations.
Important deposits have formed over
carbonatites in some parts of the
world. (Material $> 10\%$ P_2O_5
worked.) The leaching of guano
deposits on some desert islands has
resulted in enriched and valuable
deposits.

4. *Pegmatites*
Concentrations of apatite occur in
some pegmatites and can sometimes
be worked.

TABLE 2.2 (*Contd.*)

PLATINUM

1. *Magmatic*
NATIVE PLATINUM, THE ARSENIDE, SPERRYLITE, TOGETHER WITH MANY OTHER LESSER KNOWN PLATINUM-GROUP MINERALS OCCUR DISSEMINATED IN BASIC AND ULTRABASIC ROCKS SUCH AS IN THE BUSHVELD COMPLEX. PLATINUM ALSO OCCURS IN MAGMATIC NICKEL SULPHIDE DEPOSITS SUCH AS SUDBURY, ONTARIO. (Grades worked range from 4–12 g/t Pt-group, production as by-product from lower grades.)

2. *Placer*
A number of the minerals of the Pt-group metals occur in alluvial deposits derived from ultrabasic bodies containing Pt-group elements. IMPORTANT DEPOSITS OCCUR IN THE USSR.

3. *Hydrothermal*
Small amounts of Pt-group elements occur at a few hydrothermal copper deposits. The origin of such mineralization may be due to the hydrothermal leaching of Pt-bearing basic rocks.

POTASSIUM

1. *Sedimentary*
Potassium minerals such as sylvite (KCl), carnallite ($KCl.MgCl_2.6H_2O$), kainite ($KCl.MgSO_4.3H_2O$) and poly-halite ($K_2SO_4.MgSO_4.2CaSO_4.2H_2O$) occur in many evaporite deposits which are the major source of potash.

2. *Hydrothermal*
Alunite ($KAl_3(SO_4)_2(OH)_6$) is formed as a hydrothermal alteration of trachytes and rhyolites. It can be used as a source of potash.

RARE EARTHS

1. *Magmatic*
Monazite and other rare earth minerals occur in carbonatites which can be an important source. ($>2\%$ RE_2O_3 usually required to be economic, much lower grades can be worked as a by-product.) Quite high rare earth concentrations can occur in apatite and can be extracted as a by-product of phosphate production from apatite.

2. *Placer*
Monazite occurs as an accessory mineral in many acid and alkaline rocks and important detrital deposits may be formed. Monazite sands are worked at a number of localities ($>3\%$ monazite is economic).

3. *Pegmatites*
Monazite occurs in some pegmatites—sometimes an economic source.

TABLE 2.2 (*Contd.*)

SILVER

1. Hydrothermal
The major part of silver production is obtained as a by-product from lead, zinc, copper and gold ores. Silver minerals including native silver, argentite (Ag_2S), stephanite (Ag_5SbS_4), pyrargyrite (Ag_3SbS_3) and proustite (Ag_3AsS_3) occur in vein deposits, some of which are important primary silver ores. Silver minerals also occur in association with tin, copper, cobalt and nickel minerals in some vein deposits.

4. Placer
Silver forms a natural alloy with gold, and argentiferous gold, known as electrum, is an important source of silver.

2. Sedimentary
Silver is often produced as a by-product from stratiform lead–zinc and copper deposits. Silver is also produced from auriferous conglomerate deposits where it may occur in natural alloy with gold.

3. Volcanogenic
Small amounts of silver sometimes occur in massive sulphide deposits and can be recovered as a by-product.

SULPHUR

1. Sedimentary
ELEMENTAL SULPHUR OCCURS IN BEDDED DEPOSITS ASSOCIATED WITH EVAPORITES AND OFTEN PETROLEUM. THESE DEPOSITS ARE THE MAIN SOURCE OF SULPHUR.

3. Other
Sulphur dioxide is recovered during the smelting of sulphide ores to produce sulphuric acid for industrial use.

2. Volcanogenic
Elemental sulphur is found in workable quantities in areas of recent volcanism and hot spring activity.
Pyrite in massive sulphide deposits is often worked for its sulphur content. (Grades > 15% S are worked, but small amounts of copper and other base metals are usually required to make the operation economic.)

TANTALUM

1. Pegmatites
Tantalite (($Fe, Mn)Ta_2O_6$) occurs in granite pegmatites, particularly those that are Li-bearing, and is the main source of tantalum. It often occurs in association with cassiterite and beryl. Tantalite forms a solid solution

1. Pegmatites (*Contd.*)
series with columbite, the mineral of intermediate composition being known as columbo-tantalite.

2. Magmatic
Tantalite occurs in some alkaline granites which are a potential source.

TABLE 2.2 (*Contd.*)

TANTALUM (*Contd.*)

3. *Weathering*
Tantalite is a resistant mineral and
small eluvial and alluvial deposits
may form.

TIN

1. *Weathering*
CASSITERITE IS A STABLE
MINERAL AND IMPORTANT
ALLUVIAL DEPOSITS MAY
FORM BOTH ON LAND AND
OFFSHORE. Of lesser importance
are eluvial deposits which form by
the weathering of veins and
pegmatites—an important source of
tin in some parts of the world.
(Grades down to 0·05% SnO_2 can be
worked.)

3. *Pneumatolytic*
Cassiterite occurs in quartz and
greisen veins closely associated with
granite intrusions. Accompanying
minerals include wolframite,
chalcopyrite, tourmaline, topaz,
fluorite and arsenopyrite. Minerals
such as sphalerite, galena, argentite,
pitchblende and stibnite may occur
in outer zones of the same deposits
away from the deeper Sn-bearing
lodes. MAY BE IMPORTANT
ORES, AS IN CORNWALL.

2. *Hydrothermal*
Tin porphyries are an important type
of deposit and are closely associated
with dacitic porphyry intrusives. The
mineralogy is complex with
significant amounts of silver in
addition to W, Pb, Zn, Cu, Sb, Bi
and Au. Sulphides are common and
many of the Sn-bearing sulphides
such as stannite (Cu_2SnFeS_4) occur
in addition to the more common
cassiterite. MAY BE IMPORTANT
ORES, AS IN BOLIVIA.
Stanniferous quartz and quartz–
muscovite veins of simple
mineralogy occur in some parts of
the world closely associated with
granites—not an important source.
(Grades $> 0·5\%$ SnO_2 to be economic
underground.)

4. *Pegmatites*
Cassiterite occurs in pegmatites
associated with columbite, tantalite,
beryl and lithium minerals. A minor
source of tin.

TITANIUM

1. *Placer*
THE MAIN SOURCE OF
TITANIUM IS FROM ILMENITE
AND RUTILE BEACH SANDS
WHICH OCCUR IN SEVERAL
PARTS OF THE WORLD. (Saleable
concentrate: $> 52\%$ TiO_2 for
ilmenite.)

2. *Magmatic*
Ilmenite occurs with magnetite in
large masses considered to be
magmatic segregations associated
with basic igneous rocks (mainly
gabbros and anorthosites)—AN
IMPORTANT SOURCE OF
TITANIUM. Rutile also occurs in

TABLE 2.2 (*Contd.*)

TITANIUM (*Contd.*)

2. *Magmatic* (*Contd.*)
similar associations with syenites,
gabbros and anorthosites and may be
an important source.

4. *Metamorphic*
Rutile occurs in a wide range of
metamorphic rocks, but rarely in
significant concentrations. Not an
economic source at present.

3. *Pegmatites*
Rutile occurs in some pegmatites
with apatite and ilmenite and may be
an economic source.

TUNGSTEN

1. *Hydrothermal*
Wolframite ($(Fe, Mn)WO_4$) occurs in
low to high temperature veins where
it is associated with cassiterite,
chalcopyrite, arsenopyrite,
bismuthinite, scheelite, quartz and
other minerals. (Ores worked at 0.5%
WO_3, saleable concentrate: $>60\%$
WO_3.)

3. *Weathering*
Wolframite is a resistant mineral and
eluvial and alluvial deposits are
formed by the weathering of primary
tungsten deposits. Scheelite is a less
resistant mineral, but does occur in
some eluvial deposits. (Grades at
0.1% WO_3 are worked.)

2. *Metamorphic*
Many skarn deposits associated with
acid intrusions contain scheelite
($CaWO_4$) and are an important
source of tungsten.

4. *Pneumatolytic*
Wolframite occurs with cassiterite in
some greisen veins closely associated
with granitic intrusions.

5. *Sedimentary*
Some tungsten deposits of possible
sedimentary origin have been
described.

URANIUM

1. *Sedimentary*
PRIMARY URANIUM MINERALS
OCCUR IN SOME ANCIENT
PRECAMBRIAN
CONGLOMERATES (BLIND
RIVER TYPE). URANIUM
MINERALS OCCUR IN SOME
SANDSTONES, GRITS,
CONGLOMERATES AND
MUDSTONES OF DIFFERENT
AGES. THE MINERALS ARE
SOMETIMES SECONDARY AND

1. *Sedimentary* (*Contd.*)
WERE PROBABLY DEPOSITED
BY CIRCULATING
GROUNDWATERS. OFTEN
ASSOCIATED WITH V AND Se
(COLORADO TYPE)
Uranium minerals may also occur in
calcretes. ($>0.1\%$ U_3O_8, grades of
0.03% U_3O_8 worked as a by-
product.)
Uranium occurs in quite high
concentrations (>500 ppm) in some

TABLE 2.2 (*Contd.*)

URANIUM (*Contd.*)

1. *Sedimentary (Contd.)*
phosphatic and carbonaceous
sediments, but extraction is not
economic at present.

3. *Pegmatites*
Numerous uranium minerals may
occur in granitic pegmatites. SOME
ARE IMPORTANT DEPOSITS.

2. *Hydrothermal*
Uranium minerals occur in some
vein deposits often associated with
Cu, Co, Ni, Ag and Au. MAY BE
IMPORTANT DEPOSITS.

4. *Magmatic*
Uranium minerals may occur
disseminated in granitic rocks—not
an economic source at present.

VANADIUM

1. *Sedimentary*
Vanadium is often associated with
petroleum and bitumen. May be an
important source as at Minasraga,
Peru, where patronite (VS_4) occurs
with asphalite.
Uranium deposits of the Colorado
type often contain vanadium in
minerals such as carnotite
($K_2O.2U_2O_3.V_2O_5.2H_2O$) and are
important sources.

2. *Weathering*
Secondary vanadium minerals such
as vanadinite ($Pb_5Cl(VO_4)_3$) and
descloisite ($Pb(Zn, Cu)(VO_4)(OH)$)
occur in the oxidized parts of some
lead–zinc deposits, sometimes in
lateritic overburden.

3. *Magmatic*
V^{3+} replaces Fe^{3+} in some early
magnetites and can be an economic
source ($0.5–2.0\%$ V_2O_5).

ZINC

1. *Hydrothermal*
ZINC OCCURS IN CLOSE
ASSOCIATION WITH LEAD IN
MANY VEIN DEPOSITS. ALSO
OFTEN ACCOMPANYING
SILVER AND COPPER. LARGE
REPLACEMENTS OF COUNTRY
ROCK, PARTICULARLY
CARBONATES ARE COMMON.
Occasionally secondary minerals in
oxidized zones are important ores.
(Ore grade $> 5\%$ Zn.)

3. *Volcanogenic*
Sphalerite occurs in massive sulphide
deposits in volcano–sedimentary
sequences. CAN BE IMPORTANT
ORES. (Ores can be worked at 2–3%
Zn if other products produced.)

2. *Sedimentary*
STRATIFORM LEAD–ZINC
DEPOSITS IN CARBONATE
ROCKS ARE IMPORTANT
SOURCES OF ZINC.

4. *Metamorphic*
Sphalerite may occur with other
sulphides in contact metamorphic
deposits—CAN BE IMPORTANT
SOURCE. Franklinite and secondary
zinc minerals such as zincite (ZnO)
and willemite (Zn_2SiO_4) occur in
unique deposits at Franklin
Furnance, New Jersey.

TABLE 2.2 *(Contd.)*

ZIRCONIUM

1. Weathering
Zircon ($ZrSiO_4$) is a resistant mineral occurring in acid and alkaline igneous rocks and certain metamorphic rocks. Both alluvial and eluvial deposits are formed. SOME BEACH SANDS ARE AN IMPORTANT SOURCE OF ZIRCON. Both zircon and baddeleyite (ZrO_2) occur in some eluvial deposits over alkaline complexes—may be economic as a by-product. (Saleable product $>65\%$ ZrO_2.)

2. Pegmatites
Both zircon and baddeleyite occur in some pegmatites and can be an economic source.

3. Magmatic
Zircon and baddeleyite may occur in alkaline rocks including some carbonatites and may be economic as a by-product.

The prospector should also be alert to any unusual vegetation changes, which might be due to the toxic effects of mineralization, or to the presence of certain indicator plants which have been recognized as valuable prospecting aids in various regions (see p. 173). Also the recognition of gossans is an important aspect of prospecting. Gossans are defined as accumulations of yellow, red, brown or black limonite with other oxides, metal oxysalts, silica and silicates formed as a result of the leaching and weathering of sulphide bodies. True gossans can be very difficult to distinguish from ironstones and other iron oxide accumulations such as laterites or ferricretes. Table 2.3 lists some important features of gossans, but it should be remembered that there are no hard and fast rules for identifying true gossans. For more detail the reader is referred to the classic study of gossans made by R. Blanchard (1968).

It might be felt that as more and more parts of the world are covered by detailed prospecting and mapping, such methods will have less and less relevance to mineral exploration. While this is largely true, it is astonishing how major discoveries with surface showings or surface evidence are still made in areas thought to have been thoroughly mapped. This is particularly so for minerals such as sphalerite which can be very difficult to recognize in small amounts in outcrop and low grade mineralization such as that which typifies large porphyry copper deposits which may have been overlooked by earlier workers or else thought to be of little significance. As the importance of conventional prospecting, geochemistry and geophy-

TABLE 2.3

SOME IMPORTANT FEATURES FOR THE IDENTIFICATION OF GOSSANS

	True gossan	False gossan	Laterite
Field relations and structural attitude	Generally prominent outcrop ridge or spine reflecting configuration of parent sulphide body, though width is often exaggerated.	Often prominent ridge or spine, but may be recognizable as a definite sedimentary formation.	Often forms flat pavements over a wide area. Sometimes as prominent hummocky spines or ridges.
Colour	Often extreme local variation in colour—red, yellow, brown, black.	Generally dull, uniform, dark brown.	Generally dull, uniform, dark brown or red-brown.
Structural and textural characters	Cellular structure often pseudo-morphs, original sulphides and gangue minerals forming 'box-works'. Sometimes cellular structure poor or absent. Some characteristic boxworks— *Pyrite*: irregular configuration, continuity and thickness *Pentlandite*: regular and widely spaced partitions expressing (111) cleavage. *Pyrrhotite*: Typical six-sided, often internal radiating partitions. *Chalcopyrite*: rectangular, often thick walled. *Bornite*: spherical, triangular. *Sphalerite*: acutely angular intersections of partitions	May have cellular structure, but it is generally fine and of even texture.	Often well-developed pisoolitic concretionary structure. Sometimes massive and blocky. Occasionally cellular.

Geochemical characteristics	Generally high base metal values. Some typical values— *Ni gossans*: 0·1–2% Ni, 0·05–1% Cu. *Cu–Zn gossans*: <50 ppm Ni, 100–3000 ppm Cu, 50–2000 ppm Zn. *Massive sulphide gossans*: (Pyrite or pyrrhotite + minor Cu, Zn). <50 ppm Ni, 1000–5000 ppm Cu, 500–3000 ppm Zn. *Pyrite gossans*: <50 ppm Ni, <100 ppm Cu, <100 ppm Zn.	May have high base metal values similar to true gossans.	Generally low base metal values (<100 ppm Cu, <200 ppm Ni, <300 ppm Zn), but lateritized ultramafic rocks may have high base metal contents (>1000 ppm Ni, >500 ppm Cu, >500 ppm Cr).
Mineralogical characteristics	Iron oxides, hydrated iron oxides, silica, various silicates. Rare residual sulphides and gangue assemblages may occur. Occasionally oxidate minerals after original sulphides occur. These may include— *Copper*: malachite, azurite, cuprite, tenorite, chrysocolla, dioptase. *Lead*: cerrusite, minium, anglesite, leadhillite. *Zinc*: smithsonite, hydrozincite, hemimorphite, willemite. *Nickel*: green nickel blooms (annabergite, morenosite, gaspeite)	Iron oxides, hydrated iron oxides, silica, various silicates.	Iron oxides, hydrated iron oxides, minor hydrated aluminium oxides, sometimes manganese oxides and occasionally minor amounts of silicates.

sics diminishes with the discovery of near surface ore bodies by the intensive application of these methods, mineral exploration will be forced to devolve more and more on geology. The deeper deposits will be discovered by 'wildcat drill-holes', the best siting of which will depend upon a thorough appreciation and understanding of the geological processes which control the mineralization being sought. Exploration costs will increase and there will have to be a greater degree of interdependence between geology and the indirect application of geophysics and geochemistry.

2.2   TRADITIONAL PROSPECTING METHODS

With the continuing development and refinement of modern geochemical prospecting there is a growing tendency to neglect old traditional methods. In certain environments, however, traditional prospecting can be as effective and, in some cases, more effective than modern methods when applied to prospecting for heavy stable minerals such as cassiterite, columbite, gold, tantalite and wolfram. In the case of diamond prospecting traditional methods have never been supplanted, though they have been refined considerably.

Probably the oldest method of prospecting is the panning of stream sediments for gold. Owing to its very high specific gravity and distinctive colour, gold is readily detected in very small amounts by panning. The tiny specks of gold found in the pan are generally referred to as 'colours' and the prospector assessed the potential of a stream by the number of colours in the pan at each sample site. The method was extended to the panning of soil samples between streams and the potential of quartz veins was assessed by crushing and panning the vein material.

Panning is not difficult, but it requires a certain amount of skill to do it well. It is not easy to describe in words and it is best to learn it from someone who is practised at it, but the basic method is as follows:

1. Fill the pan nearly full and remove large stones by hand and throw away after briefly examining for anything of interest.
2. Fill with water, swirl around with the hand and make sure all material is wet. Further large fragments can be removed by hand at this stage and rejected after examining and feeling for heft.
3. Rock the pan from side to side with a slight rotational movement and without tipping. This gives heavier particles a chance to settle.

4. Continue the motion above and tip slightly away from you. Scrape a thin layer of the lighter material out of the pan with the hand.
5. Continue rocking motion while tipping the pan slightly forward allowing the water to wash the lighter particles out of the pan. It will be necessary to add more water from time to time.
6. Continue above process until only a small quantity of material is left.
7. Carefully pour off the remaining water until there is only sufficient to cover the small amount of material left in the pan. Swirl this water around to wash over the concentrate once and the remaining lighter particles will be swept off the 'heavies' which can be scraped to one side and collected.

Panning should be carried out with a proper prospecting pan. This is a circular thin sheet-steel pan 25–45 cm in diameter and about 10 cm deep with straight sides sloping at about 40° to the base and a slight groove running about half-way round the pan on the inside 2–3 cm below the rim. Aluminium prospecting pans are available, but they are not very satisfactory as they are not as strong as the steel ones.

Prospecting by collecting and panning soil samples is generally referred to as *loaming* and the samples as *loam samples*. The method is widely used in diamond prospecting in which the heavy kimberlite 'indicator minerals' high-Mg ilmenite, pyrope garnet and chrome diopside are sought. To illustrate the general procedure of a loaming survey an example of a tin prospecting programme in south-west Uganda is given below.

Tin prospecting by loaming

Soil samples were taken from depths of 30–50 cm filling a 20 cm × 30 cm calico cloth sample bag so that it could just be tied securely making sure to exclude any rock fragments greater than 2 cm. Each sample, which weighed approximately five kilograms, was carefully panned and the concentrate split into the following size fractions: $+\frac{1}{4}$ in, $-\frac{1}{4}$ in $+\frac{1}{8}$ in, $-\frac{1}{8}$ in $+8$ mesh, -8 mesh $+16$ mesh and -16 mesh. Magnetic impurities were removed with a strong hand magnet and the presence of cassiterite tested by tinning in zinc cups with dilute hydrochloric acid. The cassiterite grains in each size fraction were counted and multiplied by an appropriate weight factor determined experimentally and a total weight of cassiterite in milligrams was obtained for each sample. The results of a survey carried out in the above manner are shown in Fig. 2.3. Instead of giving cassiterite content in terms of weight, results could have been expressed simply

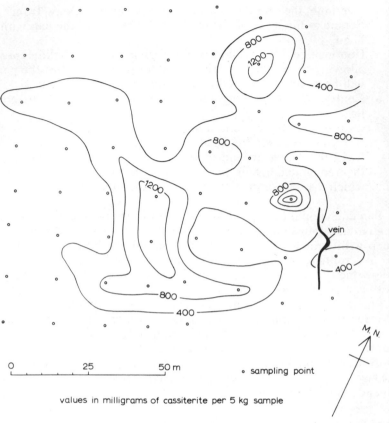

FIG. 2.3. Example of loam sampling for tin from part of a survey at Burama Ridge, Ankole, Uganda.

as a grain count for the different size fractions, which is the type of presentation that is normally used in diamond prospecting.

The area in which the work was carried out was particularly suited to the method owing to the coarse nature of the soil and the concentration of tin in the coarser size fractions (Fig. 2.4). The efficiency of the panners was tested with 10 artificial samples made up by placing varying amounts of different sized cassiterite grains in soils known to be barren. The results (Fig. 2.5) show that the panners' recovery in the coarser size fractions is extremely good and in the 10 test samples, which range from ppm Sn equivalents of 17–1432, the average loss by the panners was only 22% (Table 2.4).

The results of this loaming survey show that cassiterite contents in

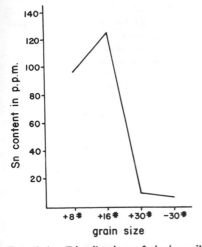

FIG. 2.4. Distribution of tin in soils from Burama Ridge, southwest Uganda (based on seven samples).

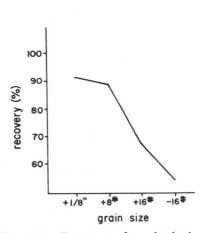

FIG. 2.5. Recovery of cassiterite by panning.

soil equivalent to 10 ppm Sn or less can be detected by panning 5-kg samples. Owing to the high stability of cassiterite and consequent low dispersal of Sn in the secondary environment, the geochemical analysis of the fine fractions is very often completely ineffective in Sn prospecting, particularly when searching for deposits of simple mineralogy such as stanniferous pegmatites and quartz-cassiterite associations. To obtain meaningful results in the area described, for instance, geochemical analysis would have had to be carried out on the −8 + 30 mesh fraction, necessitating grinding of the samples. In regions such as Tasmania or parts of Cornwall where a considerable proportion of the Sn occurs in the finer size fractions, geochemical sampling has definite advantages and may prove to be the only practical method. If the cassiterite predominates in the +30 mesh fraction, however, traditional panning methods may be much more effective than modern geochemical techniques.

Summary of the loaming method
The advantages of loaming surveys are that unskilled and semi-skilled personnel can be used, results are immediately available in the field and equipment costs are low. The main disadvantages are the large size of the samples, particularly if they have to be carried any distance and the relatively low output per man-day in treating the samples (a skilled panner should be able to process up to two samples

TABLE 2.4

ACTUAL CASSITERITE CONTENT IN TEN 5-KG TEST SOIL SAMPLES COMPARED TO CONTENT FOUND BY PANNERS

Actual content		Content found		% Lost
ppm Sn*	Number of grains	ppm Sn*	Number of grains	
17	1@8 mesh 3@16 mesh	10	1@8 mesh	41
61	3@8 mesh 13@16 mesh	29	3@8 mesh	52
83	6@8 mesh 10@16 mesh	56	4@8 mesh 7@16 mesh	32
222	2@1/8 in 11@8 mesh 6@16 mesh	209	2@1/8 in 10@8 mesh 5@16 mesh	6
280	118@16 mesh	154	65@16 mesh	45
461	3@1/8 in 27@8 mesh 19@16 mesh	417	3@1/8 in 24@8 mesh 13@16 mesh	10
789	7@1/8 in 36@8 mesh 37@16 mesh	612	4@1/8 in 33@8 mesh 37@16 mesh	22
816	13@1/8 in 18@8 mesh	796	13@1/8 in 16@8 mesh	2
956	11@1/8 in 36@8 mesh 25@16 mesh	940	11@1/8 in 36@8 mesh 18@16 mesh	2
1432	11@1/8 in 84@8 mesh 24@16 mesh	1328	10@1/8 in 79@8 mesh 22@16 mesh	7

*ppm Sn equivalent

per hour). In countries where labour is cheap, however, these disadvantages are not really a problem. Also in regions with a tradition of small-worker mining it is generally quite easy to find skilled panners.

Loaming surveys are particularly suited to gold and tin prospecting owing to the ease with which gold and cassiterite can be identified in a concentrate. Scheelite is also easy to identify as it fluoresces under short wavelength ultra-violet light. Wolfram is more difficult to identify, but it can be stained a canary yellow by boiling in concentrated nitric acid for 15 min (this also applies to scheelite). Ilmenite grains take on a dull grey or brownish coating if boiled in concentrated hydrochloric acid for about 20 min. Other minerals may be more difficult to identify quickly, but they can often be recognized under a binocular microscope with practice, and chemical spot tests can be used to aid identification. In addition geochemical analyses of the panned concentrates can be undertaken.

REFERENCES AND BIBLIOGRAPHY

Bancroft, J. A. (arranged by Guernsey, T. D.) (1960). Mining in Northern Rhodesia (a chronicle of mineral exploration and mining development). Unpublished Company Report, British South Africa Co.

Bateman, A. M. (1950). *Economic Mineral Deposits*, John Wiley and Sons Inc., New York, 916 pp.

Blain, C. F. and Andrew, R. L. (1977). Sulphide weathering and the evaluation of gossans in mineral exploration, *Minerals Sci. Engng.*, 9(3), 119–150.

Blanchard, R. (1968). Interpretation of leached outcrops, *Nevada Bureau of Mines Bull.*, **66**, Reno, Nevada, 196 pp.

Buhlmann, E., Philpott, E. E., Scott, M. J. and Sanders, R. N. (1974). The status of exploration geochemistry in southern Africa. *Geochemical Exploration* 1974, Elsevier Scientific Publishing Co., Amsterdam, 51–64.

Compton, R. R. (1962). *Manual of Field Geology*, John Wiley and Sons Inc., New York, 378 pp.

Derry, D. R. (1968). Exploration. *C.I.M. Bull.*, **61**, 200–205.

Flinter, B. F. (1955). A brief guide to the identification of the dark opaque and semi-opaque minerals known to occur in Malayan cassiterite concentrate. *Geol. Surv. Malaya*, Note No. 8.

Grip, E. (1953). Tracing of glacial boulders as an aid to ore prospecting in Sweden. *Econ. Geol.*, **48**, 715–725.

Hosking, K. F. G. (1957). The identification—essentially by staining technique—of white and near white mineral grains in composite samples, *The Camborne School of Mines Mag.*, **37**, 5–16.

Hyvarinen, L., Kauranne, K. and Yletyinen V. (1973). Modern boulder tracing in prospecting. In *Prospecting in Areas of Glacial Terrain*, Instn. Min. Metall. Lond., 87–95.

Jones, M. P. and Fleming, M. G. (1965). *Identification of Mineral Grains*. Elsevier Publishing Co., Amsterdam, 102 pp.

Kreiter, V. M. (1968). *Geological Prospecting and Exploration* (Russian translation), Foreign Languages Publishing House, Moscow, 383 pp.

Lamey, C. A. (1966). *Metallic and Industrial Mineral Deposits*, McGraw-Hill Book Co., New York, 567 pp.

Lindgren, W. (1933). *Mineral Deposits*, McGraw-Hill Book Co., New York, 930 pp.

Park, C. F. Jr. (1964). Is geologic field work obsolete? *Econ. Geol.*, **59**, 527–537.

Park, C. F. Jr. and MacDiarmid, R. A. (1975). *Ore Deposits*, W. H. Freeman and Co., San Francisco, 530 pp.

Raguin, E. (1961). *Geologie des Gites Mineraux,* Masson & Cie., Paris, 613 pp

Robinson, W. B., Stock, E. C. and Wright, R. (1973). The discovery and evaluation of the Windarra nickel deposits, Western Australia, paper presented at Western Australia Conference 1973, Aus. I.M.M.

Smith, O. C. (1953). *Identification and Qualitative Chemical Analysis of Minerals,* Van Nostrand, New York, 357 pp.

Zeschke, G. (1970). *Mineral Lagerstatten und Exploration,* Band 1, Ferdinand Enke Verlag, Stuttgart, 351 pp.

CHAPTER 3

Photogeology and Remote Sensing

Remote sensing includes any detecting or mapping techniques carried out from aircraft or spacecraft. Thus, all airborne geophysical methods are included together with aerial photography, imaging systems and air sampling methods. In this chapter, however, airborne geophysical methods will be excluded since they are described in Chapter 6.

3.1 PHOTOGEOLOGY

Photogeology can be defined simply as the interpretation of geology from photographs. As this interpretation is based mainly on aerial photography, most people would restrict the definition of photogeology to the use of aerial photographs. The earliest attempts at photogeological interpretation in the late 1800's, however, were made from land-based photography and even today photographs taken on land may be employed usefully in specialised cases such as the geological interpretation of inaccessible mountain slopes, cliff faces or open pit walls.

Aerial photographs
The first aerial photographs were taken in France in 1858 from a balloon and some years later use was made of balloon photography for intelligence purposes in the American Civil War. Intermittent use of aerial photography was made throughout the late 1800's and early 1900's, but it was not until the First World War that aerial photography became really important. After the First World War, the science of *photogrammetry* revolutionised map making and made it possible to produce accurate surveys of little known and remote areas of the world. Strictly speaking photogrammetry is simply the art of making measurements from photographs, but in general usage it is restricted to the making of maps from aerial photographs.

Today aerial photographs are usually taken from an aircraft flying at altitudes between 800 and 9000 m depending upon the amount of detail required. Photographs may be taken at different angles varying from *vertical* to *low oblique* (excluding horizon) to *high oblique* (including horizon), but for geological purposes vertical photographs are by far the most important. The scale of the photographs depends upon the terrain clearance and focal length of the camera used and is given by the expression:

$$\text{Scale} = \frac{\text{focal length}}{\text{terrain clearance}}$$

For example, a survey flown at an altitude of 5000 m over country with a mean elevation of 1000 m using a camera of focal length 20 cm would give photographs at an average scale of:

$$\frac{0 \cdot 2}{4000} = 1 : 20\ 000$$

It is immediately apparent from this that the scale of the photograph will vary with topographic relief, areas of higher elevation being at a larger scale than lower lying areas. The formula gives the scale of the negative and the scale of the final prints can be increased by making enlargements. A wide range of cameras have been used for aerial photography with negative formats ranging from $5 \cdot 7 \times 5 \cdot 7$ cm ($2 \cdot 25 \times 2 \cdot 25$ in) to $45 \cdot 7 \times 45 \cdot 7$ cm (18×18 in), but for precise survey work large formats are preferred and the standard aerial survey camera used in Britain and North America produces a negative $22 \cdot 5$ cm square (9×9 in.) from which contact prints are made. This is the standard aerial photograph (Fig. 3.1) from which most photogeological interpretation is made. The centre point on the photograph is known as the *principal point* and is located at the intersection point of lines joining opposite *fiducial marks* which are images produced on the edges of each negative by little markers fixed to the camera body.

One of the most important features of aerial photographs is that they can be viewed in pairs to produce a stereoscopic effect. This is due to parallax differences caused by photographing the same object from two different positions. To achieve this the series of photographs taken along a flight line have to overlap each other by at least 50%. If the overlap is less than 50%, there will be a gap on each photograph which cannot be viewed stereoscopically. To ensure complete stereoscopic cover an overlap of 60% is aimed at. If this is achieved, points on the ground under the principal points on each photograph will appear near the edges of adjacent photographs as shown in Fig. 3.2. A line joining all the principal points along a flight line is known as the *base line*. It will be appreciated that the base line

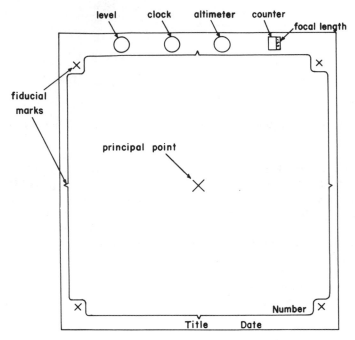

FIG. 3.1.   Standard aerial photograph.

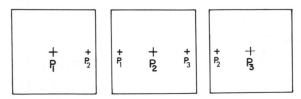

FIG. 3.2.   Three consecutive photographs along a flight line to illustrate how the principal point on a photograph should appear near the edges of adjacent photographs.

is very rarely a straight line due to the aircraft deviating from an absolutely straight flight path. For normal stereoscopic vision adjacent photographs on the base line are viewed with a stereoscope as shown in Fig. 3.3. In addition to overlap along the flight path overlap between adjacent flight lines is also necessary to achieve full photographic coverage; side-lap, as this is known, usually varies between 20 and 30%.

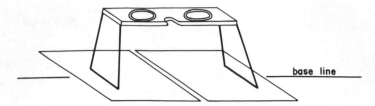

Fig. 3.3. Two photographs set up for correct stereoscopic viewing.

Taking good aerial photographs can be quite difficult as it is necessary to fly very accurately and as smoothly as possible along straight lines at a constant altitude. Weather conditions have to be very good with minimal cloud cover and atmospheric haze. To avoid long shadows the photographs should not be taken too early or too late in the day. In certain instances and particularly in areas of very low relief, however, low sun-angle photography (LSAP) can be extremely useful to bring out subdued features. In addition it is often necessary to adjust the position of the camera in the aircract to ensure that the photographs are parallel to the flight line (Fig. 3.4).

a) Camera position adjusted for true flight path

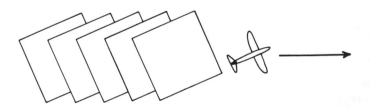

b) Camera position not adjusted for true flight path

Fig. 3.4. Effects of a cross-wind on the survey aircraft. (a) Camera position adjusted for true flight path and (b) camera position not adjusted for true flight path.

Photogrammetry

In many parts of the world one of the most important functions of the aerial photograph is to provide a substitute for a good topographic map. Even where good base maps are available, it is generally advantageous to use aerial photographs for locating oneself accurately on the ground as the photographs generally have more detail than all but the best maps. Photogrammetry has become highly specialized using expensive precision instruments for producing extremely accurate and detailed maps from aerial photographs.

Latest techniques, using a process known as differential rectification, result in the production of an *orthophoto*, which is essentially a topographic map on which the various features are shown photographically. Although the exploration geologist cannot be expected to have a detailed knowledge of photogrammetry, it is important for him to know some of the basic principles in order to produce adequate base maps if required.

The simplest type of 'map' produced from aerial photographs is the uncontrolled *print laydown*. To produce a print laydown a strip of photographs across the central part of the area is selected and alternate photographs laid down in order on a table. The edges of the photographs are trimmed and they are carefully joined together with rubber-base gum in their correct relative positions by matching photographic detail from one photograph to another. The strip can be turned over and the backs of the photographs joined more securely with cellulose tape. Next the alternate photographs of an adjacent strip are trimmed and carefully joined to the first strip in their correct relative positions. It may be found that it is more difficult to match up this strip as these photographs have to coincide not only with each other, but also with the photographs of the first strip. If any discrepancies are found, it is best to ensure that there is a good match up along the outer part of the strip as it will become progressively more difficult to add further strips if this is not done. The print laydown is then built up in this manner from the central strip by adding further strips along either side. It should be noted that it is only necessary to use half the photographs to produce the print laydown if the survey has been flown with 60% overlap. Sometimes there are places where the overlap is poor and in such cases it may be necessary to use some of the other photographs to complete the print laydown. Survey departments commonly make up print laydowns consisting of 30 or more photographs covering rectangular blocks of ground. These are numbered and photographed and single prints are made available for sale to the public. Sometimes a much more elaborate form of print laydown is made by cutting the corners and edges off the photographs, which are then glued down on a board in a mosaic. This is known as a *photomosaic* and can produce a very even print if the photographs are carefully processed to ensure that the tones match up evenly. The production of a photomosaic is a very laborious process, however, and an ordinary print laydown is adequate for most purposes.

Uncontrolled print laydowns may be quite inaccurate and difficult to join together if there is much variation in scale and displacement caused by tilt and topographic relief. It is possible to make corrections for these displacements so that accurate photogrammetric maps can be produced. Figure 3.5 shows how displacements are caused by topographic relief. Points A and B are equidistant in plan from the point vertically below the camera lens, but point B is at a higher elevation than point A. The result in the camera is that the image of point B is further away from the centre of the photograph

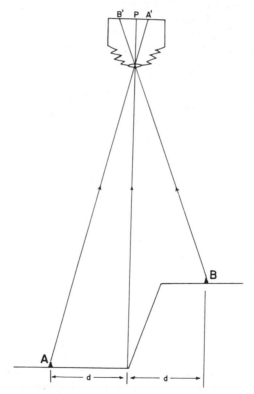

FIG. 3.5. Diagram to show how elevation differences cause displacement of the images in an aerial photograph.

than that of point A. When the camera is level, the image of the point vertically below the camera lens, known as the *plumb point*, coincides with the principal point. Displacements caused by relief are all radial from the plumb point. Figure 3.6 shows the effect of tilt and it can be seen that the plumb point is displaced from the principal point. Displacements caused by tilt are radial from the *isocentre*, a point midway between the plumb point and the principal point. Tilt displacements are much more difficult to correct than relief displacements and require elaborate and expensive equipment. Fortunately, on most aerial photographs tilt is only a minor problem and distances between plumb point, isocentre and principal point are generally very small. For this reason displacement can be considered as being radial from the principal point and quite accurate maps can be prepared by the method known as a *radial line plot*.

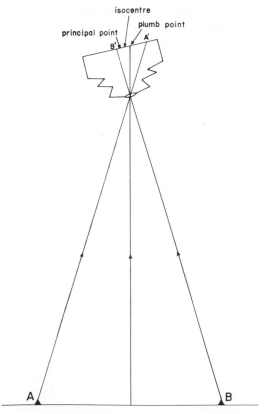

FIG. 3.6.   Diagram to show the basic effects of tilt on an aerial photograph.

To produce a radial line plot the principal point on each photograph is marked accurately with a pin prick and a cross carefully scratched on the photograph with the pin. If a yellow or red grease pencil is rubbed across the pin prick and scratches, they will show up very clearly. Then the adjacent photographs are examined under a stereoscope and the positions of the principal points carefully transferred to the adjacent photographs by pricking with a pin. The next step is to select a number of *pass points* near the edges and corners of the photographs. These are any points with sufficient photographic detail to be identified accurately on the adjacent photographs of the same strip and photographs of adjacent strips. Using a stereoscope these pass points are transferred carefully from one photograph to another within the same strips and onto photographs of the adjacent strips and marked with a pin prick. A short radial line from the principal point on each photograph is scratched through each pass point and a grease pencil rubbed across the scratches so that they show up clearly. The procedure is illustrated in Fig. 3.7 which shows pass points marked on six photographs from two adjacent strips.

A piece of plastic tracing medium is cut for each strip of photographs. Then starting with the central strip of photographs a point is marked on the plastic sheet near one end and a line drawn through it along the centre of the plastic sheet. The first photograph on the central strip is then positioned under the plastic sheet with its principal point coincident with the point already marked on the plastic sheet and the principal point of the adjacent photograph coinciding

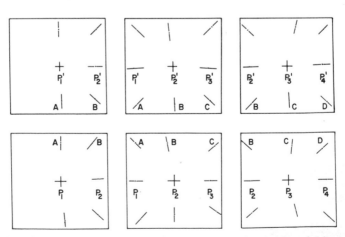

Fig. 3.7. Six photographs from two adjacent strips with principal points ($P_1 - P_4$ and $P_1' - P_4'$) and pass points (A, B, C, D) marked for a radial line plot.

with the line drawn on the sheet. The position of this second principal point is marked on the line and all the radial lines marked on the photograph are traced on the plastic sheet. The photograph is put to one side and the second photograph in the strip is positioned under the plastic sheet with its principal point coincident with the second marked point and the principal point of the previous photograph coinciding with the central line. It will probably be found that this point does not coincide with the position of the first principal point marked on the tracing medium. This does not matter as long as the principal point of the second photograph coincides with its position as marked from the first photograph. The radial lines are traced as in the case of the first photograph and the position of the third principal point marked on the strip. This photograph is put to one side and a line drawn between the second and third principal points. It will probably be found that these three principal points do not lie on a straight line, but this is quite normal. The third photograph is positioned under the sheet with its principal point coincident with the third principal point marked on the tracing medium and the principal point of the previous photograph coinciding with the line drawn between the second and third principal points. The procedure used with the first two photographs is repeated and the whole process is continued with all the photographs of the strip. The photograph numbers used should be written on the strip for identification and further plots on strips of plastic tracing medium are made up in this manner for all strips of photography being used in the compilation. An example of a short strip made up for a radial line plot is shown in Fig. 3.8.

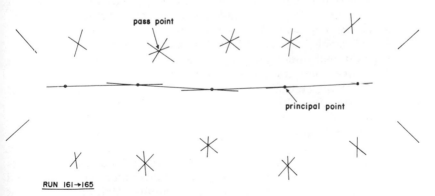

pass point

principal point

RUN 161→165

FIG. 3.8. Example of a radial line plot made up for a strip of five photographs.

At this stage the scale of the compilation is decided upon and adjustments can be made to reduce or enlarge the various strips to the required scale. The simplest procedure, however, is to make a map with the same scale as the photographs and in this case the actual scale of the completed map will be the scale of the first photograph in the central strip. It is only possible to make a map to an accurate scale if two or more survey points appear on the photographs and this is easiest if two points appear on the central strip. To produce a final map to the scale of the photographs it is now necessary to join all the strips which have to be brought to the scale of the central strip. (The scale of each strip is fixed by the first photograph used in each plot and it is generally found that the strips vary slightly in scale.) The procedure adopted is illustrated in Fig. 3.9. The central strip is placed under the middle of a sheet of plastic tracing medium, large enough for the final map and all the principal points and pass points marked. Then two pass points A and B are selected and the distance between them measured accurately. It will be found that the corresponding pass points A' and B' on the adjacent strip are either further apart or closer together than the points A and B on the first strip. All principal points and pass points on the adjacent strip have to be adjusted to the same scale as the central strip. An easy way to do this is to fasten a piece of paper to a drawing board and draw a straight line with the points A and B marked at their correct spacing. The plot of the adjacent strip is placed on the board with A' coinciding with A and fastened with a drawing pin so that the strip can be rotated around the points A and A'. The strip is rotated so that each point on the plot lies on the line AB in turn and corrected points are marked in direct proportion to their distance from A either towards or away from A depending whether A'B' is less or greater than AB. For example, if AB is 50 cm and A'B' 45 cm and a specific pass point 27 cm away from A, B' will be moved 5 cm away from A and the pass point 3 cm (5/45 = 3/27) away from A. When this is completed, the second plot can be positioned under the large sheet of tracing medium and the corrected pass points placed in coincidence with the corresponding pass points of the central strip and all corrected principal points and pass points transferred to the final compilation sheet. This process is continued with successive strips on both sides of the central strip until the compilation is complete. If a scale different to the photographs is used, the central strip has to be adjusted to the correct scale in the manner described above using ground control points appearing in the photographs and then all subsequent strips adjusted to that scale. A simpler procedure is to complete the compilation at the scale of the photographs and then photographically reduce or enlarge the compilation to the scale required.

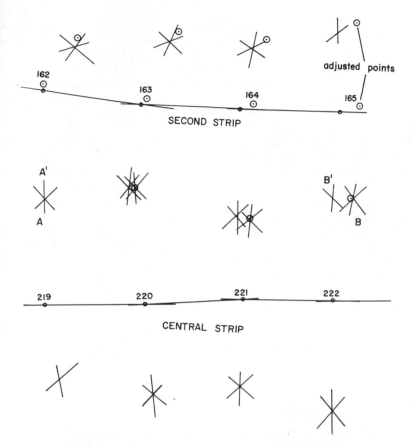

FIG. 3.9.   Method of joining adjacent radial line plots. The pass points A' and B' on the second strip have been adjusted to fit the corresponding pass points A and B on the central strip. All the other pass points and principal points on the second strip have to be adjusted accordingly.

Detail can now be plotted on the final compilation by placing the photographs under the tracing medium in turn and continuously adjusting their positions so that there is a best fit between three adjacent control points on the photographs and on the compilation. If necessary, additional control points can be added accurately by choosing points of detail appearing on two adjacent photographs in a strip, placing them in turn under the compilation sheet so that their principal points coincide with the principal points on the compilation sheet and tracing radial rays from the principal points through the control points. The points of intersection of rays from the adjacent

photographs will fix the control points accurately. If there is a
considerable difference in scale between the compilation and photographs, a plotting instrument known as a *sketchmaster* can be used.
Figure 3.10 shows the basic principle of a vertical sketchmaster. The
instrument allows the photograph being used to be viewed at the same
time as the compilation map. By raising or lowering the height of the
photograph above the table level and changing the viewing lens if
necessary the photograph image can be adjusted to match the scale of
the compilation. It is then a simple matter to trace detail onto the final
map. Detail can be transferred more accurately by a more expensive
but still relatively simple instrument known as a *radial line plotter*. The
production of a radial line plot is a laborious process and it is not
practical to make a compilation of more than six strips of ten
photographs each by the above procedure.

If required in the compilation, it is possible to measure elevation
differences from the photographs with a simple instrument known as
a *parallax bar*. This consists of two horizontal glass plates with a
small dot engraved in the centre of each at either end of an adjustable
bar. The separation between the plates can be altered by a coarse
adjustment screw and a fine adjustment screw operating a micrometer
which can be read accurately to a hundredth of a millimetre. To
operate the parallax bar, set the micrometer to mid-point and place it
on two photographs being viewed with a stereoscope so that the bar is

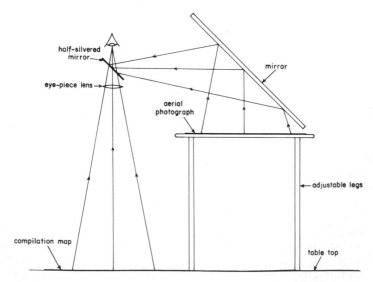

FIG. 3.10.   Basic principle of the vertical sketchmaster.

parallel to the photographic base line. With the coarse adjustment alter the separation of the plates until the dots lie over a point of known elevation. When viewed through the stereoscope, the dots should fuse and appear to float above the stereomodel. With the fine adjustment alter the plate separation and the dot will appear to float up or down depending on the direction of adjustment. Carefully float the dot up or down using the fine adjustment until the dot rests on the point of known elevation. Read the micrometer. Now float the dot up or down with the fine adjustment until it rests on the point whose elevation is to be determined and read the micrometer again. The height difference Δh between the two points is given by the relation:

$$\Delta h = \frac{(H - h)\Delta p}{W}$$

where H is the altitude of the photography, h the approximate mean elevation of the terrain covered by the photographs, Δp the parallax difference and W the mean difference in millimetres between the principal point and transferred principal point of the adjacent photograph measured on both photographs. This formula is adequate if elevation differences are small, but it becomes inaccurate with large elevation differences. A more accurate determination can be made if there are two control points with known elevation by calculating a correction factor C which is equal to $\Delta h/\Delta p$ where Δh is the elevation difference and Δp the parallax difference. For example, if two points at elevations 520 m and 710 m respectively produce a parallax difference of 5·15 mm, the correction factor C becomes 36·9 m/mm. Thus, if two points of unknown elevation produce parallax differences of 3·21 and −1·85 (negative because the micrometer reading was less than the reading for the point of elevation 520 m), the correct elevations become:

$$520 + 36·9(3·21) = 638 \text{ m}$$
$$520 - 36·9(1·85) = 452 \text{ m}$$

This method is not absolutely accurate as the correction factor C will vary with topography. For most accurate work it is necessary to determine C for a number of different elevation ranges.

The difference in parallax is due to displacement caused by variations in topographic relief. For example, if a hill appears on an aerial photograph, the image of its top point will have a greater radial displacement from the principal point than the image of a point near its base. On two consecutive photographs the two points should be the same distance apart, but because the higher point has a greater radial displacement relative to the lower point, the distance between the higher points on the two photographs will be greater than the distance between the lower points. This difference in separation measured parallel to the base line is known as the *absolute parallax* of the two points. Points of higher elevation will produce a greater absolute parallax than points of lower elevation and it is the absolute parallax we measure with the parallax bar.

Geological interpretation

In areas of good exposure a considerable amount of geological information can be obtained from aerial photographs. Different rock types may produce strong tonal and/or textural contrasts because of differences in erosion levels and patterns, colour and reflectivity, differences in vegetation cover, variations in depth of weathering and amount of soil cover and structure. Even in areas of poor exposure the underlying geology is often reflected in vegetation patterns and may show up very clearly on the aerial photographs. Photogeological interpretation can be made from print laydowns, but it is always best to view the photographs in stereo pairs. This is best done by placing a piece of clear tracing medium over one of the photographs and fastening it down along one edge with tape so that it can be lifted up. Annotations can then be made on the tracing medium while viewing the photographs with a stereoscope. The standard photogeological symbols that are commonly used are shown in Fig. 3.11.

Both *relief* and *tone* are important properties in photogeological interpretation, though neither are absolute quantities. Relief depends very much on the relative resistance of rocks to erosion and the amount of erosion that has taken place. Variations in erosion level may stand out clearly when the photographs are viewed stereoscopically since even small differences in topographic relief appear greatly exaggerated on the stereomodel. Tone is important as subtle variations may show up different rock types. However, tone is very variable as it is affected by light conditions, the season of the year, the time of day, amount of haze in the atmosphere, effects of processing, etc. Although tonal contrasts are more important than absolute tones, it can be stated as a general rule that acid igneous rocks have lighter tones than basic ones and sediments such as quartzites, limestones and sandstones have light tones while mudstones and shales have darker tones.

Lineaments may be defined as any alignment of features on an aerial photograph. The various types recognized include topographic, drainage, vegetation and colour alignments and it is generally useful to distinguish the different types in the initial interpretation. It is extremely important to mark only the natural ones and to exclude artificial alignments such as cut lines, tracks, old field boundaries,

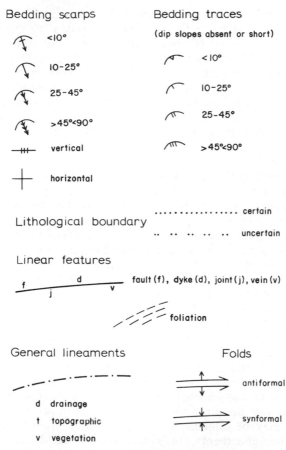

FIG. 3.11. Standard photogeological symbols.

animal paths, etc. This is generally fairly easy to do, but it is sometimes difficult to differentiate them. For example, old burn patterns caused by bush fires travelling with the prevailing wind may be confused with natural vegetation alignments.

Bedding is shown by lineaments which are generally few in number and occur in parallel groups. If particular beds are more resistant than the rocks on either side, they will show up as clear topographic lineaments. One feature of bedding lineaments is that, even if they are broken by streams, they are generally persistent and can be traced through the disruptive features. If a bed forms a ridge, it will usually have a dip slope from which the dip can be estimated. It must always be remembered, however, that dips will appear much steeper on the stereomodel than they really are due to vertical exaggeration. For example, dips of 50° or 60° may appear almost vertical and shallow dips of 15° to 20° will seem to be 45°. With practice dips can be estimated reliably in ranges of <10°, 10–25°, 25–45° and >45° < 90°. If there are no dip slopes, it may still be possible to estimate dips from bedding traces. Vertical beds cross streams and hills without any deviation, but dipping beds will show bedding traces with V-shaped deviations as they cross valleys. The pointed end of the V always indicates the direction of dip and the sharper the angle of the V, the flatter the dip.

Foliation may be indicated by lineaments and it can be generally distinguished from bedding by the fact that the parallel lineaments reflecting foliation tend to be very numerous and impersistent. Bedding lineaments are much more common and more easily recognized, however, as lineaments due to foliation are often very subdued and indistinct. In fact, it is worth remembering that in areas where bedding is not discernible in the field and only foliation can be observed, the bedding can often be recognized on an aerial photograph which takes in a large area of view and shows long subtle, topographic lineaments which often characterize bedding traces and cannot readily be seen on the ground.

Straight lineaments which appear as slight negative features generally represent faults or joints. If there is no evidence of displacement, they should normally be recorded as joints. An exception to this rule would be a major lineament which can be traced for some length. Even in the absence of evidence of displacement, such a lineament almost certainly represents a fault. Joints or faults parallel to bedding may be difficult to recognize, but, when they cut across clear bedding lineaments, they are usually unmistakable. Jointing patterns may also assist in the identification of certain rock types. For example, the characteristic jointing in granites often shows up through dense vegetation cover.

Folds are easily recognized on aerial photographs and it is often possible to plot the positions of the axial traces and to make estimates of the amount of plunge. Steeply plunging folds have well rounded noses and the bedding can be traced around in a continuous curve, but gently plunging folds will show up as two bedding lineaments meeting at an acute angle (the nose) to form a single lineament. Structurally complex areas with more than one phase of folding may also be readily discernible on aerial photographs.

Veins and dykes produce lineaments very similar to joints and faults and may be indistinguishable from them. If the veins or dykes are wide enough, however, the vein material or dyke rock may produce a tonal or relief contrast with the country rock and they are then quite distinctive. Quartz veins and acid dykes often produce light-coloured lineaments and basic dykes dark ones, but, as the relative tone depends very much on the nature of the country rock, positive identifications cannot be made purely from the photographs. There are cases on record of 'definite dolerite dykes' which turned out to be large quartz veins in the field! Pegmatites and amphibolite bands are similar to veins and dykes in appearance, but once again positive identifications can only be made after carrying out checks in the field.

In the absence of marked relief changes and distinctive topographic lineaments different rock types may be shown by changes in vegetation patterns. For example, soils underlain by calcareous rocks in the tropics often support a much thicker and greener vegetation than soils underlain by other rock types and this shows up very clearly on the photographs. The poor, shallow soils which often develop over shales support a thinner vegetation than soils over other rock types with dwarfed trees which are often quite conspicuous on the photographs. In central Africa the large mounds formed by termites are readily apparent in areas of thinner vegetation which results in a characteristic 'peppercorn' texture over areas underlain by shales. The pale, brown, sandy soils which develop over granites in the tropics vary in thickness and trees tend to grow in clusters giving a characteristic spotted or mottled texture. There are numerous other examples which vary according to the rock type and climatic environment. In addition to vegetation changes which may reflect major lithological differences, minor variations in lithology across the bedding are often reflected by subtle vegetation variations which show up as vegetation alignments paralleling the bedding.

It must always be remembered that photogeology is an aid to field mapping and not a substitute for it. Regional geological structures are often much clearer on photographs, which take in a wide synoptic view, than they are on the ground where only detailed features can be seen, but identification of rock types cannot reliably be made from photographs alone. Broad classifications can be made and lithologies on a photogeological map, for example, should be designated as: intrusive A, sedimentary unit B, metamorphic unit C, etc. It is acceptable to add: probably granite, probably limestone, probably mica schist, etc. depending upon the confidence of the observer, but definite rock types should only be given after verification in the field.

It is generally the aim in photogeology to determine the solid geology, but occasionally it is useful to map transported overburden. This is particularly important in areas where geochemical prospecting is being undertaken and it is valuable to be able to distinguish areas with residual superficial cover from areas with transported cover such as glacial drift or alluvium. This is generally very easy to do from aerial photographs as transported overburden obliterates the underlying geology, it is associated with diagnostic features (drumlins, moraines, river terraces, sand dunes, etc.) related to the mode of transport and it usually has sharp boundaries.

In addition to providing general geological information aerial photographs may contain clues directly related to mineralization. The recognition of veins and pegmatites has already been mentioned. Another important feature is the zone of inhibited vegetation growth which often occurs over shallow ore bodies where the ground is 'poisoned' by the anomalous metal content. Good examples of this are the so-called 'copper clearings' in Zambia associated with copper deposits. These are often difficult to distinguish from other natural clearings, but in areas of virgin bush the investigation of all unusual clearings recognized from the careful examination of aerial photographs has proved successful in a number of cases. J. W. Norman (1969) has given a useful summary of features related to different types of targets (Table 3.1). Some examples of photogeological interpretation are shown in Figs. 3.12–3.17.

Colour photographs
Since the eye can distinguish many more subtle changes in colour than it can in grey tones, colour photographs would appear to offer a distinct advantage over black and white ones in photogeological interpretation. To some extent this is true, but there are a number of disadvantages which have prevented the widespread use of colour aerial photography. One obvious reason is cost since colour photographs are several times more expensive than black and white ones. Another factor is the difficulty in reproducing slight variations in shade consistently in the processing of prints. There is also a considerable attenuation of colour in the atmosphere, with the blue end of the spectrum suffering greater loss than the red end. At the altitudes at which aerial photographic surveys are usually flown (4000–6000 m) the colour differentiation on ordinary colour photographs is greatly reduced. This is not true for the infra-red end of the spectrum, however, and false colour (usually shades of blue and red) infra-red photography has proved extremely useful in photogeological work. Lineaments, variations in water content and vegetation changes which are not readily apparent on ordinary black and white photographs are often very clear on these false colour photographs.

In photogeological interpretation colour photographs are more difficult to annotate than black and white ones and, although more detail is recorded by the colour variations, this can be a distraction

TABLE 3.1

SUMMARY OF DIFFERENT FEATURES ON AERIAL PHOTOGRAPHS
RELATED TO DIFFERENT TARGETS IN MINERAL EXPLORATION*

Type of target	Some air photograph characteristics
Diamond pipe	Circular depression. Stunted or reduced vegetation. Drainage towards centre with a main escape stream.
Sulphides	Vegetation inhibited over suitable host rock or contact. Gossan colours in bare soil.
Veins, including pegmatites	Linear expression—usually obvious when discordant with strike in stratified rocks. If thinner than theoretically detectable at photograph scale, they may still be detected by a linear interruption of relief (especially fault-filled veins), by a local scattering of obvious minerals (e.g. white quartz), by the interruption of drainage, and local changes in the wallrock.
Steam	Fracture traces in vicinity of current or recent igneous activity.
Bedded deposits	Traced by position in strata sequence. May be detected by colour (e.g. iron ores) or vegetation changes (e.g. phosphatic ores) or be related to a more easily detected horizon above or below.
Placer deposits	Abandoned channels shown by old meander patterns, the dark tones of organic silts, and a contrasting vegetation pattern. Barred rivers—a sudden linear change of tone that is light in colour over the impounded alluvium. Riverine plains—micro-relief related to proportions of coarse granular materials and fine silt in column.
Contact deposits	Possible environments mapped by tracing igneous intrusions cutting suitable host rocks (e.g. limestone).
Bauxite	Shown as old erosional surfaces with dense drainage pattern

*From J. W. Norman (1969).

unless the extra data recorded are related to geology. In mineral exploration, however, colour photography may be extremely valuable. Alteration zones and weathering effects often produce colour anomalies which are readily apparent on colour photographs. Probably the most spectacular examples of this are associated with some of the porphyry copper deposits in Arizona in the southwest United

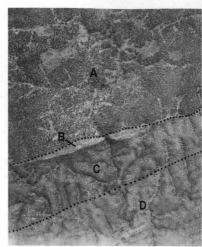

FIG. 3.12. An area in central Zambia. Four lithologies can be recognized by the distinctive tones and textures due to differences in vegetation and type of topography. The regional strike is also apparent from the clear 'grain'. Feldspathic quartzites and meta-arkoses (A) have a more subdued topography and support larger trees than the mica schists (C) and metavolcanics (D), which have thinner vegetation and a more rugged dissected topography. The quartzite (B) stands out very clearly as a major topographic lineament with a much lighter tone. The dip appears almost vertical owing to the exaggeration of the stereomodel, but it is approximately 70°N. Note the 'V' in the outcrop caused by the stream cutting through the dipping bed with the pointed end of the 'V' indicating the direction of dip.

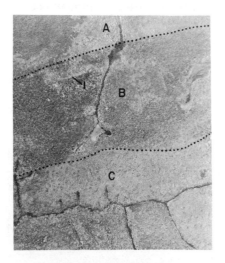

 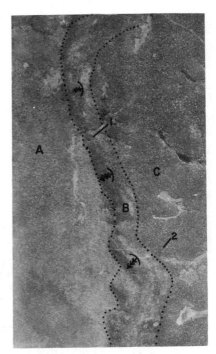

FIG. 3.14. An area in northwestern Zambia. Three distinct lithologies can be recognized: quartzites (B) dip off basement gneisses (C) which support vegetation with a darker tone than the vegetation over a sequence of schists and carbonate rocks (A). The quartzites form very distinct dip scarps with dips of 30–60°. Note the 'V' pointing in the direction of dip at 1 where the streams cut through the quartzite ridge, possibly following fault lines although no offset can be seen. A cut line can be seen at 2.

FIG. 3.13. An area in northwestern Zambia. This is a good example of different lithologies causing recognizable and distinct vegetation changes. Carbonate rocks (B) support a thicker, more luxuriant and greener vegetation which has a darker tone than the vegetation over pelitic rocks (A and C). With the exception of the prominent outcrop ridge at 1 which shows a dip of approximately 60°N, there are no other distinct lineaments on this stereo pair to indicate strike and dip. Numerous termite mounds, typical of this part of Africa, can also be seen.

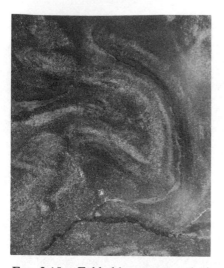 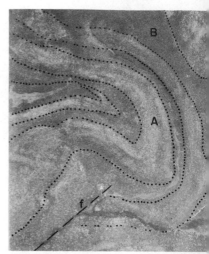

FIG. 3.15. Folded basement rocks in central Zambia. Differential weathering of a sequence of quartzites (A) and muscovite–kyanite schists (B) has resulted in the more resistant quartzites forming prominent, sinuous ridges which are very distinct on the photograph and stand out clearly on the stereomodel.

 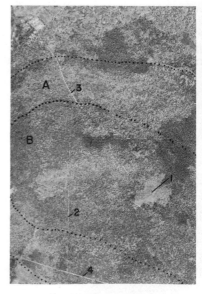

 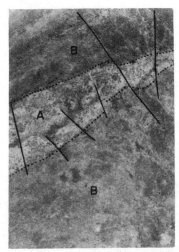

FIG. 3.17.   An area of basement rocks in eastern Zambia. The light-toned area (A) is underlain by granite gneiss, while the darker-toned areas (B) are underlain by quartz–mica schists. Faulting is marked by strong lineaments and some offsetting of the contacts. Dip scarps in the schists are due to more resistant quartz-rich layers.

States. Weathering of both the mineralized and surrounding alteration zones under the arid conditions there has produced very clear, large reddish areas around the ore bodies.

Multi-band photography

A certain amount of work has been done with multi-band photography where separate photographs are taken simultaneously through a number of lenses each fitted with different narrow band filters. The reflectance of different types of soils, rocks and vegetation often

FIG. 3.16.   An area just to the northwest of an important copper deposit in northwestern Zambia. Schists and graphitic phyllites (A) form slight topographic relief which is quite clear on the stereomodel. Carbonate rocks (B) support slightly darker and thicker vegetation. The clearing at 1 is a natural copper clearing due to a very high copper content (thousands of ppm) in the soils. Rows of prospecting pits put down to follow up the geochemical anomaly in the clearing can be seen as faint E–W lines. The cut line (2) marks a licence boundary and roads can be seen at 3 and 4. The lighter ground with slightly higher relief immediately to the northwest of the copper clearing indicates the presence of schists and phyllites.

varies slightly with different spectral bands and the study and comparison of the various spectral band photographs offers a new dimension in photogeological interpretation. However, differences are generally too small to be readily discerned by eye and machine-assisted techniques are required to emphasize the differences. Since it is rather cumbersome and expensive, multi-band photography is not widely used.

3.2   SIDE-LOOKING AIRBORNE RADAR (SLAR)

This powerful remote imaging system was originally developed for military purposes, but it has been available for civilian use for some time now. A fan-like radar beam is projected down and sideways from an aircraft and the reflected signals used to produce an image whose density is related to variations in terrain. Although the image is produced as a series of scans on a radar screen, instruments are used which enable these images to be recorded photochemically on photographic paper. The net result is a rather stark photograph not unlike a conventional aerial photograph (Fig. 3.18). The swath covered by typical SLAR imagery varies from 2 to 50 km and, although the scanning is oblique, the instrumentation converts it into an image which is essentially planimetric. The image is scanned to the right or left of the flight line so there is a gap in the imagery immediately below the aircraft.

Although SLAR images superficially resemble aerial photographs, there are some major differences. Vegetation variations do produce slightly different radar responses, but a SLAR image essentially shows the ground as it might appear on a conventional photograph stripped of vegetation. Displacements caused by relief are to the side towards the imaging aircraft and are not radial about the centre as in the case of an aerial photograph. In addition all radar shadows fall away from and are normal to the flight line. These shadows are complete black areas with no information on SLAR images, whereas most areas of shadow on aerial photographs are partially illuminated by diffused lighting. Subtle changes in colour and texture, which are readily apparent on aerial photographs, are not discernible on SLAR images.

SLAR produces excellent images which can be taken at any time of the day or night and are unaffected by cloud cover. In a relatively short time a single aircraft can provide good coverage of uniform images over wide areas that would require thousands of conventional aerial photographs. Although SLAR imagery can be used for geological interpretation, the information that can be obtained is much

FIG. 3.18. An example of SLAR imagery clearly showing San Cristobal volcano, field boundaries, roads and a small town in Nicaragua. Taken through cloud cover from 6,000 m. The direction of imaging (top to bottom) can be seen from the highlighted slopes which face the aircraft. (Courtesy of Hunting Geology and Geophysics Ltd.)

more limited than that obtained from conventional aerial photography. For this reason the most important use of SLAR imagery is to
provide good base maps in areas where persistent cloud cover is a
major problem.

Since SLAR instrumentation is extremely complex, the equipment
is very costly and it is not as widely used as it might be. However, its
value has been reduced by satellite imagery which has provided a
virtually complete catalogue of cloud-free photographs for almost all
parts of the world. Nevertheless, SLAR imagery has proved to be and
continues to be extremely useful for providing good base maps for
areas of the world such as Central America and Papua New Guinea.

3.3 SATELLITE PHOTOGRAPHS AND IMAGERY

Ever since the first high altitude photographs of the earth were
brought back from rockets fired into and above the upper atmosphere,
it has been realized that apart from its obvious military value space
photography would offer a new and exciting dimension in remote
sensing with applications in cartography, agriculture, meteorology,
forestry and geology. This was emphasized by the spectacular colour
photographs that were brought back by the United States' Gemini
space missions in the 1960's. With the advances in electronics and
communications technology attendant on the United States' space
programme of the 1960's, it became possible for orbiting satellites to
monitor the earth by continuously transmitting remarkably clear
electronically scanned images. The first civilian use of this technical
achievement was one of the Nimbus series of weather satellites
launched in 1966. In July 1972 NASA launched a satellite known as
the Earth Resources Technology Satellite (ERTS-1). This was followed by ERTS-2 in 1973. In 1975 NASA changed the names to
LANDSAT-1 and LANDSAT-2. The satellites were designed to provide virtually complete 'photographic' coverage of the entire world.
They circle the earth in a sun synchronous polar orbit at an altitude
of more than 900 km so that they cover the same ground on an 18-day
cycle. The images are scanned by a multi-spectral scanner (MSS) on
four spectral bands, visible green (5000–6000 Å), visible red (6000–
7000 Å) and two invisible infra-red bands (7000–8000 Å and 8000–
11 000 Å). The images are reproduced on photographic paper and are
available for the four spectral bands plus two false colour composites.
Band 6 (7000–8000 Å) is probably best for geological purposes. Each
standard image covers an area of 32 500 km^2 at a scale of approximately 1/1 000 000. The photographs, which are available for

almost all parts of the world below a latitude of 80°, can be ordered from the Eros Data Center, Sioux Falls, South Dakota, USA. In addition to the standard 1/1 000 000 scale photographs, transparencies (positive and negative) and enlargements at scales of 1/500 000 and 1/250 000 are available. False colour composite images can also be obtained. These often show up features not readily apparent on the ordinary black and white images. Viljoen et al (1975) give some superb examples of LANDSAT false colour imagery. LANDSAT-1 images, which have a high sun angle, are preferable for most geological purposes, but low sun angle LANDSAT-2 images can be useful to aid interpretation in structurally complex areas.

LANDSAT images may be interpreted by standard photogeological techniques, though the images do not come in stereo pairs. A pseudo-stereoscopic effect may be obtained by viewing two different spectral bands (band-lap stereo) of the same image or by examining the images of the same view taken at different times (time-lap stereo). In addition there is a certain amount of side-lap which improves with latitude. This gives a true stereo image over a narrow strip of the photograph, but only the largest topographic features produce appreciable effects. Although object resolution is generally of the order of 90–100 m, it may be slightly better if there is a strong contrast. These satellite images are ideal for outlining regional geological features and, by the same token that features not readily apparent on the ground are often very clear on aerial photographs, the satellite images, which take in a much wider view, often outline geological features not readily apparent on aerial photographs. Figure 3.19 shows a LANDSAT image of part of northern Zambia and it is remarkable how clearly it shows the regional geology (Fig. 3.20).

A lot of work has been done with LANDSAT imagery and one of the main avenues of approach has been the use of the different spectral bands to enhance features of interest. With computer-assisted techniques small contrasts in spectral reflectance may be enhanced and compared. In mineral exploration this technique has been used to try and outline areas of hydrothermal alteration associated with porphyry copper deposits (Schmidt, 1976), but only limited success can be claimed for it so far.

Satellite imagery is unlikely to play a direct role in mineral discovery as mineral deposits are such small targets in relation to the areas scanned. As aids in geological mapping and in defining favourable areas for mineralization, however, satellite images are extremely valuable. Developments in the field are so rapid that with improvements that can be expected in satellite imagery in the near future much more useful data will soon be available when new series of satellites are launched.

FIG. 3.19. LANDSAT–1 image of an area in northern Zambia shown in Fig. 3.20. Various beds and structures in the Plateau Series sediments are very clear on the satellite image.

3.4 THERMAL IMAGERY

This is a technique that was first developed for military purposes to enable observers to 'see' in the dark using infra-red wavelengths in the range 8–13 μm. A rotating mirror assembly systematically scans the ground and reflects infra-red rays to an infra-red sensor. The output from the sensor is amplified and printed as a variable density image very similar in appearance to an ordinary aerial photograph. Aircraft fitted with a thermal imaging device can fly a survey to produce coverage in the same manner as for a conventional photographic survey. Since thermal emission varies greatly with the time of day or night, however, surveys are normally flown just before dawn.

Geological features not apparent on ordinary photographs often show up quite clearly on thermal imagery (Fig. 3.21). The thermal

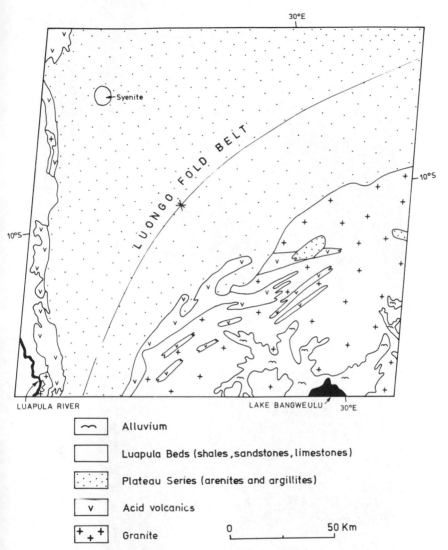

Alluvium

Luapula Beds (shales, sandstones, limestones)

Plateau Series (arenites and argillites)

Acid volcanics

Granite

0   50 Km

FIG. 3.20. Geology of part of northern Zambia (after Geological Survey of Zambia).

capacity of the ground is strongly affected by moisture content and features such as fault zones, for example, can often be recognized on thermal images by virtue of the moisture contents being higher than the surrounding area. Any areas of abnormal heat flow may also show

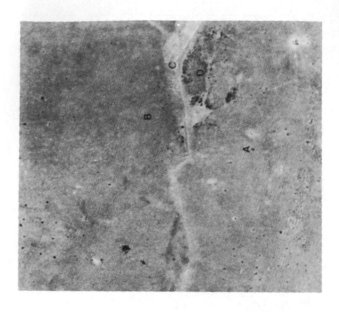

FIG. 3.21.   Pre-dawn thermal infra-red scanner image (left) compared to a conventional aerial photograph (right) of the same area. The geology consists of poorly exposed, flat-lying dolomites in the Transvaal and it is impressive how the thermal imagery shows up the jointing pattern (A) and small faults (B), most of which are not visible on the aerial photograph. (Courtesy of Spectral Africa Pty. Ltd., Randfontein, South Africa.)

up and can be useful in searching for potential targets for geothermal energy. There is often an increased heat flow due to ore bodies and some workers have suggested that thermal imagery might be useful for outlining potential mineralized areas. However, the heat flow is generally too small to be detectable at the surface and can only be measured below ground level.

A new development in thermal imagery is the planned launching of a satellite by NASA in 1979 to be known as the Heat Capacity Mapping Mission (HCMM). This satellite will provide thermal imagery of the earth in the same manner as the more conventional imagery provided by LANDSAT. These images will no doubt be a useful complement to LANDSAT imagery as an aid to mineral exploration.

3.5 OTHER REMOTE DETECTION METHODS

In the last ten years considerable attention has been paid to the detection of vapour in the atmosphere from aircraft and spacecraft by detecting absorption spectra in reflected solar radiation. The biggest use so far is probably monitoring pollution over and in the vicinity of industrial areas by measuring the amount of NO_2 and SO_2 in the air. The method may have some value in prospecting as SO_2 and other gases may be associated with certain sulphide deposits. The most useful vapour from the prospecting point of view is probably mercury as it is known that mercury vapour is emitted from many mineral deposits. The absorption wavelengths for mercury, however, are heavily absorbed by the ozone layer in the upper atmosphere and very little solar radiation in this waveband reaches the earth's surface. This may be unfortunate from a prospecting point of view, but it is hardly unfortunate for other reasons as this radiation in the ultraviolet part of the spectrum is highly dangerous to most living things.

A common gas with an absorption spectrum in the visible region between 5000 and 6000 Å is iodine vapour. Barringer Research in Canada has devised instrumentation for the detection of iodine vapour in the atmosphere (Barringer, 1969) and tests have been carried out to see if it might have some application in prospecting. It is known that chloride brines are associated with some forms of ore deposition and it is highly likely that other halogens will also be present. Iodine is also associated with oil-field brines.

The basic instrumentation developed is known as *correlation spectrometry*. Solar radiation is passed through a spectrometer from a telescope looking down on the earth and is made to vibrate across the diffraction grating by an oscillating refractor plate. The desired wavelengths selected by the diffraction grating are passed on to an optical correlation mask placed at the usual exit slit of the spectrometer. This mask is a photographic replica of the iodine spectrum. If iodine vapour is present in the light-path, then the absorption spectrum will correlate with the optical mask. As this spectrum vibrates across the mask, a beat will be produced which can be detected on a photomultiplier tube. The principle of the correlation spectrometer can be applied to many other gases including NO_2 and SO_2. To date the method has shown little promise in practical mineral exploration. The remote detection of iodine shows up concentrations of kelp seaweed along coastlines and pollution control by the monitoring of SO_2 and NO_2 has already been mentioned.

More recent developments include the extension of spectral correlation techniques into the infra-red region of the spectrum and the development by the US Geological Survey of a technique for measuring solar stimulated luminescence of materials on the earth's surface. Another interesting technique, first developed by R. J. P. Lyon (1965), measures the infra-red emission spectra of rocks and soils. Present instrumentation can roughly differentiate broad classes of silicate rocks from an aircraft with a reasonable degree of reliability. Development work in these techniques is continuing and, although they are of dubious value in mineral exploration at the present time, it is possible that further research may result in remote detection methods which could make contributions to mineral discovery.

3.6   AIR SAMPLING METHODS

Minute mineral particles from rocks and soil are carried up into the atmosphere by air currents. In addition metallo-organic compounds are dispersed by humus and vegetation and are present in the atmosphere in minute amounts. Above and in the vicinity of mineral deposits there can be expected to be a greater concentration in the atmosphere of particles with a metal content related to the mineral deposit. Weiss (1971) has described a technique patented by him for collecting these aerosol particles from an aircraft and subsequently analyzing them for their trace element content. The sample collector consists of an aluminium frame 30 cm square around which are wound 450 turns of 0·1 mm nylon thread. This frame is towed on a rope behind the aircraft with a ground clearance of 60–90 m. After a short exposure, the sample collector is hauled back on board, placed in a plastic bag and another sample collector passed out from the aircraft. The nylon threads are removed in the laboratory, ashed and analyzed spectrographically. More detailed analyses can be carried out by drawing the nylon threads through a small nylon micro-web pad. It is possible to analyze the particles concentrated on the pad by an electron probe. Test flights have been flown over a number of areas with positive results. It is claimed that mineralized outcrops can be located within an area of 2–3 km^2 in survey areas of many tens of thousands of square kilometres.

Another air sampling system has been developed by Barringer Research in Canada and is known under the trade name, AIRTRACE®. Details of the equipment have not been published, but is is far less cumbersome than Weiss's method and positive results are claimed for it. Air is drawn into the AIRTRACE® sampler on board an aircraft flying at an altitude of 60 m; the metals detected directly include Cu, Pb, Zn, Ni and Ag in addition to Hg vapour. The AIRTRACE® system placed on board a survey aircraft carrying geophysical equipment can provide additional information which may be useful in assessing the geophysical anomalies. For instance, an EM conductor associated with trace elements in the atmosphere might rate a higher priority than one that showed little or no response on AIRTRACE®.

Although more research is being undertaken in air sampling methods and improvements can be expected, it is highly unlikely that they will ever supplant more conventional work on the ground. As ore bodies become more difficult to detect and discover, it is hard to conceive how a method which at best locates targets to areas of several square kilometres can play a significant role in future discovery.

REFERENCES AND BIBLIOGRAPHY

Allum, J. A. E. (1966). *Photogeology and Regional Mapping*, Pergamon Press Ltd., Oxford, 107 pp.

Allum, J. A. E. (1970). Consideration of the relative values of true and infrared colour aerial photography for geological purposes, *Trans. Instn. Min. Metall.*, Lond., 79, B76–87.

American Society of Photogrammetry. (1960). *Manual of Photographic Interpretation*, Washington D.C., 868 pp.

Arnold, C. R., Rolls, P. J. and Stewart, J. C. J. (1971). *Applied Photography*, The Focal Press, London and New York, 510 pp.

Barringer, A. R. (1969). Remote sensing techniques for mineral discovery, Paper 20, 9th CMM Conference, London.

Eardley, A. J. (1942). *Aerial Photographs: Their Use and Interpretation*, Harper and Brothers Publishers, New York, 203 pp.

Greenwood, J. E. G. W. (1965). Air photographs in economic mineral exploration, *Geol. Surv. Canada*, Paper 65–6, 43–49.

Gregory, A. F. and Moore, H. D. (1975). The role of remote sensing in mineral exploration with special reference to ERTS-1, *C.I.M. Bull.*, May 1–6.

Kilford, W. (1973). *Elementary Air Survey*. Pitman and Sons, London, 363 pp.

Lueder, D. R. (1959). *Aerial Photographic Interpretation*, McGraw-Hill Book Co., New York, 462 pp.

Lyon, R. J. P. (1965). Analysis of rocks by spectral infrared emission (8–25 microns), *Econ. Geol.*, **60**, 715–736.

Lyon, R. J. P. and Lee, K. (1970). Remote sensing in exploration for mineral deposits, *Econ. Geol.*, **65**, 785–800.

Miller, V. C. and Miller, C. F. (1961). *Photogeology*, McGraw-Hill Book Co., New York, 248 pp.

Moffit, F. H. (1964). *Photogrammetry*, Instructional Textbook Co., Scranton, Pa., USA, 455 pp.

Norman, J. W. (1969). The role of photogeology in mineral exploration, *Trans. Instn. Min. Metall.*, Lond., **78**, B101–107.

Reeves, R. G. (ed.) (1968). *Introduction to Electromagnetic Remote Sensing*, American Geol. Institute, Washington.

Schmidt, R. G. (1976). Exploration for porphyry copper deposits in Pakistan using digital processing of LANDSAT-1 data, *Jour. Research U.S. Geol. Surv.*, **4**, 27–34.

Smith, J. T. (ed.) (1968). *Manual of Color Aerial Photography*, American Society of Photogrammetry, Falls Church, Virginia, 550 pp.

Talvitie, J. and Paarma, H. (1973). Reconnaissance prospecting by photogeology in northern Finland. In *Prospecting in Areas of Glacial Terrain*. Instn. Min. Metall., London., 73–81.

Viljoen, R. P., Viljoen, M. J., Grootenboer, J. and Longshaw, T. G. (1975). ERTS-1 imagery: an appraisal of applications in geology and mineral exploration, *Minerals Sci. Engng.*, **7**(2), 132–168.

Weiss, O. (1971). Airborne geochemical prospecting, *Geochem. Explor.* C.I.M. Spec. **11**, 502–514.

CHAPTER 4

Geochemical Prospecting

Geochemistry is simply defined as a study of the chemistry of the earth and its component parts. The main tasks of the pure geochemist are: (i) to determine the abundances of the elements and their isotopes in the earth and (ii) to study the distribution and migration of different elements in the various components of the earth (rocks, minerals, air, oceans, etc.). The science of geochemistry is very young and has largely developed during the present century. The application of geochemistry to prospecting is even younger, dating from pioneering work carried out in the USSR in the 1930's. Further developments took place in North America and elsewhere in the 1940's, but it was not until the 1950's that geochemistry really became the common and important prospecting tool it is today.

4.1 DISTRIBUTION OF ELEMENTS

If one looks at a table of the average amounts of the various elements in the crustal rocks of the earth (Table 4.1), only eight elements have abundances greater than 1% and, of those eight, oxygen and silicon together make up almost 75% of the earth's crust. Aluminium, iron, calcium, sodium, potassium and magnesium follow in decreasing order with aluminium at approximately 8% and magnesium at 2%. These elements combine in varying proportions to form the various common silicate minerals of which the vast majority of rocks are composed. If we look at the table of crustal abundances again, we see that the common base metals such as copper, lead, zinc and nickel are quite rare, each forming less than 100 ppm and in the case of lead as little as 15 ppm of the earth's crust. Tin forms only 3 ppm, tungsten 1 ppm and gold is much rarer than we might think at 0·005 ppm. It is quite clear that ores of some of these metals, which might have grades of many per cent, are highly anomalous features in the crust. For instance, a sulphide ore containing 15% lead and a rich gold ore

TABLE 4.1

AVERAGE CRUSTAL ABUNDANCES OF VARIOUS ELEMENTS IN PARTS PER MILLION (AFTER MASON, 1958)

Element	ppm	Element	ppm	Element	ppm
Oxygen	466 000	Yttrium	40	Caesium	1
Silicon	277 200	Lithium	30	Holmium	1
Aluminium	81 300	Neodymium	24	Europium	1
Iron	50 000	Niobium	24	Thallium	1
Calcium	36 300	Cobalt	23	Terbium	0·9
Sodium	28 300	Lanthanum	18	Lutetium	0·8
Potassium	25 900	Lead	15	Mercury	0·5
Magnesium	20 900	Gallium	15	Iodine	0·3
Titanium	4 400	Thorium	10	Antimony	0·2
Hydrogen	1 400	Samarium	7	Bismuth	0·2
Phosphorus	1 180	Gadolinium	6	Thulium	0·2
Manganese	1 000	Praseodymium	6	Cadmium	0·2
Fluorine	700	Scandium	5	Silver	0·1
Sulphur	520	Hafnium	5	Indium	0·1
Strontium	450	Dysprosium	5	Selenium	0·09
Barium	400	Tin	3	Argon	0·04
Carbon	320	Boron	3	Palladium	0·01
Chlorine	200	Ytterbium	3	Platinum	0·005
Chromium	200	Erbium	3	Gold	0·005
Zirconium	160	Bromine	3	Helium	0·003
Rubidium	120	Germanium	2	Tellurium	0·002
Vanadium	110	Beryllium	2	Rhodium	0·001
Nickel	80	Arsenic	2	Rhenium	0·001
Zinc	65	Uranium	2	Iridium	0·001
Nitrogen	46	Tantalum	2	Osmium	0·001
Cerium	46	Tungsten	1	Ruthenium	0·001
Copper	45	Molybdenum	1		

grading 50 g/tonne both represent a concentration of 10 000 times over the average crustal abundances. On the other hand, the ore of a common metal such as iron grading 60% represents a concentration of only 12 times the average crustal abundance. These concentration processes take place because of the differing chemical and physical properties of the various elements which permit segregation into distinct mineralogical phases or associations.

In 1922 Goldschmidt proposed a geochemical classification of the elements (Table 4.2) under four headings: *siderophile* (association with metallic iron); *chalcophile* (affinity for sulphur); *lithophile* (affinity for oxygen); and *atomophile* (affinity for elemental state). The

TABLE 4.2

GOLDSCHMIDT'S GEOCHEMICAL CLASSIFICATION OF THE ELEMENTS

Siderophile			Chalcophile			Lithophile					Atomophile
Fe	Co	Ni	Cu	Ag		Li	Na	K	Rb	Cs	H N (C) (O)
Ru	Rh	Pd	Zn	Cd	Hg	Be	Mg	Ca	Sr	Ba	Inert gases
Os	Ir	Pt	Ga	In	Tl	B	Al	Sc	Y		
Au	Re	Mo	(Ge)	(Sn)	Pb	Rare earths					
Ge	Sn		As	Sb	Bi	(C)	Si	Ti	Zr	Hf Th	
C	P		S	Se	Te	(P)	V	Nb	Ta		
(Pb)	(As)	(W)	(Fe)	(Mo)	(Cr)	O	Cr	W	U		
						(H)	F	Cl	Br	I	
						(Tl)	(Ga)	(Ge)			
						(Fe)	Mn				

classification is essentially empirical, but it does have a firm theoretical basis. For instance, the ionization potential of an element is a rough guide to its geochemical classification. Elements with low ionization potentials such as Na, K, Ca and Mg are lithophile, elements with high ionization potentials such as Ni, Co and Au are siderophile and elements with intermediate ionization potentials such as Cu, Zn, Pb and Ag are chalcophile. Some elements are borderline cases and appear in more than one group. When this occurs, the element is shown in brackets under the group(s) of secondary affinity. This system was mainly a result of the study of the distribution of elements in meteorites, which are presumed to have solidified from a liquid state, and the classification really refers to the behaviour in liquid–liquid equilibria between melts. For this reason the classification is not always a true guide to the elemental association in minerals that have formed in the earth's crust by processes such as metamorphism, sedimentation or weathering. In addition to the effects of pressure and temperature the affinity of an element is also dependent upon the chemical environment. For example, metallic iron is a major constituent of many meteorites, but in the earth's crust where abundant oxygen is present iron is strongly lithophile and native iron is only known to occur at a few sites, one of the better known being at Disko Island, Greenland. Chromium is strongly lithophile, but in iron meteorites, where there is little oxygen available, chromium enters the sulphide phase.

Although many minerals exhibit covalent bonding or partial covalent bonding between the constituent atoms, the majority of the common minerals can be considered as ionic structures. For this

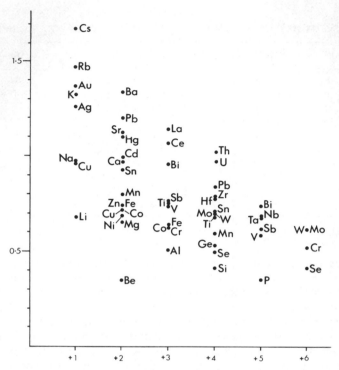

FIG. 4.1. Diagram showing the ionic radii of various elements. Ordinate: ionic radius (Å); abscissa: valency.

reason ionic size or radius is of great importance in atomic substitution and elemental associations in different minerals. Figure 4.1 shows the relationship between ionic radius and valency for a wide range of cations. Substitution is quite common in minerals with ionic structures between ions whose ionic radii do not differ by more than 20%. Indeed, some rarer elements are accepted so readily into the structures of the minerals of more abundant elements that they never form distinct minerals of their own. For example the substitution of Rb for K and Hf for Zr is so widespread that no discrete Rb or Hf minerals have ever been described. The valencies do not have to be the same provided electrical balance is maintained by substitution of an additional ion or ions. For example Ca^{2+} substitutes for Na^+ in plagioclase feldspar because Al^{3+} substitutes for Si^{4+} thus maintaining electrical balance. If the valencies differ by more than one, substitution hardly ever takes place due to the difficulty in achieving electrical balance. In addition little or no substitution takes place between elements which form covalent bonds and those which form ionic bonds. For example, Au, Ag, Cd, Cu and Hg show a strong preference for covalent bonding and for this reason no substitution takes place between Au or Ag and K, between Cu and Na, between Cd and Ca or between Hg and Sr as one might expect from the similarities in ionic radii shown in Fig. 4.1. A certain amount of anomalous behaviour in substitution is also exhibited by the transition elements which include Ti, V, Cr, Mn, Fe, Co and Ni and are defined as those elements whose atoms have an incompletely filled inner shell with more than 8 electrons. For example, some substitution does take place between Ni^{2+}, Mg^{2+} and Fe^{2+} and between Cr^{3+} and Fe^{3+} as might be expected from Fig. 4.1, but both Ni and Cr show a marked enrichment in ultrabasic igneous rocks relative to Mg and Fe. The reasons why Ni and Cr are largely removed from a magma in the earliest stages of crystallization are not fully understood.

Goldschmidt formulated some empirical rules about ionic substitution which are valid for purely ionic bonding:

1. If two ions have the same radius and the same charge, they will enter a given crystal lattice with equal facility.
2. If two ions have similar radii and the same charge, the smaller ion will enter a given crystal lattice more readily.
3. If two ions have similar radii and different charges, the ion with the higher charge will enter a given crystal lattice more readily.

The substitution of a minor element for a major one with the same charge and similar ionic radius is known as *camouflage* (e.g. Hf^{4+} for Zr^{4+}), the substitution of a minor element for a major element with a similar ionic radius and lower charge is known as *capture* (e.g. Ba^{2+} is captured by K minerals), and the substitution of a minor element for a major element with a similar ionic radius and higher charge is known as *admittance* (e.g. Li^+ is admitted into Mg minerals).

Although substitution and association among the elements is a complex subject and is not fully explained by current theories, some appreciation and understanding of it is important in exploration geochemistry. It explains why a degree of caution must be exercised in the interpretation of anomalous concentrations of elements in soils and stream sediments. For example, Cu and Ni contained in the ferromagnesian minerals of basic rocks, Zn in magnetite and spinel, and Nb in Ti minerals often give rise to soil anomalies similar to those over mineralized rocks. In addition elemental associations are often of assistance where certain elements can be used as pathfinders or indicators of the mineralization of associated elements, e.g. Hg in Cu, Pb, Zn sulphide deposits, As in certain Au deposits, Rb in porphyry Cu deposits and Mo in some W occurrences.

4.2 PRIMARY DISPERSION

Primary dispersion was originally defined as the distribution of elements in that part of the geochemical cycle (Fig. 4.2) concerned with the deep-seated processes of magmatic differentiation and metamorphism. In exploration geochemistry, however, primary dispersion has gradually come to have a looser definition and has become synonymous with the distribution of elements in unweathered rocks and minerals whatever their origin. A knowledge of primary dispersion in an area is often of assistance in the interpretation of both stream sediment and soil sampling surveys. It indicates what background ranges might be expected over specific rock types and assists in distinguishing between anomalies due to possible mineralization and those due to high-background unmineralized rocks. Table 4.3 gives the mean values for a number of elements in some of the major igneous and sedimentary rock types.

Primary dispersion can also be used directly in geochemical

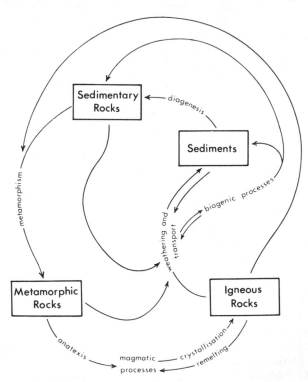

FIG. 4.2.   The geochemical cycle.

TABLE 4.5

MEAN VALUES (ppm) FOR SOME IMPORTANT ELEMENTS IN MAJOR IGNEOUS AND SEDIMENTARY ROCK TYPES. NUMBERS IN BRACKETS REFER TO SOURCES

Element	Igneous rocks				Sedimentary rocks			
	Ultrabasic	Basic	Acid	Alkaline	Sandstone	Limestone	Shale	Black shale
Antimony	0·1 (3)	0·2 (3)	0·2 (3)	—	1 (3)	—	1-3 (3)	—
Arsenic	1-2·8 (3, 4)	2 (3, 4)	1·5 (3, 4)	—	—	2·5 (3)	4-15 (2, 3, 4)	75-225 (2)
Barium	2-15 (3, 4)	250-270 (3, 4)	600-830 (3, 4)	—	100-500 (2)	20-200 (2, 3)	300-800 (2, 3, 4)	450-700 (2)
Beryllium	0·2 (1, 4)	0·1-1·5 (1, 3, 4)	3-5 (1, 3, 4)	2-12 (1)	1 (2)	<1-1 (2, 3)	1-7 (2, 3, 4)	1 (2)
Bismuth	0·02 (3)	0·15 (1, 3, 4)	0·1 (3)	—	0·3 (2)	—	0·2-1 (2, 3, 4)	—
Boron	5 (3)	5-10 (3, 4, 6)	15 (3, 4)	9 (6)	—	9-10 (3, 6)	10-100 (3, 4)	—
Cadmium	0·1 (1)	0·2 (1, 3, 4)	0·1-0·2 (1, 3, 4)	0·1 (1)	—	0·1 (3)	0·2-0·3 (2, 3, 4)	—
Chromium	2 000-3 400 (3, 4, 6)	200-340 (3, 4, 6)	2-4 (3, 6)	1 (6)	10-100 (2)	5-10 (2, 3)	100-160 (3, 4, 7)	10-500 (2)
Cobalt	150-240 (3, 4, 6)	25-75 (3, 4, 6)	1-8 (3, 6)	8 (6)	1-10 (2)	0·2-4 (2, 3)	10-50 (2, 3, 4, 7)	5-50 (2)

TABLE 4.3 (Contd.)

Element	Igneous rocks				Sedimentary rocks			
	Ultrabasic	Basic	Acid	Alkaline	Sandstone	Limestone	Shale	Black shale
Copper	10–80 (3, 4)	100–150 (3, 4, 6)	10–30 (3, 4, 6)	—	10–40 (2)	5–20 (2, 3)	20–150 (2, 3, 4, 7)	20–300 (2)
Fluorine	100 (4, 8)	340–500 (4, 5, 8)	480–810 (5, 8)	570–1 000 (5)	180–200 (5)	220–330 (5, 8)	500–940 (5, 8)	—
Gold	0·1 (4)	0·035 (4)	0·01 (4)	—	—	—	—	—
Lanthanum	3·3 (3)	10–27 (3, 4)	25–46 (3, 4)	—	—	6 (3)	20–40 (3, 4)	25–100 (2)
Lead	0·1 (3)	5–9 (3, 4)	10–30 (3, 6)	—	10–40 (2)	5–10 (2, 3)	16–20 (2, 3, 4, 7)	20–400 (2)
Lithium	2 (1, 4)	10–15 (1, 3, 4)	30–70 (1, 3, 4)	28 (1)	7–29 (2)	2–20 (2, 3)	50–60 (2, 3, 4, 7)	17 (2)
Manganese	1 100–1 300 (4, 6)	2 200 (4)	600–965 (4, 6)	—	—	385 (6)	670–890 (4, 6)	—
Mercury	—	0·08–0·09 (3, 4)	0·04–0·08 (3, 4)	—	0·03–0·1 (2)	0·03–0·05 (2, 3)	0·4–0·5 (2, 3)	—
Molybdenum	0·3–0·4 (3, 4)	1–1·4 (3, 4)	2 (3, 4)	—	0·1–1 (2)	0·1–1 (2, 3)	1–3 (2, 3, 4)	10–300 (2)
Nickel	800–3 000 (3, 4, 6)	50–160 (3, 4, 6)	2–8 (4, 6)	2–4 (6)	2–10 (2)	3–12 (2, 3)	20–100 (2, 3, 4, 7)	20–300 (2)

Niobium	15 (1,3,4)	20 (1,3,4)	20 (3,4,6)	30–900 (1,6)	—	—	—	20 (2,3,4)	—
Silver	0·3 (4)	0·3 (4)	0·15 (4)	—	0·4 (2)	—	0·2 (2)	0·9 (2)	—
Tantalum	<1–1 (1,3,4,6)	0·5–1 (1,3,4,6)	3–4 (1,3,4,6)	1–2 (1,6)	—	—	—	2–3·5 (3,4)	—
Tin	0·5 (3)	1 (3)	3 (3)	—	—	—	—	—	—
Titanium	3 000 (4)	9 000 (4)	2 300 (4)	4 400 (6)	—	—	—	4 300–4 500 (4,6)	—
Tungsten	0·5 (3)	1 (3)	2 (3)	—	—	—	0·5 (3)	2 (3)	—
Uranium	0·001–0·03 (3,4)	0·6–0·8 (3,4,6)	3·5–4·8 (3,4,6)	—	—	—	2 (3)	3·2–4 (3,4)	—
Vanadium	50–140 (3,4)	200–250 (3,4)	20–25 (3,4)	34 (6)	10–60 (2)	—	2–20 (2,3)	50–300 (2,3,4,7)	50–2 000 (2)
Zinc	50 (3,4)	90–130 (3,4,6)	40–60 (3,4)	—	5–20 (2)	—	4–25 (2,3)	50–300 (2,3,4)	100–1 000 (2)
Zirconium	20–70 (1,3,4,6)	100–150 (1,3,4)	170–200 (1,3,4)	300–680 (1)	—	—	20 (3)	120–200 (3,4)	10–20 (2)

Sources: 1 (*Vlasov*, 1966); 2 (*Krauskopf*, 1955); 3 (*Taylor*, 1964, 1966); 4 (*Vinogradov*, 1959); 5 (*Fleischer and Robinson*, 1963); 6 (*Rankama and Sahama*, 1950); 7 (*Shaw*, 1954); 8 (*Turekian and Wedepohl*, 1961).—*No data.*

exploration. Rock sampling surveys, often referred to as lithogeochemical surveys, can be carried out in a similar manner to soil and stream sediment sampling surveys to search for anomalous metal concentrations. Rocks are a poor sampling medium, however, since small samples are less likely to be representative of the surrounding area than equivalent soil or stream sediment samples and rocks have to be crushed for analysis making the surveys more costly. Nevertheless, lithogeochemical surveys may be undertaken on a regional scale to define metallogenic provinces or belts by locating rock types likely to be associated with mineralization, or on a much more local scale to locate blind ore bodies or extensions to known deposits by delineating the primary dispersion haloes that are sometimes associated with mineralization. Although lithogeochemistry is widely used in the USSR and much success is claimed for it, the use of primary dispersion in exploration geochemistry is a relatively new development and at present it is difficult to assess its full value.

The search for metallogenic provinces or belts

Probably one of the oldest attempts in the application of lithogeochemistry has been in the search for tin. It is well known that tin mineralization is spatially and genetically associated with certain granites and it would be extremely valuable to be able to distinguish 'tin' granites from barren ones. In a study of granites in the USSR Barsukov (1957) concluded that, although 'tin' granites often contained four to five times as much tin as the 3–5 ppm contained in barren granites, there is very little difference between them. Jedwab (1955) compared a barren granite with a 'tin' granite in France by using the Li content of feldspar and the Sn content of biotite. The 'tin' granite contained an average of 110 ppm Sn in the biotite compared to only 67 ppm in the barren granite and the Li content of the feldspar from the 'tin' granite was 141 ppm, contrasting sharply with only 36 ppm from the barren granite. Although this would appear to offer a simple method for distinguishing barren granites from 'tin' granites, more detailed and thorough investigations by other workers have failed to find any unambiguous correlation between tin content and mineral potential. Flinter (1971) and Hesp (1971) both made comprehensive studies of tin in granites in Australia and concluded that the tin content is not suitable for calculating the degree of mineralization associated with a particular granite. Beus and Sitnin (1972) in a study of granites in the USSR, on the other hand, concluded that a Sn content of 20 ppm or more is a good indicator of the mineralizing potential of a granite. Thus, there is a considerable degree of conflict in the data collected from different parts of the world and, although much more research needs to be undertaken before firm conclusions can be drawn, it would appear likely that there is no geochemical criterion that can be universally applied to distinguish 'tin' granites from barren ones.

Studies have been applied to granitic rocks for mineralization other than tin. Brabec and White (1971) analyzed over 300 samples from the Guichon Creek batholith in British Columbia for copper and zinc. A number of important porphyry copper deposits are contained within the batholith and, although copper values ranged from 1–1600 ppm, some of the major copper deposits are located in an area of the batholith with low copper contents and they concluded that a high copper content of a particular intrusive phase does not necessarily indicate its superior ore potential. In a study of granitic rocks in the Yukon, Garrett (1973) was able to distinguish plutons known to be associated with mineralization and likely to be associated with mineralization from unfavourable plutons, using advanced statistical techniques of principal-component and multiple-regression analysis. The results are well shown by zinc in which 57 plutons gave low principal-component scores and were considered as unfavourable and 15 gave high scores. Of those 15, nine were either associated with mineralization or had mineral showings in their vicinity. Lawrence (1974) used Rb/Sr ratios as a guide to mineralization in the Galway granite, Ireland. He showed that the Murvey granite, which contains sub-economic traces of molybdenite and chalcopyrite mineralization, has Rb/Sr ratios of 4/43, well above the values for unmineralized granite.

Regional bedrock geochemical studies have been applied to rocks other than granites. Davenport and Nichol (1973) made a study of some of the vulcano–sedimentary belts in the Canadian Shield in which some important massive zinc–copper sulphide deposits occur. They concluded that the zinc content of felsic volcanic rocks could be used as a guide to potentially favourable areas since economically significant mineralization occurs in volcanics with high zinc contents. Cameron et al (1971) collected 1079 rock samples from 61 ultramafic bodies in the Canadian Shield, some of which contain nickel mineralization and some of which are barren. They divided the occurrences into three categories: *ore*, *minore* (weakly mineralized to minor ore) and *barren* (may contain minor sulphides). An analytical technique was used which preferentially leached the sulphides and Ni, Cu, Co and S were determined for each sample. It was found that Cu and S were best to distinguish the three categories with Ni of lesser importance and Co of little value. Discriminant function analysis was used to treat the data and it was found that discriminant scores for Cu, Ni and Co of 5·5 and over indicated *minore* and *ore*, though a few *barren* localities gave scores over 5·5. Mean values for the different categories gave 0.059% S for *barren* and 0.582% S for *ore* and Cu/Ni ratios of 0.234 for *ore*, 0.070 for *minore* and 0.045 for *barren*.

The search for localised deposits

Varying degrees of rock alteration are commonly associated with vein deposits and large alteration zones generally occur around porphyry copper deposits. Even when these alteration effects appear to be absent or weakly developed, they are often manifest as small changes in the trace element content of the country rock in the immediate vicinity of mineralization. In addition ore bodies may also be surrounded by an aureole or primary dispersion halo of weak mineralization or small concentrations of associated trace elements. Thus, if either country rock alteration or a primary dispersion halo or both occur with a particular ore body, it may be possible to locate a concealed ore body by finding the outer alteration zone or primary dispersion halo. Considerable success has been claimed for this technique in evaluating and forecasting the ore potentials of the deeper parts and flanks of known deposits in the USSR (Ovchinnikov and Grigoryan, 1971). Figure 4.3 shows an example of a primary dispersion halo around a lead–zinc ore body in skarn rocks in a district of Middle Asia in the Soviet Union. In Chile Rb and Sr have been used as guides to copper mineralization emplaced in andesites (Oyarzum, 1974). Normal unmineralized andesites have a Rb/Sr ratio of 0.2 which falls to 0.01 in the epidotized zone surrounding mineralization and rises to 2.6 within the mineralized zone. Pantazis and Govett (1973) analyzed a large number of rock samples for Cu, Zn, Ni and Co around the Mathiati cupriferous pyrite mine in Cyprus and used discriminant function and determinative function analysis to treat the data. The results were somewhat inconclusive, but by using a combination of statistical techniques it was possible to define the mine area as a primary target.

Summary

With a few exceptions the results of primary dispersion surveys have not proved particularly encouraging. Regional studies may work in one area and not in another apparently similar one. In addition, studies have shown that a number of major ore bodies such as the Broken Hill lead–zinc deposit in Zambia have little or no primary dispersion haloes. Far more research needs to be done in this relatively new field and, although primary dispersion surveys may have limited application in exploration geochemistry at the present time, they may well increase in importance as our knowledge expands and more indirect methods have to be applied with the discovery of the shallower targets using more conventional techniques.

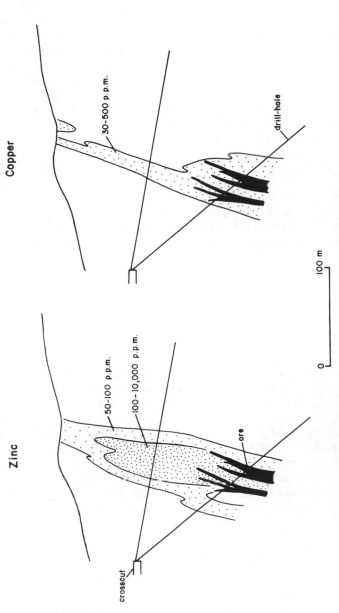

FIG. 4.3. Example of primary dispersion haloes around a lead–zinc ore body in skarn rocks in the USSR (after Ovchinnikov and Grigoryan, 1971).

4.3 SECONDARY DISPERSION

During weathering, rocks break down by physical and chemical processes which disperse the various elements contained in the original rocks into soils, stream sediments, ground waters, river waters, the sea, the air, plants and animals. The degree of secondary dispersion an element may undergo is expressed by its *mobility*. Some elements such as Be, Au, Sn, Si and Ti occur in or as stable minerals which are very resistant to the effects of normal chemical weathering. These elements have a low mobility and are generally dispersed as clastic fragments by slow mechanical weathering into soils and stream sediments. At the other extreme are the highly mobile elements such as Na, K and Mg which readily enter the water-soluble phase and are widely dispersed by ground and surface waters. Many of the elements of interest in geochemical exploration fall between these two extremes, but may display a wide range in mobility depending on the environment. Secondary dispersion is of prime importance in exploration geochemistry as it results in the various elements present in an ore body being dispersed over a much wider 'target area', thus enabling the presence of mineralization to be detected as an anomalous metal content in soils, stream sediments, plants and surface and ground waters.

The importance of pH and Eh

The pH of a solution is defined as the negative logarithm to the base 10 of the hydrogen ion activity in gram-ions/litre. It varies from 0 for strongly acidic solutions to 14 for strongly basic or alkaline ones. Neutral solutions, i.e. ones in which the concentrations of hydrogen and hydroxyl ions are equal, have a pH of 7. Natural waters with pH's less than $4\cdot0$ or greater than $9\cdot0$ are rare and most fall within the range of $5\cdot5$ to $8\cdot5$. pH can be determined by special indicator papers or solutions, but the most accurate measurements are made with pH meters which use special glass electrodes sensitive to hydrogen ion concentrations. For calibration purposes there are a number of standard buffer solutions which can be made up. For example, a $0\cdot05$M solution of potassium hydrogen phthalate has a pH of $4\cdot0$ at 20°C and a $0\cdot01$M solution of borax has a pH of $9\cdot2$ at 20°C. The pH of a solution varies slightly with temperature and corrections need to be applied if very precise measurements are being made. Soil pH is determined by measuring the pH of an aqueous slurry made up by adding distilled water to soil in a container (50 g of soil in 50 ml of water is commonly used).

pH is extremely important in determining the mobility of an element in the weathering environment. Of the metallic elements only

he alkalis and alkaline earths are soluble over virtually the full pH
ange. Most metals are soluble in acid solutions, but are precipitated
s the alkalinity increases. The pH at which the hydroxide of an
lement precipitates is known as the *pH of hydrolysis*. Table 4.4 lists
he pH's of hydrolysis for some common elements and it gives a
ough indication of the solubility to be expected in nature. In additon
o pH, however, other factors, which can affect the solubility, need to
e taken into account. For instance, although the table shows that
itanium is soluble below a pH of 5, this is only so if it is released
rom a weathered mineral as $Ti(OH)_4$. Since this rarely happens and
itanium is almost always released as the insoluble dioxide, titanium is
highly immobile element even under strongly acid conditions. In
addition the presence of certain anions may restrict or increase the
solubility range. For example, in the presence of chloride ions, both

TABLE 4.4

THE pH'S OF HYDROLYSIS OF SOME COMMON ELEMENTS
(FROM BRITTON, 1955)

pH	Element	Environment
11		
	Magnesium	Strongly alkaline soils
10		
9		
	Divalent manganese	
	Lanthanum	Sea water
8		
	Silver	
	Divalent mercury	
7	Zinc	River water
	Cobalt	Rain water
	Nickel	
6	Lead	
	Divalent iron	
	Copper	
	Titanium	
5		
	Hexavalent uranium	Peat water
4	Aluminium	
3		Waters from oxidizing
		sulphide deposits
	Trivalent iron	
2	Tin	

lead and silver remain insoluble at low pH's. In the case of uranium the formation of uranyl carbonate complexes extends the solubility of uranium well above the pH of hydrolysis shown in the table.

In addition to pH the *redox potential* or Eh, which is a measure of the reducing or oxidizing potential of an environment, is extremely important. The oxidation potential is a relative figure and the standard reference is the reaction:

$$H_2 = 2H^+ + 2e$$

which is arbitrarily fixed as $0.00\,V$ at a pressure of 1 atm and at pH = 0. The Eh values extend on either side of the zero, positive values indicating an oxidizing potential and negative values a reducing potential relative to the standard hydrogen half cell reaction. Oxidation potentials are dependent on the concentrations of the reacting substances and reactions involving hydrogen or hydroxyl ions are strongly affected by pH. The range of oxidation potentials in natural environments is restricted by the reactions:

$$H_2 = 2H^+ + 2e \qquad\qquad Eh = 0.00\ V$$

and

$$2H_2O = O_2 + 4H^+ + 4e \qquad Eh = 1.23\ V$$

which define the Eh range over which water is stable. Figure 4.4 shows an Eh–pH diagram on which the normal weathering environment is outlined and a number of natural environments are shown. Eh can be measured on some pH meters, but Eh measurements are influenced by many factors and field measurements often show a big variation and interpretation is not always satisfactory. In the natural environment redox potentials are largely dependent on the amount of oxygen available and the amount of organic matter present. Hot, well-drained environments to which atmospheric oxygen has ready access have a high oxidation potential, whereas waterlogged environments with a high organic content are strongly reducing. In addition to removing oxygen by its oxidation to produce reducing conditions, organic matter also has a strong tendency to adsorb hydrogen ions and produce conditions of low pH. Thus, organic-rich environments, such as peat bogs, tend to be acid and reducing. Waterlogged, organic-free environments are only oxidizing where they are in free contact with the atmosphere. Confined waters rapidly lose their oxygen and by hydrolysis of silicates become alkaline so that below the water table environments tend to be alkaline and reducing.

The Eh and pH conditions of an environment are extremely important in determining the mobility of most elements of interest in

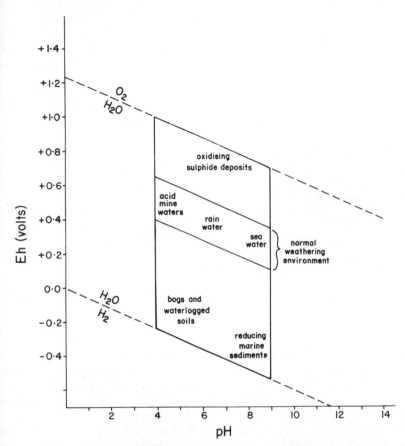

FIG. 4.4. Eh–pH diagram showing various natural environments.

exploration geochemistry and Eh–pH diagrams have been determined and plotted by different workers for many chemical reactions of interest in geochemistry. Garrels (1960) and Garrels and Christ (1965) have dealt with the fundamentals of the subject and Hansuld (1966) has discussed the application of pH and Eh in interpretation of geochemical data. Figure 4.5 shows the results of a soil sampling survey over copper mineralization along a fault zone in Puerto Rico. A Mo anomaly coincides with the Cu anomaly at the eastern end where pH conditions are acidic. In the central area where pH conditions are mixed the Mo anomaly becomes more diffuse and at the western end of the fault zone where conditions are alkaline and Mo is highly mobile there is no Mo anomaly accompanying the strong Cu anomaly.

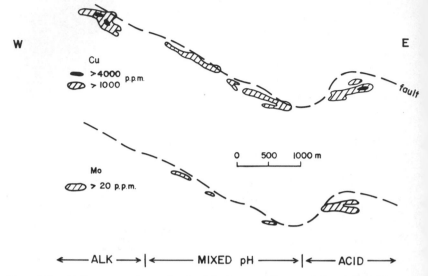

FIG. 4.5. Soil survey from the La Muda area, Puerto Rico (after Hansuld. 1966).

Adsorption

Clays, organic matter and hydrous oxides of iron and manganese all have the ability to adsorb metallic ions and are thus very important in determining mobility of many elements in the secondary dispersion environment. The scavenging effects of both iron and manganese hydrous oxides are well known and the presence of either in soils or stream sediments can result in a marked build up of trace metal values. Cd, Co, Ni and Zn are strongly scavenged by Mn oxides, As is strongly scavenged by Fe oxides and elements weakly scavenged include Cu, Mo and Pb (Nowlan, 1976). In Table 4.5 the percentages of Cu, Pb and Zn in solution adsorbed by goethite are related to pH and it is interesting to note that significant percentages are adsorbed at pH values below the pH's of hydrolysis shown in Table 4.4. This generally has a detrimental effect in exploration geochemistry as both iron and manganese hydrous oxides may significantly restrict the mobility of many metals with a consequent reduction in the size of secondary dispersion haloes. Also false anomalies caused by hydrous oxides of iron and manganese adsorbing and concentrating trace elements from normal background sources can be confused with significant anomalies derived from mineralized sources. On the other hand the scavenging effects of Mn–Fe oxides have been put to practical use. In drainage surveys collection and analysis of Mn–Fe

TABLE 4.5
ADSORPTION OF Cu, Pb AND Zn BY GOETHITE
FROM 3.2×10^{-5}M SOLUTIONS (DATA FROM
FORBES *ET AL*, 1976)

pH	Metal adsorbed (%)		
	Cu	Pb	Zn
4·7	8	8	
5·2	44	29	
5·5	75	37	
5·9	90	63	11
6·4			19
7·2			61

oxide concretions and coatings for their trace element content may result in better anomaly–background contrast than that obtained from conventional sediment surveys.

Nowlan (1976) showed in a survey in Maine that anomaly contrast may be significantly increased by analyzing Fe–Mn oxides. This is particularly true for the weakly scavenged elements such as Cu. It would appear that scavenging of the weakly scavenged elements only takes place when they are present in above-normal amounts, whereas elements such as Zn are so readily scavenged that there is little or no enhancement of contrast. Carpenter *et al* (1975) showed in test surveys over mineralized areas in the south-west United States that anomaly–background contrast was enhanced for both Zn and Cu in Fe–Mn oxide coatings, but in the case of Pb anomaly–background contrast was much better in −80 mesh sediments. In addition the ratios Zn/Mn and Cu/Mn in the Fe–Mn oxide coatings enhanced the downstream detectability of mineralization.

Clay minerals are strong adsorbents and will hold various cations by virtue of unsatisfied electric charges both at crystal edges and within the lattice layers. The cation exchange capacity is measured in milliequivalents per 100 g (me/100 g) and varies from 3–15 for kaolin to 10–40 for illite to over 100 for montmorillonite and vermiculite (Grim, 1953). In addition to the clay minerals many other fine-grained minerals or colloidal particles have the ability to adsorb cations. The amount of adsorption increases as the grain size of the adsorbent decreases and the amount of a substance adsorbed from solution increases with the concentration of the adsorbate in solution. In addition highly charged ions are adsorbed more readily than those of lower charge.

Organic matter can adsorb considerable quantities of trace elements and expressed in terms of cation exchange capacity organic carbon may reach values up to 500 me/100 g which is considerably higher than the cation exchange capacity of montmorillonite. Govett (1960) has shown that there is strong positive correlation between the organic carbon content of soils in Zambia and their cation exchange capacity. There is also positive correlation between organic carbon content and total Cu in background areas. In both anomalous and background areas there is a very strong positive correlation between readily extractable Cu and organic carbon content. Chowdhury and Bose (1971) undertook some experimental studies on the adsorption of various metals by 'humic acid' at pH 5. They showed that the maximum metal holding capacity of 'humic acid' is of the order of 40 me/100 g and that Pb attains equilibrium with a minimum concentration in solution followed by Cu, Zn, Ni and Co. The release of metals retained by the 'humic acid' was also determined at pH's ranging from 5 down to 0. This showed that the Pb is most strongly held with 60% retained at pH 3·5 compared to 32% Cu, 15% Zn, 11% Ni and 7% Co. At pH 2·5, 35% of the Pb is still held, but virtually all the Cu, Zn, Ni and Co has been released; Pb is only totally released into solution at pH 1. The manner in which organic matter adsorbs is only poorly understood, though the formation of metallo-organic complexes such as porphyrins is known to play a part. Nevertheless, from a practical point of view the role of organic matter in secondary dispersion is extremely important. Marked concentrations of trace metals can occur in poorly drained, organic-rich soils and it is very difficult to decide whether anomalies in such environments are due to mineralized sources or an unusual build-up from high background rocks.

Dispersion processes
Many different agencies are responsible for dispersion of elements in the secondary environment, but dispersion processes can be discussed under three main headings: mechanical, solution and biogenic.

Mechanical processes
Clastic fragments derived from the weathering of rocks are dispersed principally through the agencies of gravity, water, wind, ice and animals. Overburden on hill slopes tends to move downwards under the influence of gravity and movement may be extremely slow as in the case of soil creep or rapid and sudden as in the case of large landslides. In mountainous terrain there may be considerable movement of material down steep slopes forming thick deposits of colluvium or talus at the foot of slopes. The use of talus as a sampling

medium in mountainous regions can be of value in reconnaissance geochemical surveys, particularly in arid areas where there may be little or no stream drainage (Maranzana, 1972). Even in relatively well-drained mountainous areas talus sampling can be a useful method for reconnaissance work (Hoffman, 1977).

Flowing surface waters carry suspended particles and move larger clastic fragments along the bottom, both eroding and transporting overburden and weathered rock. Sheetwash or surface run-off carries material over the land surface, generally only eroding and transporting the finest particles, but on steep hillsides considerable amounts of coarse material may be carried and deposited as alluvial fans at the base of slopes. Much of the surface water eventually finds its way into gulleys, streams and rivers where erosion and transportation are often accelerated. Fast flowing, turbulent waters erode and transport material more effectively than slower moving waters and, as the velocity decreases, transported sediment is deposited, often at considerable distances from its place of origin. Very sluggish and slow moving water in swamps and lakes carries very little sediment in suspension and such environments are mainly centres of deposition from inflowing waters.

Glaciers can be important agents of erosion and transportation. During the Pleistocene large areas of Eurasia and North America were covered by continental ice sheets, which has resulted in the development of dispersion patterns very different from unglaciated areas. Boulders may be carried for hundreds of kilometres by glaciers, but most of the basal till formed during the glaciation and deposited under the ice sheet consists of clay-sized particles (rock flour), cobbles and boulders of local origin transported no more than one or two kilometres and generally very much less. Sometimes several basal till layers occur and are due to the cyclic advance and retreat of the ice sheet. Mineral fragments and particles from ore deposits are often well preserved in the basal till just above bedrock and may form wide dispersion trains or fans in the direction of ice movement. Figure 4.6 shows the dispersion of copper values in the basal till 'down-ice' from an ore deposit in Quebec. In glaciated areas thick ablation till deposits formed by the final melting and retreat of the ice sheet together with fluvio-glacial sediments deposited by melt waters often cover large areas and render ordinary geochemical methods ineffective. In such areas deep sampling methods have to be used to obtain samples from the basal till in which dispersion patterns of interest may occur.

Wind is another agent of mechanical dispersion, but is of very little practical value in geochemistry. In fact, thick deposits of wind-blown sand and loess in arid regions generally render geochemical prospecting methods totally ineffective. In the vicinity of mining or industrial

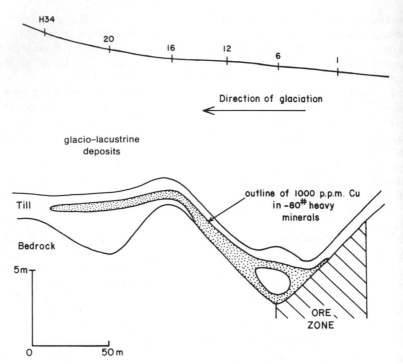

FIG. 4.6. Dispersion 'down-ice' in basal till from the Louvem deposit, Va^l d'Or, Quebec (after Garrett, 1971).

centres wind-borne particles may result in widespread contamination and make the application of geochemical methods very difficult.

Various burrowing animals move soil and overburden, but are generally insignificant as agents of mechanical dispersion. Notable exceptions to this are termites which form large mounds in tropical regions and are known to bring mineral particles up from the water table to which they require access to survive. Some gold deposits in Rhodesia have been discovered as a result of prospecting termite mounds (West, 1970) and d'Orey (1975) gives an interesting example of the location of copper mineralization in Mozambique concealed beneath transported overburden by sampling and analyzing material from termite mounds.

Solution process

Under the influence of weathering forces numerous constituents of the original rocks and minerals pass into solution and are carried away by surface and ground waters to be eventually precipitated or redeposited because of adsorption or changes in pH, Eh and chemical environment. This results in the elements of interest in ore deposits being dispersed over a wider area which can vary from a narrow zone little larger than the sub-outcropping mineralization to extensive lateral dispersion characterized by hydromorphic anomalies which may be considerably removed from the source areas. In stream sediments, which are the products of mechanical processes of dispersion, the trace element content of interest in geochemical prospecting may be largely due to the transport of metals in solution and subsequent adsorption by the fine-grained sediments and coprecipitation with Fe and Mn oxides. It is only in the case of stable minerals such as gold, cassiterite and monazite that the elements of interest are mainly present in discrete clastic mineral grains.

Biogenic processes

Plants take up numerous trace elements together with their normal nutrients from the soil. When the plants die and decay, the more soluble constituents are removed by groundwaters and the less soluble constituents may accumulate to an appreciable extent in the humus layer. Thus, plants can play an important role in secondary dispersion. Certain bacteria, algae and other micro-organisms have a big influence on dispersion in the secondary environment. Some anaerobic bacteria derive their energy from oxidising iron compounds and there are others that can reduce sulphates to sulphides (Baas Becking and Moore, 1961). Marked concentrations of heavy metals have been found in certain algae (Cannon, 1955) and there is evidence that some micro-organisms can precipitate metals (Lovering, 1927). Some higher organisms also appear capable of concentrating heavy metals. The oxygenating compound in the blood of some arthropods and molluscs is based on the copper-bearing substance haemocyanin analogous to the iron-bearing haemoglobin. Up to 2·3% Zn and 0·6% Cu has been found in the soft parts of oysters living in seawater slightly contaminated with Zn and Cu from old mine tailings off the coast of British Columbia and up to 3% Zn and 660 ppm Cu has been found in oysters taken from the Atlantic Ocean (Boyle and Lynch, 1968). Both coal ash and petroleum contain concentrations of many different elements and it is not unlikely that they were accumulated by the original organisms, particularly in the case of petroleum.

Thus, there are numerous dispersion processes in the secondary environment which result in the formation of many different types of anomalies. The most important of these are: residual anomalies, mechanically transported anomalies, hydromorphic anomalies, stream sediment anomalies, lake sediment anomalies, vegetation anomalies and water (stream, lake and groundwater) anomalies.

Residual anomalies form wherever rocks weather *in situ* and solution processes have not removed the elements of interest so that anomalies in the overburden lie directly over their sources. Such anomalies occur in all parts of the world, but are particularly common in non-glaciated regions where geochemical methods have proved highly effective because of the simple relationship between a residual anomaly and buried ore. Nevertheless, there are many complicating factors and interpretation may not always be straightforward. Both mechanical and chemical processes can result in strong concentrations of trace elements in the overburden that may be due to weakly mineralized or even unmineralized sources. The reverse may also occur and the partial leaching out of elements of interest from the overburden may result in a weak anomaly overlying good mineralization. Some of the problems are illustrated by Fig. 5.12, in the next chapter, which shows a marked build up of Zn values in the overburden owing to the effects of tropical weathering and the accumulation of secondary Zn minerals derived from thin, sporadic willemite veinlets in dolomitic limestone. Table 4.6 compares the Zr and Nb content of residual soils overlying the Sukulu carbonatite in Uganda with the Zr and Nb values in bedrock obtained from the same pits as the soil samples. The higher Zr and Nb values in the soil are due to a mechanical enrichment of the zirconiferous minerals, zircon and baddeleyite, and niobium-bearing mineral, pyrochlore, which are resistant to weathering and have become concentrated in the soils formed by a process of leaching of the carbonates and weathering and consequent breakdown of the various silicate minerals which made up the parent carbonatite rock.

The dangers and pitfalls in evaluating the mineral potential from a residual anomaly are illustrated in Fig. 4.7 which shows profiles across two occurrences of copper mineralization in different parts of Zambia discovered as a result of geochemical prospecting. Deposit B consists of good copper mineralization totalling about 8 million tonnes at 2·8% Cu in porous sandstones and conglomerates. Leaching has penetrated to depths of 30 m and more resulting in a marked depletion of copper in the weathered overburden. As a result there is only a poor soil anomaly with maximum values of 280 ppm Cu and low background to anomaly contrast. Deposit A consists of low-grade mineralization with very poor continuity at different stratigraphic levels in dolomite, tillite and shale. Accumulations of malachite occur near the base of the weathered overburden and there is a very good soil anomaly with maximum values well over 1000 ppm Cu and marked background to anomaly contrast.

Mechanically transported anomalies occur wherever clastic fragments of the source rocks or minerals have been moved by agents of

TABLE 4.6

COMPARISON OF Nb AND Zr CONTENT IN ppm OF
BEDROCK AND SOIL SAMPLES TAKEN FROM PITS
IN THE WEST VALLEY, SUKULU CARBONATITE,
UGANDA

Pit no.	Zr content		Nb content	
	Bedrock	Soils	Bedrock	Soils
O4X	20	595	50	933
H7	270	315	520	1050
O4Z	45	455	180	770
B9	375	326	250	1237
O6V	130	705	100	1080
L1	20	403	300	1610
O6X	50	1033	320	1595
D7	55	444	350	1555
B1	10	580	20	2830
O4B	50	773	2100	3370
O6B	500	348	2000	1367
D1	10	352	210	2408
D3	130	510	500	1848
O6J	10	678	100	985
O2J	200	366	185	1392
Average	125	526	479	1602

mechanical transport. The most widely documented of these occur in glacial overburden, but soil creep, landslip, sheetwash and wind can all result in transported anomalies. If an anomaly has been transported, the location of the primary source may involve a long and detailed search in the reverse direction of transport. In the case of gravity transport the sense of movement is usually obvious from the direction of ground slope, but in the case of glacial transport the local direction of ice movement is not always so obvious, though it is generally known in regional terms. Glacial boulder trains may be many kilometres long, but it is unusual for geochemical dispersion trains in glaciated regions to be more than 1 or 2 km long owing to the effects of dilution which make the odd erratic mineralized boulders at some distance from a mineralized source virtually undetectable by geochemical sampling of the overburden. A good example of this is given by Kauranne (1959).

Hydromorphic anomalies are due to solution processes, hence the name 'water-formed'. Elements carried away in solution may be

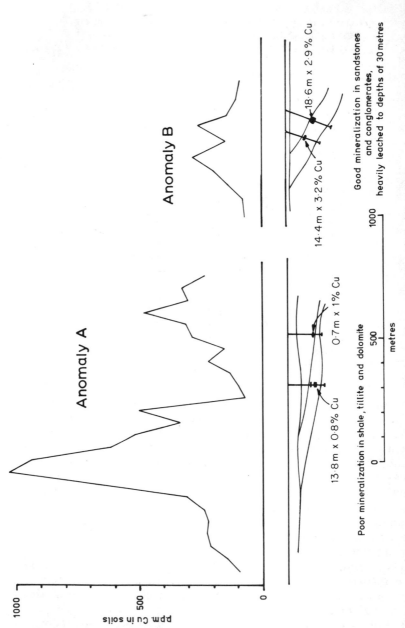

FIG. 4.7. Residual anomalies over two copper deposits in Zambia both discovered as a result of geochemical prospecting.

precipitated or deposited again when the environment changes, resulting in lateral dispersion from the parent source. The amount of transport may vary considerably from short distances which result in a marginal widening of a residual anomaly due to hydromorphic processes to long distances where the anomaly may be considerably removed from its parent source. Hydromorphic anomalies are often characterized by a marked build up of metal content so that the anomalous values may be misleadingly high. The cations are also generally loosely held by simple adsorption on organic matter, clays and iron oxides so that hydromorphic anomalies are characterized by a high proportion of cold-extractable metal (cxMe).

Figure 4.8 gives an example of a hydromorphic anomaly in central Ireland. The zinc values have been hydromorphically dispersed in the near-surface soils by seepage and drainage down gentle slopes over a distance of 800 m from the source, whereas lead, which has a much lower mobility, has suffered very little hydromorphic dispersion. Figure 4.9 shows a hydromorphic copper anomaly associated with a residual anomaly in central Zambia. The residual soil anomaly falls off downslope towards seasonal swampy ground with a high organic content where there is a sudden and spectacular build up of copper values in the soils some 900 m from the source. There is also a good cold-extractable Cu anomaly over the swampy ground which is completely absent over the mineralization. Another example of a hydromorphic anomaly is given in Fig. 4.10 which shows an extremely long hydromorphic Cu anomaly along a watercourse in northwestern Zambia. The area is covered by thick unconsolidated desert sands, though present climatic conditions are far from arid with an average annual seasonal rainfall of 950 mm. There are few rock exposures and the area has little topographical relief with drainage typified by long swampy watercourses (locally known as dambos) up to 2 km wide in which a central channel is usually defined, though flow is generally sluggish. In addition to anomalous Cu values (up to 8000 ppm) Zn and Ni values are also anomalous with maximum values of 440 and 555 ppm respectively. All metals show an increase in concentration downstream in reverse order of normal drainage dispersion patterns with a gradual build up of values throughout the length of the dambo. The soils have a high organic content and there is extensive precipitation of iron oxides with a considerable development of laterite down the centre of the dambo. Insufficient follow-up work was done to decide whether the anomaly might be due to a mineralized or unmineralized source, though it is considered likely that the source could be unmineralized basic igneous rocks. Such an anomaly is an extremely ill-defined target for follow-up work and illustrates the problems in dealing with some hydromorphic anomalies.

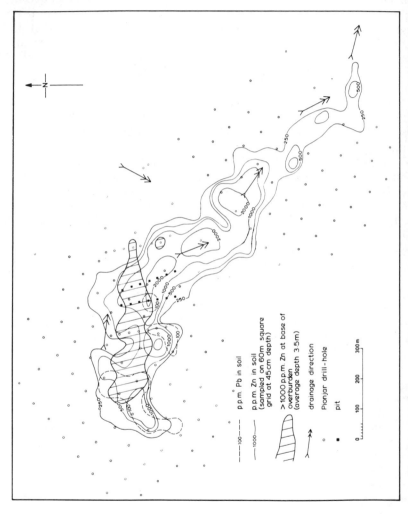

Fig. 4.8. Example of a hydromorphic zinc anomaly from central Ireland.

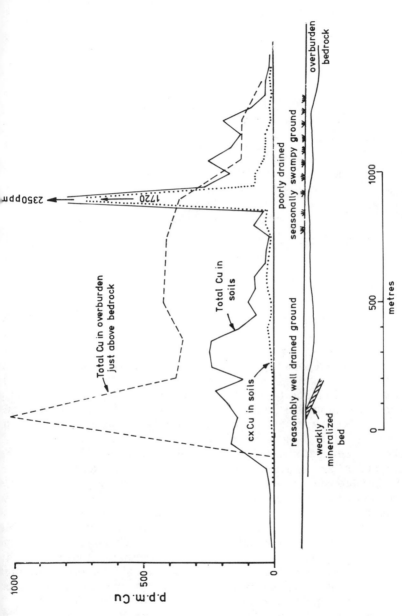

FIG. 4.9.   Example of a hydromorphic copper anomaly associated with a residual anomaly over a copper occurrence in central Zambia.

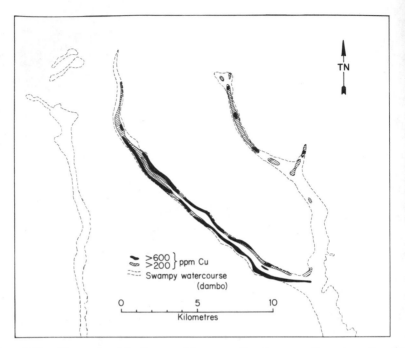

FIG. 4.10. Example of hydromorphic anomalies from northwestern Zambia.

It must be emphasized that it is erroneous to assume that if a geochemical anomaly is shown to be hydromorphic, it indicates an unmineralized or poorly mineralized source. Some hydromorphically transported anomalies are associated with good ore deposits, but equally, there are many others that have been shown to be due to unmineralized, high-background rocks.

Stream sediment anomalies form when mineral deposits or source rocks are eroded and the erosional products are swept into stream beds and deposited as sediment. In the case of resistant ore minerals such as cassiterite or beryl the elements of interest are contained in clastic mineral grains and such a stream sediment anomaly is really a type of mechanically transported anomaly. In the case of less stable source minerals, the elements of interest may be contained in clay sized particles derived from the chemically weathered products of the source deposit or they may be due to adsorbed ions on organic matter and fine mineral particles. In addition much of the anomalous metal content may be held with precipitated Fe and/or Mn oxides.

In the typical stream sediment anomaly the concentrations of the elements of interest decrease downstream away from the source as

more and more barren material enters the stream bed and contributes
to the total sediment. In well dissected areas undergoing active
erosion mineralized sources often give rise to long dispersion trains
making stream sediments a good sampling medium for reconnaissance
surveys. Figure 4.11 shows a stream sediment anomaly in eastern
Uganda which led to the discovery of some bodies of carbonatite.

Lake sediment anomalies are similar to stream sediment anomalies
in as much as both clastic ore grains and adsorbed and chemically
precipitated metals may occur in the sediments. Unlike many streams,
however, lakes are mainly centres of deposition. Materials brought in

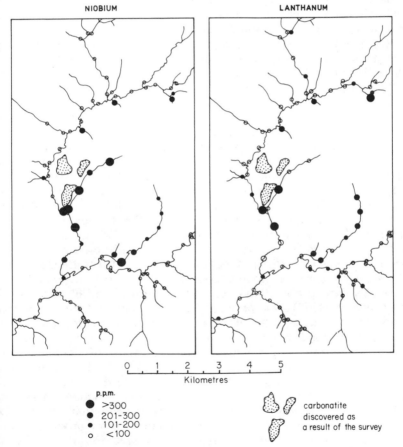

FIG. 4.11. Stream sediment anomalies which led to the discovery of
carbonatite in eastern Uganda.

by running water are mostly deposited as deltaic sediments near the point of entry and only the finest particles are carried into the centres of lakes. Thus, lake bottom sediments tend to be muds and, if water circulation is poor, there may also be an accumulation of organic matter.

Lake environments vary considerably depending upon many factors such as size, depth, climatic conditions, nature and volume of incoming water, etc. In temperate regions with hot summers and cold winters lake waters become stratified in winter and summer owing to temperature differentials between surface and bottom waters. When this happens, water circulation is restricted to the near-surface waters. In the spring and autumn the temperature differences even out and water circulation may extend to the bottom of the lake providing it is not exceptionally deep; in very deep lakes the bottom waters may be permanently stagnant. Lakes in tropical regions often have a poor and irregular circulation except for those at high altitudes where there is enough heat loss to prevent stratification so that waters in these lakes may circulate continually.

Although lake sediments are affected by many factors so that anomaly patterns may not only vary considerably from lake to lake but even within a single lake, lake sediment sampling may be a useful reconnaissance exploration method in certain regions of the world. This is particularly true for glaciated areas with poor drainage and numerous lakes. Both near-shore and centre-lake sediments can be used, though centre sediments are preferred as near-shore sediments are often largely subaqueous bank soils. Nevertheless, Minatidis and Slatt (1976) found in a survey of part of the Kaipokok region of Labrador that near-shore lake sediments can be a guide to anomalous areas. In addition organic-rich muds are generally the best sampling medium. Lakes with rocky, sandy bottoms do not lend themselves to lake sediment surveys. Good accounts of lake sediment sampling are given by Allan (1971), Allan *et al* (1973), Hoffman and Fletcher (1976) and Cameron (1977). One problem with lake sediment surveys is that interpretation may be difficult because of all the variables that can be involved. A good knowledge of the drainage, water circulation, chemical environment, regional geology, influx of sediment and general limnology (study of lakes) is required if the source of a lake sediment anomaly is to be tracked down. As a word of caution it must be remembered that spurious anomalies are not uncommon in lake sediments. For example, Cameron (1977) showed that Ni and Co derived from unmineralized metasedimentary rocks produced very strong anomalies in lake sediments in an area in the Northwest Territories, Canada.

Vegetation anomalies are caused by plants taking up trace elements from the soil and concentrating them in leaves, stems, twigs, bark, etc. Whether or not a vegetation anomaly will be present over a mineral deposit depends on a number of factors. Different plant species vary widely in their uptake of elements and an anomaly may be shown by some plants and not by others growing on the same site. Trace element contents in plants may vary seasonally and different organs of the plant generally vary in their ability to concentrate metals. In addition it is important for the element of interest to be in a soluble form available to the plant. In certain circumstances the detection of vegetation anomalies can be a useful method in exploration geochemistry and vegetation surveys are described and discussed in Section 4.8.

Water anomalies in wells, springs, rivers and lakes form whenever chemical weathering of mineral deposits or source rocks results in elements of interest passing into solution. The metal content of most natural waters is present mainly as simple cations, but a certain amount may also be present as ions adsorbed on suspended mineral particles or organic matter or as suspensions of very fine insoluble mineral particles. The concentrations of trace elements are normally very low with most being present at a few ppb (where 1 billion = 10^9) or tens of ppb. The detection of natural water anomalies (hydrogeochemistry) can be a useful method in exploration geochemistry. One of the main problems is that anomalies can be extremely variable and dispersion erratic owing to loss of metals at so-called precipitation barriers which may occur whenever the Eh, pH or chemical environment changes. Hydrogeochemical surveys are briefly described in Section 4.8.

Pathfinder elements

In geochemical exploration it is sometimes advantageous to use an associated element as an indicator of the element sought. Such an indicator element is known as a pathfinder. There may be a number of different reasons for deciding to use a pathfinder in an exploration programme. In the case of rare elements a more common and abundant associated element may not only be easier to analyze, but may result in a much better definition of possible targets. An example of this is As which has been used as a pathfinder for Au in various parts of the world. For example, James (1957) carried out surveys over gold mineralization in Rhodesia and found that the As content of soils was on average 400 times the Au content over known gold mineralization which was well defined by As anomalies. Another type of pathfinder is one that is chosen simply on the grounds that it is easier and cheaper to analyze. An example of this is the use of Pb or Zn as pathfinders for fluorite. Before the specific-ion electrode technique for the analysis of F was developed, F analyses were slow and costly and, since small amounts of galena and other minerals generally accompany fluorite mineralization, geochemical exploration for fluorite could be undertaken by analysing for Pb and other base metals. Pathfinders may also be chosen on the grounds of their wider dispersion than the main element thus presenting a larger 'target'. Obvious candidates for this type of pathfinder are gases which may diffuse upwards through rocks and overburden and thus provide a means of detecting buried and concealed mineralization. Gaseous pathfinders that have been used include Rn for U, He for U and hydrocarbons, Hg vapour for various base metal ores and SO_2 for sulphide deposits in general (McCarthy, 1972). These pathfinders probably have the best potential for future work and are briefly described below.

Radon gas is part of the decay series of both uranium and thorium, but whereas Rn-220 derived from thorium only has a half-life of 54·5 s, Rn-222 derived from uranium has a half-life of 3·8 days and thus any radon detected in groundwater or soil gas will be largely due to the presence of uranium. Since Rn is an α-emitter, its detection in a gas sample simply involves the detection of α-particles. One method of carrying out Rn soil gas surveys is to use an instrument known as a *radon emanometer*. To carry out a measurement a perforated tube is pushed into the soil to a depth of 1–1·5 m and a sample of gas is drawn into a chamber containing a zinc sulphide α-particle detector. If any Rn-222 is present in the gas sample, α-particles given off as it decays cause a fluorescence of the zinc sulphide which is detected by a photocell. Radon in water samples can be measured by drawing air through the water into the detection chamber of the instrument. Another technique of alpha detection involves the use of special cellulose nitrate films which are sensitive to α-particles and are unaffected by β-particles, γ-radiation or light (Basham and Easterbrook, 1977). In the TRACK ETCH® method patented by the Terradex Corporation of the United States α-sensitive films are fastened in the bottom of plastic cups which are placed in an inverted position in holes in the ground. Left undisturbed for a period of 3–4 weeks any Rn that diffuses into the cup will be recorded on the film when it decays and releases α-particles. Any other α-emitters that may be present in the soil will not register on the film as they are unable to penetrate the plastic cup or traverse the full air space in the cup. The films are sent to the laboratory where they are treated and the etched tracks counted. Results are quoted in terms of tracks per square millimetre (T/mm^2). TRACK ETCH® has an advantage over the radon emanometer in that the Rn is measured over a long period of time and is thus much more likely to give consistent results. Radon emanometers measure the Rn content of a small volume of gas at one moment in time and results are often extremely erratic and may show poor repeatability. Beck and Gingrich (1976) give an account of a TRACK ETCH® orientation survey over uranium mineralization in northern Saskatchewan clearly showing that the method can detect uranium mineralization at depth (Fig. 4.12).

Helium atoms are formed whenever α-particles, given off during radioactive decay, lose their charges. There are eleven α-emitters in the U-238 decay series, nine in the U-235 decay series and eight in the Th-232 decay series. Thus, He can be expected to accumulate to an appreciable extent in minerals, groundwaters and soil gases over U and Th deposits. Helium is also known to accumulate in some petroleum

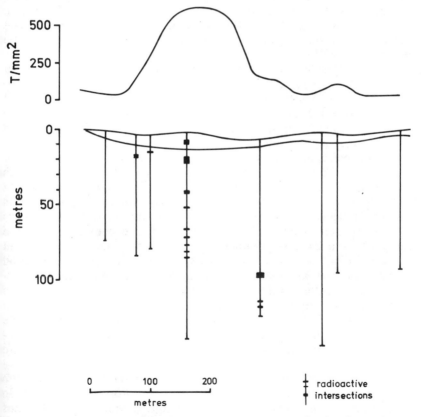

FIG. 4.12. Track etch anomaly over uranium mineralization in Saskatchewan (after Beck and Gingrich, 1976).

and natural gas traps and so He has potential as a pathfinder for both U (Dyck, 1976) and hydrocarbons. Being a noble gas, however, He is not easy to analyze and requires a mass spectrograph. Until recently these were costly pieces of equipment, but less expensive field instruments for the detection of He are now available. A number of surveys using He as a pathfinder for U have been conducted with varying degrees of success in various parts of the world, but owing to its widespread occurrence, derivation from both Th and U and its inert properties, it is not nearly as useful as Rn-222 for U exploration.

Numerous and different types of base metal deposits may contain Hg either as discrete Hg minerals or in trace amounts within the lattices of various ore minerals, though complex Pb–Zn–Ag ores usually contain the highest levels. Portable spectrometers are available (Robbins, 1973) to measure the quantity of Hg in a gas sample and can be used for carrying out surveys of Hg content in soil gas in the same manner as the radon emanometer is used for Rn. The instrument is extremely sensitive and very low levels of Hg can be detected, though in wet, organic-rich soils the response may be poor as such soils tend to retain Hg. In addition to detecting Hg vapour in soil gas, ordinary soil sampling surveys can be carried out for Hg, but the samples should not be force-dried before analysis as this may result in loss of Hg. Friedrich and Hawkes (1966) give an example of the use of Hg as a pathfinder in an area in Mexico. Although Hg would appear to offer a good tool for detecting concealed ore deposits, only limited success can be claimed for it to date. In many instances when it has been demonstrated that a base metal deposit is associated with a Hg anomaly, the ordinary base metals in the deposit give equally good and often better anomalies. Also the Hg content of ore minerals in different deposits can be extremely variable as shown by Sears (1971), who determined the Hg content of 600 ore samples from three gold mines and 24 base metal mines in Quebec, and Jonasson and Sangster (1974) who determined the Hg content of sphalerite from 66 sulphide deposits all over Canada. In addition elemental Hg may remain locked in the sulphides and not be released unless the deposits are undergoing active weathering. Nevertheless, the detection of Hg vapour does have potential as a method for locating blind sulphide deposits (McNerney and Buseck, 1973) and it is probable that successful discoveries based on Hg vapour detection will be made in the future.

The oxidation of sulphide minerals results in the formation of SO_2 which, therefore, has potential as a pathfinder for shallow sulphide deposits in general. Various methods are available for the analysis of SO_2, one of the more commonly employed being correlation spectrometry (see Chapter 3). The use of SO_2 is unlikely to be as important a method as Hg 'sniffing', however, since SO_2 will only be released by oxidizing sulphides. Meyer and Peters (1973) give an example of an SO_2 anomaly over massive sulphides in Newfoundland.

Although it is not strictly a geochemical method, it is worth mentioning that dogs have been successfully trained to 'sniff out' sulphides concealed by overburden. It is not known what gases the dogs detect but it is thought not to be SO_2 or H_2S. Nilsson (1973) and Hyvarinen et al (1973) give some interesting examples of the use of dogs for prospecting in glaciated regions of Scandinavia.

In addition to the use of gases for detecting metallic ore deposits, the detection of various hydrocarbons is used in geochemical exploration for petroleum. These are not strictly pathfinders as they represent major components of the numerous hydrocarbons contained in oil or gas deposits. Although this technique has been used for surface prospecting for petroleum, it is of most use in offshore work and in evaluating potential oil-bearing strata from studies of the various hydrocarbon gases in boreholes.

In addition to the more common pathfinders, research has been undertaken by various workers on a wide range of indicator elements. Iodine vapour has been proposed as a possible pathfinder for both metallic ore deposits and petroleum (see Chapter 3), but no success can be claimed for it to date. Palladium and iridium have been used to evaluate Ni gossans (Travis et al, 1976) and rhenium has been proposed as a pathfinder for porphyry coppers (Coope, 1973), though there is no evidence yet that it is a feasible method. Some workers can be criticized for completely losing sight of the purpose of a pathfinder. There is no point in using a pathfinder if the major element of interest is easier to analyze and produces anomalies as good or nearly as good as the proposed pathfinder. Learned and Boissen (1973) used Au as a pathfinder for porphyry Cu mineralization in Puerto Rico and, although the Au anomalies could be claimed to define one of the sites of Cu mineralization somewhat better than the Cu anomalies, it does seem to defeat the main aims in using a pathfinder element.

Summary

The dispersion of elements in the secondary environment is dependent on a great number of variables and a wide variety of anomalous patterns can develop, some of which may be related to mineralization and some of which may not be. There is an extremely large literature on geochemical exploration and countless case histories for numerous different elements have been documented. Basically, exploration geochemistry is a simple technique, but interpretation may not be so easy as there are numerous variables and few general

TABLE 4.7

SUMMARY OF THE DISPERSION OF VARIOUS ELEMENTS IN THE SECONDARY ENVIRONMENT AND APPLICATIONS IN EXPLORATION

ANTIMONY

Soils:	5 ppm.
Waters:	1 ppb.
Mobility:	Low.
Uses:	Geochemical prospecting for Sb has been undertaken, but is not very important. Both Sainsbury (1957) and Chakrabarti and Solomon (1971) describe surveys for Sb. It has also been used as a pathfinder for gold (James, 1957) and may produce coincident anomalies over some base metal deposits (Hawkes, 1954).

ARSENIC

Stream sediments:	1–50 ppm.
Soils:	1–50 ppm.
Waters:	1–30 ppb.
Plant ash:	1–2 ppm, >10 ppm may indicate mineralization. Concentrations up to 1% observed in certain plants growing over mineralized zones.
Mobility:	Fairly low, readily scavenged by iron oxides.
Uses:	Has been mainly used as a pathfinder for Au and Ag vein-type deposits.

BARIUM

Soils:	100–3000 ppm. Anomalous concentrations over barite mineralization >5000 ppm. Peaks at many percent.
Waters:	10 ppb.
Mobility:	Low.
Uses:	Has been used in geochemical prospecting for barite, but dispersion limited by low mobility.

BERYLLIUM

Stream sediments:	<2 ppm. Values >2 ppm may delineate areas of beryl mineralization.
Soils:	<2–6 ppm. Values >10 ppm may define beryl-bearing pegmatites. Peak values >100 ppm over rich zones.
Mobility:	Low to moderate.
Uses:	Be has been used in geochemical exploration for beryl deposits (Debnam and Webb, 1960; Reedman, 1973). Similar anomalous values may occur over unmineralized alkaline rocks (Reedman, 1974).

BISMUTH

Soils:	<1 ppm. Values >10 ppm may define Bi mineralization.
Mobility:	Low.

TABLE 4.7 (*Contd.*)

BISMUTH (*Contd.*)

Uses: Little work has been done with geochemical prospecting for Bi. Most Bi is produced as a by-product of other ores and there are only a few very small deposits that have been worked for Bi alone. Surveys in Zambia show peak values of 200 ppm over Bi-bearing vein deposits. May also have value as a pathfinder for certain vein Au deposits.

CADMIUM

Soils: <1–1 ppm. Values over a few ppm are anomalous and may be due to mineralization containing traces of Cd.

Mobility: High—closely follows Zn.

Uses: As in the case of Bi, Cd is produced as a by-product of other ores (lead–zinc) so that there has been little work done on prospecting for Cd. It has been used as an aid in lead–zinc prospecting to distinguish between anomalies likely to be due to mineralization (Zn + Cd) from those unlikely to be due to mineralization (Zn only). Surveys in Ireland have shown that this can be misleading since very high Cd values (>200 ppm) have been found with a Zn anomaly apparently unrelated to mineralization and low Cd values (a few ppm) are associated with a strong Zn anomaly related to good mineralization.

CHROMIUM

Stream sediments: 5–1000 ppm.

Soils: 5–1000 ppm. Values >1000 ppm may be due to chromite mineralization, but can also be due to un-mineralized ultrabasic rocks.

Mobility: Low, but may be high under high Eh and pH conditions if released as the chromate ion.

Uses: Has been used in chromite prospecting where values >1% Cr in residual soil overlie chromite-bearing rocks, but chromite is so readily identified in heavy mineral concentrates that geochemical prospecting for chromite is rarely necessary. Has been used as a pathfinder for ultrabasic rocks which may contain Pt-group metals.

COBALT

Stream sediments: 5–50 ppm.

Soils: 5–40 ppm. Anomalous concentrations over mineralization >100–500 ppm.

TABLE 4.7 (*Contd.*)

COBALT (*Contd.*)

Waters:	0·2 ppb.
Plant ash:	9 ppm.
Mobility:	Moderately high, but readily scavenged and held by Fe–Mn oxides.
Uses:	Has been used for Co prospecting, but, since Co is generally produced as a by-product of other metals, surveys are rarely conducted for Co alone. Useful as an ancillary element in surveys for other base metals which may be accompanied by Co mineralization.

COPPER

Stream sediments:	5–80 ppm. >80 ppm may be anomalous.
Soils:	5–100 ppm. Anomalies >150 ppm may indicate mineralization. High background basic rocks can give rise to values of many hundreds of ppm.
Waters:	8 ppb. >20 ppb may be anomalous, but hydrogeochemistry rarely used for Cu owing to limited mobility.
Plant ash:	90 ppm. Values >140 ppm may be anomalous.
Mobility:	High at pH's below 5·5, low at neutral or alkaline pH. Also may be adsorbed by organic matter and coprecipitated with Fe–Mn oxides, but Cu is less readily scavenged by Fe–Mn oxides than other base metals (e.g. Co, Zn, Ni).
Uses:	Stream sediment and soil sampling surveys have been widely used in all parts of the world in Cu prospecting and there is a large literature on the subject. Biogeochemical methods have also been used with some success. To help distinguish anomalies due to unmineralized basic rocks from anomalies likely to result from mineralization the Co/Ni ratio has been used in soil surveys. A high Co/Ni ratio (>1) indicates that anomalous Cu values are more likely to be due to mineralization than Cu anomalies accompanied by low Co/Ni ratios.

FLUORINE

Soils:	200–300 ppm. Anomalies over mineralization >1000 ppm with peaks at many thousands of ppm.
Waters:	50–500 ppb. Values >1000 ppb in river waters may be due to mineralization.
Mobility:	Fairly low.
Uses:	Geochemical surveys have been undertaken for fluorite in various parts of the world using soils, ground-

TABLE 4.7 (*Contd.*)

FLUORINE (*Contd.*)

waters and river waters as sampling media. F now commonly used as a direct indicator, but Pb and/or Zn generally used as pathfinders before advent of specific-ion electrode analytical technique (Farrell, 1974, Friedrich and Pluger, 1971).

GOLD
Soils: <10–50 ppb. Values >100 ppb may indicate mineralization.
Waters: 0·002 ppb
Mobility: Generally extremely low under neutral, alkaline and reducing conditions, but may be moderately high with formation of complex ions under oxidizing conditions in both acid and alkaline environments (Lankin *et al*, 1971).
Uses: A number of soil surveys using Au as a direct indicator of Au mineralization have been conducted in various parts of the world with considerable success. Before cheap and sensitive AAS (see p. 185) analytical method for Au was available, the use of pathfinders such as As and Sb was common, but not used so widely nowadays (Brown and Hilchey, 1974).

HELIUM
Atmosphere: 5·2 ppm by volume.
Waters: $4·76 \times 10^{-8}$ cm^3 STP/g (Clarke and Kugler, 1973).
Mobility: Extremely high as an inert gas dissolved in waters and diffusing through overburden and fractures in rock.
Uses: Pathfinder for U and hydrocarbons using both soil gas and He dissolved in groundwaters.

LEAD
Stream sediments: 5–50 ppm.
Soils: 5–80 ppm. Values >100 ppm may indicate Pb mineralization.
Waters: 3 ppb.
Plant ash: 70 ppm.
Mobility: Low.
Uses: Geochemical surveys for Pb using soils and stream sediments have been successfully employed all over the world. Biogeochemical and hydrogeochemical surveys have also been used with a certain amount of success. Owing to the low mobility of Pb, Zn is often a better indicator of Pb or Pb–Zn mineralization. Pb has been used as a pathfinder for barite and fluorite mineralization.

TABLE 4.7 (*Contd.*)

LITHIUM

Stream sediments:	10–40 ppm.
Soils:	5–200 ppm.
Waters:	3 ppb.
Mobility:	Moderate to high.
Uses:	Stream sediment and soil surveys have been used in regional reconnaissance prospecting for various pegmatite deposits since complex Li-bearing pegmatites generally contain minerals of interest such as beryl, cassiterite, pollucite, columbite, in addition to the Li minerals which are of potential economic value. Rarely used.

MANGANESE

Stream sediments:	100–5000 ppm.
Soils:	200–3000 ppm.
Waters:	<1–300 ppb.
Plant ash:	4800 ppm.
Mobility:	Usually very low, may become mobile under acid, reducing conditions as divalent ion.
Uses:	Soil and vegetation surveys have been conducted in prospecting for Mn ores, but Mn is more commonly used as an ancillary element in geochemical surveys to aid interpretation.

MERCURY

Stream sediments:	<10–100 ppb
Soils:	<10–300 ppb. Values >50 ppb may indicate mineralization such as Pb–Zn–Ag ores.
Soil gas:	10–100 ng/m^3, >200 ng/m^3 over base metal ores.
Waters:	0·01–0·05 ppb. Values >0·1 ppb may be due to Hg mineralization. Hg in waters readily adsorbed by solids, so waters are not good prospecting medium.
Mobility:	Generally low, but high as vapour phase.
Uses:	Has been used successfully in prospecting for Hg ores using stream sediments and waters (Dall'Aglio, 1971) and soils. Also used as a pathfinder of base metal ores (Friedrich and Hawkes, 1966). The vapour phase which can be detected in very small amounts in soil gas or the atmosphere has potential as a pathfinder of many ores (McNerney and Buseck, 1973). However, this is only true if Hg is present in elemental state. Many ores which contain Hg in sulphides may not release any Hg vapour unless undergoing weathering.

TABLE 4.7 (*Contd.*)

MOLYBDENUM

Stream sediments:	< 1–5 ppm. > 10 ppm may indicate Mo mineralization.
Soils:	< 1–5 ppm. > 10 ppm may indicate Mo mineralization.
Waters:	< 1–3 ppb.
Plant ash:	13 ppm. Very high Mo concentrations (> 1%) have been found in the ash of certain plants growing over Mo deposits.
Mobility:	Generally high, but is low under acid and reducing conditions when it is readily adsorbed by iron oxides and clay minerals.
Uses:	Stream sediment, soil and vegetation surveys have all been successfully employed in prospecting for Mo deposits. Mo is also used as a pathfinder for porphyry Cu deposits.

NICKEL

Stream sediments:	5–150 ppm.
Soils:	5–500 ppm. > 500–several thousand ppm may indicate mineralization.
Waters:	<1–10 ppb.
Plant ash:	65 ppm.
Mobility:	Fairly high.
Uses:	Stream sediment, soil and vegetation surveys have all been successfully employed in prospecting for Ni (e.g. Philpott, 1974). The Ni/Cr ratio has been used to distinguish between soil anomalies likely to be due to mineralization from those due to unmineralized ultrabasics. A Ni/Cr ratio <1 indicates no mineralization (Cox, 1974). Cu anomalies >200 ppm usually accompany Ni anomalies in soil over mineralization. High Ni values similar to those over mineralization may occur over unmineralized basic or ultrabasic rocks. Ni has been used as a pathfinder for kimberlites in diamond prospecting where country rocks are low in Ni.

NIOBIUM

Stream sediments:	5–200 ppm. Values > 200 ppm may indicate Nb-bearing minerals.
Soils:	5–200 ppm. Values > 200 ppm may indicate Nb-bearing minerals.
Mobility:	Low.
Uses:	Both stream sediment and soil surveys have been successfully employed to locate pyrochlore-bearing carbonatites (Bloomfield *et al*, 1971; Reedman, 1974) and columbite-bearing pegmatites. Unmineralized or poorly mineralized alkaline rocks may give high values in stream sediments and soils.

TABLE 4.7 (*Contd.*)

PHOSPHORUS

Stream sediments:	100–3000 ppm.
Soils:	100–3000 ppm. Values >5000 ppm may indicate phosphate-rich rocks.
Mobility:	Despite the fact that P is essential to life and is taken up by plants from soils, P generally occurs only in sparingly soluble compounds and overall mobility is low.
Uses:	Geochemical prospecting for P has only been used rarely, but it works extremely well in locating phosphate-rich rocks (Bloomfield *et al*, 1971; Reedman, 1974).

RADIUM

Stream sediments:	Measured in terms of radioactivity, usually picocuries/gram (pCi/g). 0·2 pCi/g. Values >1·0 pCi/g may indicate U mineralization.
Mobility:	Fairly low, adsorbed by organic matter.
Uses:	Can be used as a pathfinder for U in stream sediments and soils (Morse, 1971; Sutton and Soonawala, 1975).

RADON

Soil gas:	Measured by α counts. Over U mineralization values may be several hundred α counts/min with short measuring time of radon emanometer.
Waters:	Measured in terms of radioactivity, usually picocuries/litre (pCi/litre). 10–30 pCi/litre. Values >100 pCi/litre may be due to U mineralization.
Mobility:	Extremely high as an inert gas dissolved in waters and diffusing through overburden and fractures in rock.
Uses:	Rn in soil gas and waters is widely used as a pathfinder for U mineralization. Extensive dispersion haloes cannot form owing to the short half-life (Morse, 1971; Stevens *et al*, 1971; Beck and Gingrich, 1976).

RARE EARTHS

Of the rare earths (RE) Ce, La and Y have been used in geochemistry most commonly and some figures for La (pathfinder of cerian sub-group) and Y (representative of yttrium sub-group) are given.

Stream sediments:	20–500 ppm La.
Soils:	20–1000 ppm La. Values several thousand ppm+ may indicate RE mineralization. <10–100 ppm Y.
Plant ash:	16 ppm (total RE).
Mobility:	Moderately low.
Uses:	La has been used successfully in stream sediment and soil surveys for locating carbonatites with which RE

TABLE 4.7 (*Contd.*)

RARE EARTHS (*Contd.*)

minerals may be associated (Reedman, 1974). RE elements may also occur replacing Ca in minerals such as apatite and perovskite and may result in soil values similar to those due to the presence of discrete RE minerals such as monazite.

SILVER

Soils:	<0·1–1 ppm. Values >0·5 ppm may indicate mineralization.
Waters:	0·01–0·7 ppb.
Mobility:	Fairly low.
Uses:	Has been used in prospecting for Ag and Ag–Au deposits. Sometimes also a useful ancillary element for surveys for complex ores which are accompanied by significant Ag contents.

TIN

Stream sediments:	<5–10 ppm. Values >20 ppm may indicate mineralized areas.
Soils:	<5–20 ppm. Values >50 ppm may indicate mineralization.
Mobility:	Low.
Uses:	Stream sediment and soil surveys have been successfully employed in Sn prospecting in various parts of the world. Owing to the ease of identifying cassiterite in heavy mineral concentrates, however, traditional prospecting methods are often better than geochemical methods if Sn is present in the coarser size fractions.

TITANIUM

Stream sediments:	500–10 000 ppm.
Soils:	500–10 000 ppm.
Waters:	3 ppb.
Mobility:	Low.
Uses:	Owing to ease of identifying ilmenite and rutile in heavy mineral concentrates, geochemical prospecting for Ti has hardly ever been undertaken. Often used as an ancillary element in regional surveys where it often has considerable value for delineating different rock types.

TUNGSTEN

Stream sediments:	<2–10 ppm. Values >10 ppm may indicate mineralized areas.

TABLE 4.7 *(Contd.)*

TUNGSTEN (Contd.)

Soils:	<2–20 ppm. Values >20 ppm may indicate mineralization and values >200 ppm observed over main ore zones.
Mobility:	Low to moderate.
Uses:	Stream sediment and soil surveys have been successfully employed in various parts of the world in prospecting for tungsten deposits.

URANIUM

Stream sediments:	<1–5 ppm. Values >5 ppm may be due to mineralization.
Soils:	<1–10 ppm. Values >10 ppm may be due to mineralization.
Waters:	<1–1 ppb. Values >2 ppb may indicate mineralization.
Plant ash:	0·6 ppm.
Mobility:	Extremely high, though readily held by organic matter.
Uses:	Stream sediment, soil, vegetation and water surveys have been successfully employed in uranium prospecting (Morse, 1971).

VANADIUM

Soils:	20–500 ppm.
Waters:	<1 ppb.
Plant ash:	22 ppm.
Mobility:	Low.
Uses:	Little use has been made of V in geochemical prospecting, though it is sometimes used as an ancillary element in regional surveys. Can be used to indicate V-rich sulphide deposits.

ZINC

Stream sediments:	10–200 ppm. Values >200 ppm may indicate mineralization.
Soils:	10–300 ppm. Values >300 ppm may indicate mineralization, but residual anomalies over good mineralization generally >1000 ppm.
Waters:	1–20 ppb. Values >20 ppb may indicate mineralization.
Plant ash:	1400 ppm.
Mobility:	High, but adsorbed by organic matter and readily scavenged by Mn oxides.
Uses:	Zn has been widely employed in stream sediment, soil, vegetation and water surveys all over the world with considerable success in prospecting for zinc, lead–zinc and complex base metal ores.

TABLE 4.7 (*Contd.*)

ZIRCONIUM

Soils:	50–600 ppm. Values > 1000 ppm indicate possible interesting concentrations of zirconiferous minerals.
Mobility:	Extremely low.
Uses:	Zr has been little used in geochemical prospecting. Owing to irregular and widespread distribution of zircon in igneous rocks and as a detrital mineral, soil values often show wide fluctuations.

Sources: data compiled from a variety of sources including Swaine (1955), Taylor (1966), Vinogradov (1959) for soils, Cannon (1960) for plant ash and Hawkes (1957) and Wedepohl (1969) for waters, in addition to a number of other sources cited in the table plus unpublished survey and company data available to the author.

rules that can be applied universally. Table 4.7 lists a number of different elements, indicating mobilities, average background contents in various sampling media and expected anomalous concentrations together with a brief summary of applications in applied geochemistry. The table can only be used as an approximate guide since it is not possible to set universal threshold levels. Concentrations that may be due to mineralization for a particular element in one area may only represent background concentrations in another area. Nevertheless, the table does serve to indicate ranges in the right order of magnitude. For example, 200 ppm Cu in a soil is not particularly high, but is is extremely high and anomalous for Be.

4.4 SOIL TYPES

The study of soils is known as *pedology* and has become a highly specialized discipline with a terminology and classification that is both bewildering and daunting to the non-specialist. This is made more confusing by the fact that usage tends to differ from country to country. Nevertheless, it is important that anyone working in exploration geochemistry should at least have a basic knowledge of soils and the soil-forming process since soil is such an important sampling medium.

Soil may be simply defined as a mixture, in varying proportions, of decayed and decaying organic matter (humus) and various mineral particles derived from the weathering of underlying rocks. The agriculturist would also add to the definition that soil is the medium in which plants grow and by which they are supplied with water and mineral nutrients. To the civil engineer soil has a somewhat looser definition and is synonymous with *regolith*, i.e. the unconsolidated material, of whatever origin, that nearly everywhere forms the surface of the land and rests on solid rock. The exploration geochemist would adhere to the stricter definition of the soil scientist.

The formation of soil (*pedogenesis*) is extremely complex, but involves two processes, weathering and profile development. In any given soil these processes may be present at various stages of advancement and, although they may, and generally do, take place simultaneously, weathering is the essential one and profile development may be weak or even absent.

Weathering

Weathering is accomplished through physical, chemical and biological agencies within the zone of influence of the atmosphere. Physical agents include frost action, alternate heating and cooling, solution and sand blasting. Chemical reactions include hydration, hydrolysis, oxidation, reduction, action of acids and cation exchange. *Hydration* involves the chemical combination of water with a substance, e.g. the hydration of haematite to form limonite. *Hydrolysis*, which is essentially the absorption of hydrogen and hydroxyl ions, is a primary factor in the breakdown of silicate minerals. The absorbed hydrogen ions release cations which are removed in solution or are available for further chemical reactions. *Oxidation* takes place under wet, aerated conditions and chiefly involves the alteration of ferrous and manganous compounds to oxides and hydroxides of higher valency. *Reduction* takes place under anaerobic, 'oxygen hungry' conditions causing compounds to give up oxygen. For example, ferric compounds are changed to ferrous ones and sulphates are converted to sulphides. The *formation of weak acids* in the weathering environment accelerates the attack on many minerals. These acids include carbonic acid, which is formed when carbon dioxide is dissolved in water, sulphuric acid formed by the oxidation of sulphides, and clays and colloidal organic matter which may adsorb hydrogen ions and thus act as weak acids. *Cation exchange*, or base exchange as it is sometimes called, involves the exchange of cations in certain minerals with other cations in solution. Clay minerals, particularly montmorillonite, have a high cation exchange capacity.

The role of biological agents in weathering is really not separable from physical and chemical agents as the action of plants and animals is to assist in both the physical and chemical processes. Plant roots penetrate and break rocks and small animals burrow into the ground to help in the destruction of soft rocks and facilitate the entry of air and water. The transpiration of plants adds carbon dioxide to the atmosphere, the acids built up around plant roots assist in the breakdown of minerals and numerous bacteria are responsible for many chemical reactions. For example, nitrifying, sulphonofying and iron bacteria derive their energy from the oxidation of simple inorganic compounds.

Thus, weathering is a complex process, strongly influenced by climate and topography, in which a number of different agencies acting in concert break down solid rocks into particles and convert the original minerals into various clays, hydrous oxides and other secondary minerals which are in closer equilibrium with their environment at the surface of the earth. Minerals vary in their resistance to the forces of weathering and many minerals such as quartz, muscovite, rutile, ilmenite, chromite, beryl and cassiterite are often found unaltered in soils, though they may have suffered a certain amount of fragmentation.

The soil profile

As the products of weathering accumulate to form the basic substance of the soil, chemical and biological agencies act on the soil in varying degrees, dependent on climatic conditions, to develop the soil profile. Soil scientists recognize four zones in the soil profile. These are:

A zone—zone of maximum biological activity and eluviation
B zone—zone of illuviation, i.e. accumulation by deposition and precipitation
C zone—weathered bedrock
D zone—parent material

The A and B zones, which are known together as the *solum*, may have the following subdivisions:

A_0—largely organic matter, partly decomposed
A_1—dark coloured horizon, organic-rich, mixed with mineral matter
A_2—light coloured horizon, zone of maximum eluviation
A_3—transitional between A and B
B_1—uppermost zone of illuviation
B_2—brown horizons, accumulation of clay minerals and sesquioxides
B_3—transitional to C

The A zone develops as rain water percolates downward removing soluble compounds and fine material such as colloidal clays and sesquioxides in suspension. Dead plant debris, which accumulates at the surface, decays to form dark coloured humus. Micro-organisms such as algae, fungi, bacteria, protozoa, nematodes, worms and insects play a major role in the decomposition of organic matter.

Below this dark, organic-rich layer is a light coloured horizon where leaching and eluviation reach a maximum. Below this A horizon, clays and sesquioxides are deposited, resulting in the characteristic brown, red–brown or yellow–brown colour of the B horizon. Other material in solution may also be precipitated in this horizon or may be carried down to the water table to pass eventually into the surface drainage. This process of leaching in the A horizon and accumulation in the B horizon is known as *podzolization*. Passing downward from the B horizon the zone of recognizable weathered bedrock, the C zone, is reached. In this zone the term *saprolite* is sometimes used for highly weathered rock which is soft and crumbly but retains the *in situ* texture and structure of the parent rock so that it may be recognized as weathered granite, schist, gabbro, etc. The C zone becomes less weathered downwards until the fresh parent material or D zone is reached.

World classification of soils
In the traditional classification of soils, which dates back to the latter part of the last century, climatic conditions were considered to be the dominant factor in pedogenesis and soil types were grouped in climatic zones (*zonal soils*). It was recognized, however, that many soils did not fit into these zonal groupings owing to sets of local conditions such as impeded drainage or the dominating influence of a particular parent material. These soils became known as *intrazonal soils*. Finally, a third group of young, immature and poorly developed soils, which did not fit into either of the above categories, was recognized. These soils were called *azonal soils*.

In the United States a new system of classification, based on measurable properties in the field, has been developed by the Department of Agriculture and has been designed to avoid some of the controversy that is inevitable in the traditional classification, which is mainly based on genesis. This new classification, known as the Comprehensive System of Soil Classification (CSSC), has a completely new terminology which is largely based on Greek and Latin roots. This has resulted in a long list of new names such as plinthaquepts, cryaquents, spodosols, nadurargids and inceptisols, which are both difficult to pronounce and remember. Although this new classification undoubtedly has its advantages, it is not used widely outside the United States and geologists prefer to use the older classification which is still the basic system used throughout the world and is simpler to the non-specialist.

Zonal soils
The zonal soils can be roughly divided into two main groups: *pedocals* and *pedalfers*. The pedocals are soils in which calcium

carbonate accumulates and the pedalfers are soils in which compounds of iron and aluminium accumulate. Pedocals require relatively
dry conditions in which to develop, whereas pedalfers are restricted
to wetter regions. An annual rainfall of 60 cm approximately
separates the pedocals from pedalfers. Although these terms are
widely used, they are not satisfactory for the purposes of
classification since some soils, such as Prairie soils, do not fit into
either category. Table 4.8 lists some of the great zonal soil groups and
the conditions under which they develop. The divisions cold, temperate and tropical imply conditions of temperature and are not
restricted to geographical zones. For instance, typical tropical red
earths develop in temperate regions if conditions are hot and wet
enough. Likewise, soils typical of temperate regions develop in up-
land areas of the tropics. Some of the important zonal soil groups are
described below:

> *Tundra soils.* These soils occur in the areas of permafrost and
> consist of brown soil at the surface, often rich in humus. Owing
> to the impeded drainage, a bluish-grey, water-logged subsoil rests
> on the permanently frozen material.
> *Podzols.* These are soils with well-developed A and B horizons.
> Main types include brown podzolic soils, grey–brown podzols
> and the red and yellow podzols of warm, humid regions.
> *Chernozems* (Russian for black earth). Very thick, black A
> horizon. Calcareous soil with calcium carbonate concretions in
> the C horizon.

TABLE 4.8
SOME OF THE GREAT ZONAL SOIL GROUPS AND THE CONDITIONS
UNDER WHICH THEY DEVELOP

Very cold zone		Tundra soils	
	arid	Brown soils Desert soils	↑ pedocals →
Temperate	semi-arid grasslands	Chestnut soils Chernozems Prairie soils	
	forest–grassland transition	Degraded chernozems Non-calcic brown soils	← pedalfers →
	timbered regions	Podzols	
Tropical	hot, humid	Tropical red earths	↓

Chestnut soils. These are similar to chernozems, but develop in areas of lower rainfall. They have a dark-brown A horizon.

Prairie soils. Deep, dark-brown or reddish-brown A horizon without lime accumulation in their profile.

Brown soils. Brown surface soils grading into a white or grey calcareous layer.

Desert soils. Poorly developed profile, light-coloured with little or no organic matter. There are grey desert soils and red desert soils and they are often overlain by calcareous material at the surface.

Tropical red earths. These include the yellow–brown and red–brown lateritic soils (latosols). They are thoroughly leached and are characterized by a thin, poorly developed A horizon and thick B horizon with marked accumulation of sesquioxides of iron and aluminium. The laterization process is really an extreme form of podzolization and, if it is very pronounced, it may result in the formation of laterites (iron rich) or bauxites (aluminium rich). Concentrations of Ti, Mn or Ni are also known in some laterites.

Intrazonal soils

Some of the main groups of intrazonal soils are listed in Table 4.9 together with the conditions required for their formation.

Bog soils. Characterized by thick accumulations of black, decaying vegetable detritus—essentially an exaggerated A_0 horizon. These may be acid (acid peat soils) or slightly alkaline (fen peat soils) if alkaline substances are dissolved in the marsh water.

TABLE 4.9

SOME OF THE MAIN INTRAZONAL SOIL GROUPS AND THE CONDITIONS REQUIRED FOR THEIR FORMATION

Conditions		Soil groups	
marshes, swamps, imperfectly drained areas		Bog soils Humic-gleys Meadow soils Ground-water podzols	
imperfectly drained arid regions		Saline soils (solonchaks) Alkali soils (solonetz) Soloths	increasing rainfall
lime-rich parent	wet dry	Brown forest soils Rendzinas	

Humic-gleys. Water-logged soils rich in organic matter and characterized by a green–grey colour due to the reducing, anaerobic conditions which keep iron in the ferrous state.

Meadow soils. They owe their character to a fluctuating water table so that the surface is very wet at certain times of the year. A thick, dark, organic-rich A horizon overlies a greyish, gleyed B horizon with rusty mottlings and spots.

Ground-water podzols. These soils are formed in areas where the water in the lower layers is constantly draining away allowing horizontal leaching and eluviation to take place. A typical profile is thin organic litter (A_0) over a grey A horizon (as opposed to the grey–green of a gleyed soil) overlying a mottled, yellowish-brown and grey B horizon.

Solonchak. These form in areas of low rainfall and are characterized by a light-grey colour and poorly developed profile with accumulations of soluble salts.

Solonetz. These soils form under somewhat wetter conditions than the solonchaks so that a certain proportion of the soluble salts are leached out.

Soloth. This is simply a solonetz with signs of podzolization due to a slightly wetter climate.

Rendzinas. Very thin soils of various colours containing fragments of limestone resting directly upon the limestone bedrock.

Azonal soils

There are three main groups of azonal soils:

Regosols. Develop on unconsolidated deposits such as moraine and sand dunes.

Lithosols. Develop on steep slopes. They are thin, stony soils containing fresh and weathered rock fragments.

Alluvial soils. Develop on recent alluvium.

Physical properties of soils

The following physical properties are commonly used in the description of soils:

Colour

This is one of the most obvious properties used in the general description of soils and is often a rough guide to the chemical and mineralogical composition. The identification of colour, however, is extremely subjective and it is very difficult to quantify. A number of attempts have been made by various workers to record colours, the best known system being the Munsell notation, but all systems suffer

from the difficulty that soil colours tend to vary with moistur
content, compaction and the nature and intensity of the illuminatin
light. In describing colour it is best to use a fundamental colour suc
as red, yellow, black, white, grey, green, brown and then modify it i
necessary using an adverb such as pale, reddish, greyish, dark, etc

Texture
To the farmer soil texture is the property that describes the ease o
difficulty of working the soil. A sandy soil is described as light and
clay soil as heavy, so that soil texture may be described according t
the particle size of the soil. The following particle sizes are used:

	very coarse	1·0–2·0 mm
	coarse	0·5–1·0 mm
sand	medium	0·25–0·5 mm
	fine	0·1–0·25 mm
	very fine	0·05–0·1 mm
silt		0·002–0·05 mm
clay		<0·002 mm

Soil textures can thus be described as silty clay, sandy loam, clay, silt
etc.

Structure
Soils *in situ* may have a definite structure which is described by
self-explanatory terms such as laminated, prismatic, columnar
blocky, etc.

Plasticity
The degree to which a moistened lump of soil can be moulded to
different shapes is known as its plasticity.

Cohesion
Used to describe the degree to which moistened lumps of soil hole
together.

Classification of local soils
On a local scale soils are classed in soil series which are defined a
being alike in: 1. geology, 2. mode of deposition (residual, water
wind, ice), 3. colour, 4. topography, 5. drainage (free, good, impeded
water-logged, etc.), 6. profile, 7. pH (generally varies from less than
5·5 for strongly acid soils to greater than 8·0 for strongly alkalin
soils) and 8. climate. A local place name may be used to identify the
series, e.g. Broadmoor Series, Lawrence Series. Slight differences in

he soils within a series may be described with textural terms, e.g.
.awrence loams, Lawrence sandy loams, etc.

4.5 STATISTICAL TREATMENT OF DATA

A considerable amount of work has been carried out on the statistical
reatment of geochemical data and there is now a large literature on
he subject. Much of the more recent research concerns the treatment
f multivariate data and involves the use of quite complex computer
rograms. To understand many of these newer techniques fully
equires a knowledge of advanced mathematics, a field in which few
xploration geologists have much training. In practical mineral
xploration, however, a simple statistical approach is all that is
enerally required and, although some of the advanced techniques
nay be of some value in certain cases, they more properly belong to
he realm of academic research.

Selection of anomalous values

The first task in looking at a set of geochemical analytical results from
. particular area or survey is to select the anomalous values which
night be indicative of a mineralized source. The level above which
alues may be considered anomalous is known as the threshold and
ias been mathematically defined as the mean value plus two standard
leviations, or simply the top $2\frac{1}{2}\%$ of values. A set of geochemical
esults may represent a statistical sample of one, two or more distinct
>opulations, it may represent an incomplete sample population, i.e.
>ne with an excess of low or high values, or it may be a combination
>f several of these. For this reason it is not really satisfactory simply
o calculate a threshold value according to the definition 'mean plus
wo standard deviations'. It is always best to make an appraisal of the
analytical results by studying a frequency distribution. One way of
loing this is to plot a histogram of values such as is shown in Fig.
..13. Although this is often adequate, it is not the clearest method of
analyzing the sample data. It is now generally recognized that most
geochemical distributions are log-normal, i.e. the logarithms of the
values form a normal distribution. One of the easiest ways of showing
his is to use log–probability graph paper on which log-normal dis-
ributions plot as straight lines. Figure 4.14 shows the same data as
iven in the histogram in Fig. 4.13 plotted as a cumulative frequency
:urve on log–probability paper. This set of data contains two distinct
>opulations, A and B, combined in the ratio 1:1 and, although the two
>opulations can be distinguished in the histogram, they are much
:learer in the curve A + B, plotted on log–probability paper. Figure

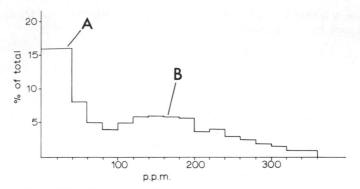

FIG. 4.13. Histogram showing two populations, A and B.

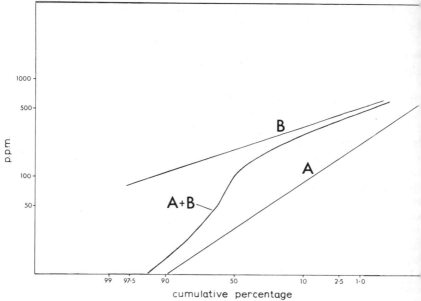

FIG. 4.14 Log–probability plot of the data in the histogram in Fig. 4.13.

4.15 shows the characteristic log–probability plots of distribution most commonly encountered.

Let us consider some actual examples. The analytical results for C and Zn from a soil sampling survey over a carbonatite plug in eastern Uganda are shown in Table 4.10. Note that the cumulative percentages have been calculated from highest to lowest. The resultant

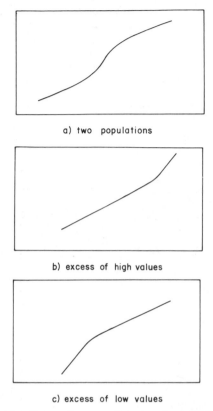

a) two populations

b) excess of high values

c) excess of low values

FIG. 4.15.   Some characteristic log–probability plots.

log–probability plots are given in Fig. 4.16. From these it is immediately apparent that Zn contains two populations whereas Cu indicates a single log-normal population. An approximate but adequate method for extracting the two Zn populations is to draw lines tangential to both tails of the curve passing through points equidistant from the point of inflexion. This has been done for the Zn curve in Fig. 4.16 and, using the definition that mean background is the 50 percentile and threshold the $2\frac{1}{2}$ percentile, the two populations have mean backgrounds of 100 and 300 ppm and thresholds of 150 and 760 ppm respectively. In this instance, this is a reflection of soils derived from two distinct rock types. In the case of Cu, the mean background of 80 ppm and threshold of 180 ppm can be read directly from the plotted line.

In cases where a population contains an excess of high values the

TABLE 4.10

DISTRIBUTION OF COPPER AND ZINC VALUES IN THE −150 MESH FRACTION OF 251 SOIL SAMPLES COLLECTED OVER THE LOLEKEK CARBONATITE, EASTERN UGANDA

Mid-class mark (ppm)	No. samples copper	Cumulative per cent	No. samples zinc	Cumulative per cent
25	19	100	—	—
50	68	92·4	6	100
75	98	65·3	69	97·6
100	41	26·3	91	70·1
125	15	10·0	33	33·9
150	8	4·0	—	—
175	—	—	1	20·7
200	—	—	6	20·3
225	1	0·8	3	17·9
250	—	—	7	16·7
275	—	—	2	13·9
300	1	0·4	—	—
350	—	—	4	13·2
375	—	—	2	11·6
400	—	—	2	10·8
425	—	—	1	10·0
450	—	—	1	9·6
500	—	—	7	9·2
550	—	—	3	6·4
600	—	—	3	5·2
650	—	—	3	4·0
700	—	—	1	2·8
775	—	—	1	2·4
900	—	—	2	1·9
1 000	—	—	3	1·2
	251		251	

threshold value should be taken at the break in the curve if this falls above the $2\frac{1}{2}$ percentile. A distribution of this type indicates an incomplete population with an excess of anomalous values and, as these might be significant, it is safest to take a threshold at a lower value than the mathematically defined 'mean plus two standard deviations'. In cases where the break occurs below the $2\frac{1}{2}$ percentile the threshold can be taken normally as for the straight line case. It is customary to divide anomalous values into two classes, possibly anomalous and probably anomalous, with the latter defined as the mean plus three standard deviations or the top $\frac{1}{2}\%$ of values. Some

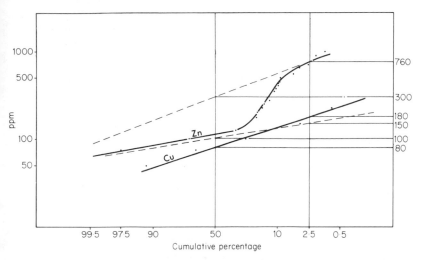

FIG. 4.16.   Log–probability plots of copper and zinc values in the − 150 mesh fraction of 251 soil samples from the Lolekek carbonatite, eastern Uganda.

workers have defined a further category, definitely anomalous, but this seems unnecessary.

Figure 4.17 shows the log–probability plot of over 20 000 zinc values from a soil sampling survey in Ireland. The curve suggests an incomplete population with an excess of high values and an abnormal number of low values. In this case the possibly and probably anomalous cut-offs were taken at the breaks in the curve at 160 and 300 ppm respectively. The excess of values below 55 ppm is due to a large number of samples from bog soils with low zinc values.

In making a statistical appraisal of geochemical data from any survey area, it is advisable to consider samples derived from different rock types or from different environments in separate groups. For example, if a soil or stream sediment sampling survey had been carried out over an area underlain by limestone and basement granite, it would be best to determine means and thresholds for two groups of sample values, limestone derived and granite derived. In areas of poor rock exposure it is not possible to do this and, in fact, a careful interpretation of the geochemical data may assist in elucidating the poorly known geology. In addition to geological differences it may be necessary to consider different sample types, e.g. low organic and high organic. This can be extremely important in stream sediment surveys; samples from marshy or boggy streams with high organic contents should always be considered separately from samples col-

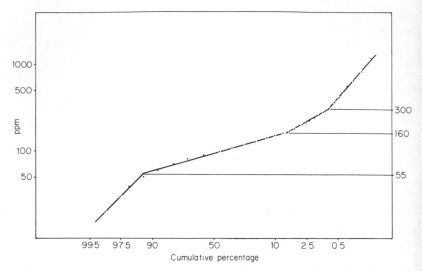

FIG. 4.17.   Log–probability plot of zinc values in the − 80 mesh fraction of over 20 000 soil samples from a survey in central Ireland.

lected from freely drained areas. A certain amount of judgement and common sense has to be exercised in appraising geochemical data and this grouping of sample types should not be taken to extremes. For instance, it is rarely necessary to divide the samples into more than three groups unless the survey area is very large.

In summary, it should be remembered that is is not necessary to plot graphs and histograms to determine the strongly anomalous values; they are obvious from a quick look at the data. A thorough statistical appraisal is important to select the low-order anomalies, which are not readily apparent and may be of more significance.

Moving average maps

The moving average or rolling mean technique is used to smooth irregularities in the survey data. Since this levels off both localized lows and highs, it tends to defeat the main aim in exploration geochemistry of defining anomalous sources. For this reason raw data maps should be used in preference to moving average maps for defining small scale exploration targets. In regional studies, however, the moving average technique is ideal for outlining broad trends and can be applied to both soil and stream sediment surveys. The technique is also extremely useful in the evaluation of eluvial and alluvial deposits where raw data maps may not show lateral grade variations very clearly.

To produce a moving average map a square template known as the 'search area' or 'window' is placed on the map and an arithmetic or geometric mean is calculated from all the values falling within the square. The search area is moved across the map in successive steps with an overlap of the previous positions and mean values are calculated for each position of the search area. The various mean values are plotted at the centre positions of each square and the results are contoured to produce the final moving average map. Two possible procedures for a soil sampling survey on a square grid are illustrated in Fig. 4.18. In the first case a search area covering four sample points is used with a 50% overlap. In the second case a search area covering nine sample points is used with a $33\frac{1}{3}$% overlap. In both examples the mean values calculated for the three consecutive positions of the search area shown are plotted at points 1, 2 and 3. This process is repeated until the whole survey area is covered. In producing a moving average map it is important to state the size of the

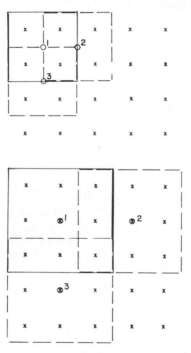

FIG. 4.18.   Diagram to show how different moving average maps can be produced for the same survey by varying the size and amount of overlap of the search area.

search area used together with the amount of overlap (Fig. 4.19). If there is no overlap, the resulting map is not strictly a moving average map, but may be referred to as a 'fixed mean map'. It is possible to use various shapes of search areas such as rectangles, circles and ellipses, but elongated search areas (rectangles, ellipses) are only used when it is desired to introduce a bias to the data. For example, they can be used to suppress or enhance a regional trend depending upon the orientation of the search area. However, these types of search areas are rarely used and should only be used with the greatest degree of caution.

Since the distribution of metal values in nature generally approximates to log-normal, many workers advocate using the geometric mean for geochemical moving average maps. While this may be adequate for outlining regional trends, it suppresses erratic 'highs' which may be significant in mineral exploration. To overcome this, anomalous areas are sometimes shown in conjunction with a moving average map by calculating standard deviations for each position of the search area. The resulting standard deviation map can be contoured and anomalous zones are those with highest values. This method does not distinguish between anomalously high or low areas, however, and raw data maps are to be preferred in mineral exploration.

Computer methods

Since geochemistry involves the collection and study of large amounts of numerical data, computers have special appeal to many geochemists. While this has resulted in many useful applications, there has been a tendency in some quarters to put too much reliance on computer methods; some workers may even be criticized for verging on the naive belief that computers can somehow extract valuable information from a set of data which contains nothing of real significance. The uses of computers in geochemistry can be placed under two main headings: (i) rapid processing of old manual methods and (ii) use of new techniques only possible with computers.

Under the first heading computers can be programmed to undertake tasks such as calculating thresholds, plotting contour maps and plotting moving average maps. In general exploration work, the cost of this is often unjustified and anomaly maps can be produced manually on a routine basis quite adequately as work progresses. The use of a computer only really becomes desirable if a large number of elements have been analyzed and if treatments such as moving averages are to be applied. Once data are on computer cards, it is a simple matter for a computer to run different programs to treat the data in various ways that would be too time-consuming by hand.

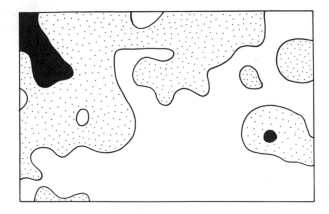

Raw Data Contoured

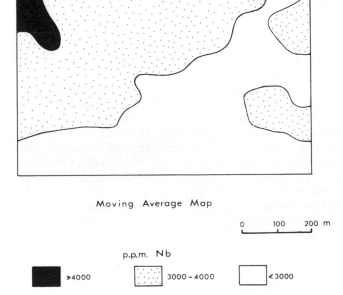

Moving Average Map

FIG. 4.19.   Raw data contours compared to moving average map for niobium values in soil over part of Sukulu carbonatite, Uganda. Sampled on a 60 m square grid to a depth of 3 m. Moving average calculated from four-sample search area with 50% overlap.

Under the second heading there is a wide range of techniques that can be applied to large sets of data. These include discriminant analysis, factor analysis, trend-surface analysis and correlation and regression analysis.

Correlation studies between different elements determined for a batch of samples from a particular area have become popular in computer work. The linear correlation coefficient (r) for two variables X and Y is defined as:

$$r = \frac{\Sigma(X_i - \bar{X})(Y_i - \bar{Y})}{\sqrt{\Sigma(X - \bar{X})^2 \Sigma(Y - \bar{Y})^2}}$$

where $r = +1$ shows perfect positive correlation, $r = 0$ shows no correlation and $r = -1$ shows perfect negative correlation. The calculations are simplified by the following relationships:

$$\Sigma(X - \bar{X})^2 = \Sigma X^2 - \frac{(\Sigma X)^2}{n}$$

$$\Sigma(Y - \bar{Y})(X - \bar{X}) = \Sigma XY - \frac{\Sigma Y \Sigma X}{n}$$

The measure is purely statistical and does not imply any causal relationship between the variables. For example, the correlation coefficient for La and Nb values from 251 soil samples taken over the Lolekek carbonatite complex in eastern Uganda is $+0.88$ indicating very good positive correlation and a sympathetic relation. However, although this is due to the fact that both La and Nb are enriched in carbonatite compared with other rock types, the elements are contained in different minerals. If correlation studies are extended to include the examination of correlation coefficients between all possible pairs of variables in a multivariate set of data, the method is known as *cluster analysis*, which involves the computation of similarity matrices and requires the use of a computer. Correlation coefficients can be used to examine the similarity or relation between samples (Q-mode) or between the variables (R-mode).

Regression quantifies the relation between variables by means of an equation. In the case of linear regression between two variables X and Y the equation is of the form:

$$Y = b_0 + b_1 X$$

where Y is said to be regressed on X and b_0 is the Y intercept and b_1 the slope of the line. For given values of X and Y the coefficients b_0 and b_1 can be calculated by the method of least squares (see Chapter 9). Both correlation and regression analysis can be carried out when causes of correlation between variables are poorly understood or not known and grouping of the variables according to their mutual correlations may increase one's understanding of geochemical relationships within a particular survey area. However, one must always be wary against too ready an acceptance of computer produced data. Mathematically defined relations do not necessarily imply geochemical relations. Chapman (1976) gives some examples of the pitfalls that may be encountered in correlation and regression analysis.

An extension of correlation analysis is *factor analysis* which is a technique to ascertain the dimensionality of multivariate observations. In other words, if N is the number of variables, factor analysis can be used to determine whether there are N independent variables or a smaller number which can therefore be contained in a smaller number of components or factors. The theory is quite complex and is fully explained in a textbook by Harman (1960) and briefly introduced in a number of books on statistical methods in geology such as Krumbein and Graybill (1965). A number of computer programs are now available to carry out the extremely involved computations and the technique has become quite popular in geochemistry. Both Q-mode and R-mode factor analyses are undertaken, but Q-mode is probably more useful as it emphasizes relations among samples. Factor analysis produces a number of different factor models and it is then necessary to decide which one or ones are most significant. For example, in a multi-element reconnaissance survey of a particular area a four-factor model may be judged to be most significant. It is then possible to calculate a factor loading or score for each sample point and maps showing the distribution of factor scores for each factor model can be produced. In some instances this may outline significant anomalous areas more clearly in terms of a particular factor model (associated elements) than a conventional map of single element concentrations. An example of this is given by Nichol (1973).

Another computer technique that has been used in geochemistry is *trend surface analysis*. This is a method of data treatment that has wide application in gravity surveying to produce residual gravity maps, but is of rather dubious value in geochemistry. Essentially the method is based on defining a surface in terms of a polynomial function and determining the best fit to a given set of geochemical data by using the method of least squares to calculate the coefficients of the polynomial. In theory polynomials up to any power can be used, but it is normal practice to restrict them to linear, quadratic or cubic ones. When a trend surface has been calculated, it is considered to represent the regional variation; the local departures or residuals from this trend surface are then taken as the anomalies. An example of this technique for a regional survey in Sierra Leone is given by Nichol et al (1969).

Discriminant analysis yields a function for assigning a number of measured variables to one of several groups on the basis of a linear function of the form:

$$aA + bB + cC + dD + \cdots + K$$

where A, B, C, etc. are the variables (geochemical analyses for different elements in this case), a, b, c, etc. are coefficients chosen to weight the variables according to their contribution to the discrimination and K is a constant. A simple example is given by:

$$D = 0.012Zn + 0.018Nb + 0.007La - 0.17Cu - 0.0006Ti$$

which is a discriminant function devised to distinguish ijolite-derived and carbonatite-derived soils from a carbonatite complex in eastern Uganda, positive values indicating carbonatite bedrock and negative values ijolite bedrock beneath the soil cover. An example of some actual results for this discriminant function is given in Table 4.11. The coefficients of a discriminant function are determined by taking into

TABLE 4.11

SOME SOIL SAMPLE VALUES FROM THE LOLEKEK CARBONATITE IN EASTERN UGANDA TO ILLUSTRATE THE DISCRIMINANT FUNCTION ABOVE

| | Values in ppm | | | | | Bedrock |
Cu	Zn	Nb	La	Ti	D	indicated
25	1 250	1 000	2 800	6 000	+45	carbonatite
150	90	50	100	20 000	−35	ijolite
70	600	230	1 000	20 000	−6	ijolite
40	1 000	450	1 800	10 000	+20	carbonatite
35	100	60	100	15 000	−12	ijolite

If $D > 0$, bedrock is carbonatite; if $D < 0$, bedrock is ijolite.

consideration the variance of the variables and maximizing the ratio of the difference between the sample means and the standard deviations within the groups. The calculations involve the determination of matrices and have to be done on a computer. The reader is referred to Krumbein and Graybill (1965) for a good introduction to discriminant functions and other statistical methods in geology. In geochemistry discriminant analysis is commonly employed to differentiate rock types, but it is also used as a means for distinguishing anomalies likely to be associated with mineralization from those not due to mineralization. Anomalies can also be grouped according to the type of mineralization that they are likely to be associated with, but dealing with more than two groups increases the complexity of the calculations considerably.

4.6 DRAINAGE SURVEYS

Geochemical drainage surveys are concerned with the search for anomalous metal contents in stream waters and sediments which might relate to possible mineral deposits. The sampling of sediments is of far greater importance than the sampling of stream waters and only stream sediment surveys will be considered here, though hydrogeochemical surveys are briefly discussed in Section 4.8.

Stream sediment surveys are ideally suited for geochemical reconnaissance of well dissected areas undergoing active erosion. In areas of mature and old drainage patterns with sluggish run-off the method is much less applicable. Even in such areas, however, stream sediment surveys can be useful in the initial reconnaissance stage, though the interpretation may be much more difficult as seepage and hydromorphic transport often play a dominant role.

In planning a stream sediment survey it is first necessary to establish the type of coverage and thus sample density required for the target(s) being sought. The sample density will be governed ultimately by the drainage density and size of catchment areas, but as a rough guide sample densities of one per $2\text{--}10 \text{ km}^2$ are used in large regional surveys and sample densities of one per $0.5\text{--}2 \text{ km}^2$ are commonly employed for more detailed reconnaissance sampling. For outlining regional geological features and metallogenic provinces sample densities as low as one per 100 km^2 or more have been used, e.g. Armour-Brown and Nichol (1970). If the area being sampled is completely new, it is advisable to undertake an orientation survey before embarking on the full scale survey. An orientation survey can be conducted over an area of known mineralization to determine not only the geochemical response of the mineralization in the drainage, but also the effective sample density and optimum size fraction for analysis. It is often difficult to convince exploration companies of the value of orientation surveys and, although they are not really necessary in a known environment, the relatively low cost and short time spent on an orientation survey may save considerable time and money later and may be a significant factor in the success or failure of a survey in an untested environment, particularly when prospecting for the less commonly sought elements.

Field procedures

Stream sediment surveys should be confined to small low-order streams. The actual limit will vary with each survey as it depends upon the targets sought, but large rivers with wide catchment areas should not be sampled. The sampling interval chosen will depend upon the degree of coverage required, but whatever interval is selected, one should make a point of sampling immediately upstream of all confluences. Each sample should be collected from the active part of the stream and, as only the fine sand and silt fraction is normally analyzed, the sample should contain as much fine material as possible. This is often quite difficult to achieve in mountainous areas with very active erosion and it may be necessary to search around behind boulders or in eddy pools to find sufficient fine material. Samples may be collected whether the stream is wet or dry and in some parts of the world with very seasonal rainfall, streams are more often dry than wet. Ample −80 mesh material for analysis is afforded by 80–120 g of sediment, placed in the standard Kraft paper sample packet. It is good practice when collecting samples to fill the sample packet at each sample site with several small amounts of sediment collected from different points within 5–10 m of each other. Stream sediment samples can be generally scooped up by hand, but in areas with steep banks and deep water this may be very difficult. In such cases a small coal shovel or garden trowel on the end of a broom handle facilitates sample collection. In all cases, the sampler must guard against collecting collapsed bank material instead of active sediment. Some workers advocate the collection of samples in duplicate from sites 15–20 m apart (e.g. Howarth and Lowenstein, 1971; Bolviken and Sinding-Larsen, 1973), but this is not really necessary and can reduce productivity in the field considerably.

The field crew should discuss and plan each day's sampling in the base camp the evening before and approximate sample sites can be marked on the maps or aerial photographs. Although careful and thorough planning is important for any survey, it is particularly important in remote areas where helicopters are commonly used for transport if field crews are to be deployed efficiently. Each sampler should carry sufficient pre-numbered sample packets for the day's work and the actual sample locations together with sample numbers can be marked on the maps as the samples are collected. It is rarely necessary to mark sample sites in the field as the precise sample location is not required when an anomalous area is revisited for follow-up work. Field notes should be taken at each sample site and entered in a notebook or on specially prepared field sheets. The use of the latter is recommended, but they should not be too involved or complicated. Some very comprehensive field sheets have been designed with a large number of items for the sampler to tick off or fill in This can result in reduced sampling productivity and, as most samplers are not trained geologists, the information is rarely recorded accurately. It is far better to have a limited amount of reliable information than a large amount of dubious data. The information recorded should include the amount of organic matter present, the nature of the stream and sediment, any outcrops present and whether there is any iron or manganese precipitate. In areas where the streams are flowing the measurement of pH of the waters can be useful. Figure 4.20 gives an example of a field sheet with recorded notes. This particular field sheet has columns for analytical results and, although this is very useful, it would not be practical if a large number of elements were being analyzed. Productivity in the field varies widely according to conditions and mode of transport, but it usually averages between 15 and 40 samples/man-day.

Project __Makondu Area__   Sampler __J.H.R.__
                                                          Date __29 - 9 - 75__

Sample No	Values in p.p.m.					Organic Content			Sediment				Remarks
	Cu	Pb	Zn	Ni	Co	H	M	L	VC	C	M	F	
2101	45	10	30					✓			✓	✓	stream 6m wide — sluggish flow
2102	50	15	35					✓				✓	granite outcrops stream 2m wide - sluggish flow
2103	30	15	55			✓						✓	swampy stream 5m wide
2104	65	10	35					✓	✓	✓			stream 3m wide — good flow brown sand
2105	55	15	40				✓	✓	✓				many granite boulders strong flow - 3m wide
2106	45	15	40				✓	✓	✓				''
2107	50	15	45					✓	✓	✓			granite outcrops stream 1m wide - strong flow
2108	35	10	35				✓				✓	✓	stream 3m wide sluggish flow - deep water
2109	90	20	120			✓						✓	swampy - much Fe hydrox precipitate
2110	60	15	95				✓		✓		✓		dry gully 2m wide - sandy

FIG. 4.20.   Example of a stream sediment survey field sheet.

Plotting and presentation of data
The first step in the presentation of the results of a drainage survey is to plot all the streams on a transparency or series of transparencies covering the area surveyed. A transparency for each element is laid over the base map on which the sample numbers are marked and the

respective values plotted. After carrying out a statistical appraisal of the data to select background and threshold values as described in Section 4.5, one of two standard presentations can be used: a dot map or worm diagram. In the first case each sample site is marked with a black dot, the size of which is related to the metal content of the sample. In the second case the thickness of the line showing the stream course between sample sites is related to the metal content of the sample immediately downstream. In neither case is it necessary to show actual values, though they can be included if desired. Although there is probably little to choose between the methods (Fig. 4.21), dot maps are preferable for low density surveys and worm diagrams are preferable for high density surveys. In the case of dot maps unsampled streams are obvious, but in worm diagrams unsampled streams can be shown with a dashed line. The divisions commonly used for exploration surveys are: background and less, background to possibly anomalous, possibly anomalous to probably anomalous, and greater than probably anomalous. For very strongly anomalous values further divisions can be used. Black and white dot maps or worm diagrams have the advantage that prints can be obtained readily for reports, but for working maps coloured worm diagrams with varying colours instead of line thicknesses are useful. The colours normally used for the divisions above are yellow, green, blue and red.

In mineral exploration stream sediment data should not normally be presented in the form of contoured maps. In regional surveys, however, which are undertaken to outline broad trends relating to regional geology and possible mineralized areas and provinces, it is normal practice to present the results of a stream sediment survey as a series of contoured maps. Owing to the irregular areal distribution of stream sediment data, this is usually done by dividing the area under consideration into a number of equal units or cells and applying a moving average technique or trend surface analysis.

Follow-up work
The type of follow-up work undertaken depends very much on the nature of the original survey. If it was a wide spaced reconnaissance survey, it would be quite usual for the first stage of the follow-up work to consist of closer interval sampling along the anomalous streams. If the sample density of the original survey was quite high, the follow-up work would normally take the form of a soil sampling grid over the area around and between the anomalous streams. As a preliminary stage to a detailed soil sampling grid, it is often useful to undertake *bank sampling*. This consists of sampling bank material from both sides of an anomalous stream which may show from which side of the river the anomaly originates and where the anomaly

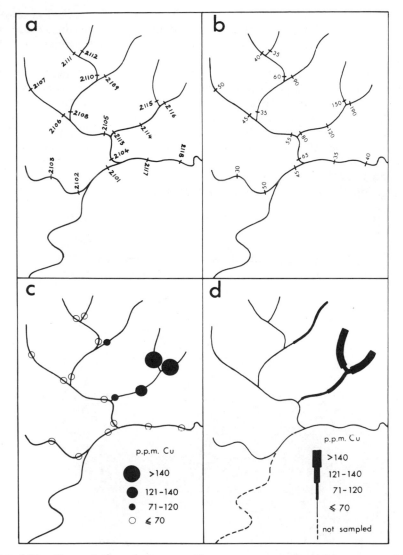

FIG. 4.21. Presentation of stream sediment survey results; (a) base map with sample numbers, (b) drainage map with copper values, (c) dot map, and (d) worm diagram.

cut-off occurs. If exposure in the area is very good, detailed geological mapping and prospecting may be sufficient to locate the source of anomalous elements. In most cases though, a soil sampling survey will have to be carried out.

If there are many anomalies and the initial survey area is very large, it will be necessary to be selective. It is normal practice to give the anomalies a priority rating, classifying them into first, second, third, etc. priority, though it is unnecessary to have more than four categories. This can be done most easily on the strength of the anomalies, but this is rarely satisfactory as other factors need to be taken into account. Some workers (for example Dubov, 1973; Hawkes, 1976) have attempted to do this on a purely mathematical basis, but there are too many variables and geochemical anomalies

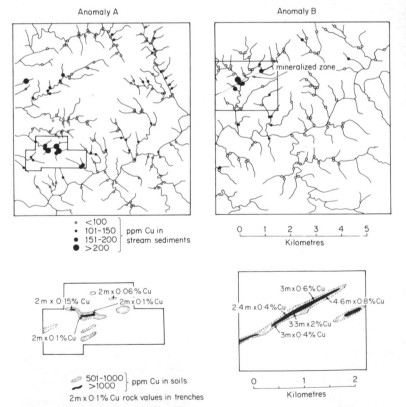

FIG. 4.22. Two stream sediment anomalies and their respective follow-up soil blocks from a belt of metavolcanic rocks in central Zambia.

cannot be quantified in this way except in the simplest cases. The factors to consider are: the strength of the anomaly, associated elements (if any), nature of the stream, local geology, amount of organic matter, presence or absence of iron or manganese oxides and pH. It should be appreciated that a comparison with results obtained over known mineralization in the same environment is the most useful measure, but such data (from orientation surveys) are all too rarely available.

Figure 4.22 illustrates some of the pitfalls in trying to assess geochemical anomalies. The area sampled covers a belt of metavolcanic rocks in central Zambia and the follow-up soil sampling to two similar stream sediment anomalies is shown. Anomaly B is due to a long zone of low-grade copper mineralization up to 60 m wide marked by conspicuous amounts of malachite exposed in trenches and surface outcrop. Anomaly A, on the other hand, which appears to be the better of the two at the stream sediment sampling stage, is due to unmineralized metavolcanic rocks as trenching failed to reveal any visible mineralization. Up to 0·1% Cu in the rocks, however, is sufficient to explain the anomalies.

4.7 SOIL SURVEYS

Soils are the most important sampling medium in geochemical prospecting and may be used for both detailed and reconnaissance work. Following a regional stream sediment survey, detailed soil sampling of the anomalous catchments is usually a first step in follow-up work. In areas with poor drainage a reconnaissance soil sampling survey will be necessary. A number of different sampling schemes or patterns can be used depending upon a number of factors. The obvious and most widely used pattern is the square or rectilinear grid. In highly dissected areas with rugged terrain it may be difficult to lay out a good sampling grid and an alternative method that can be used in such cases is *ridge and spur sampling*. As the name implies, this consists of taking soil samples along the tops of watershed ridges and spurs. Used in conjunction with stream sediments, this method gives very good coverage of areas with difficult terrain. Very often in reconnaissance work it is desirable to undertake a preliminary assessment by running a few soil sampling traverses across the area of interest instead of becoming involved in the major task of laying out a grid. Existing roads, tracks or boundary fences can be used for locating the traverse lines and a high productivity can be achieved in terms of line-kilometres, particularly if a motor vehicle is being used. This method has been successfully employed in first stage reconnaissance work despite the fact that sampling coverage is very uneven.

Notwithstanding the success of other methods, the carefully laid out sampling grid is the most satisfactory method of collecting soil samples. The normal procedure is to lay out an accurate base line and then run perpendicular cross lines from it. A surveying chain or tape can be used for measuring distance and a theodolite for setting off the lines, but a compass is usually adequate unless there is undue magnetic disturbance. In relatively flat, open areas a bicycle wheel with an odometer attached can be used very successfully for measuring distance. If the vegetation is thick, it will be necessary to cut the lines. In addition it is important to mark the positions of lines and sampling points with pegs, blazes or flagging tape. Although it is necessary to lay out a sampling grid carefully so that the sampling points are located accurately, it is important not to waste a great deal of time in precision surveying. In geochemical prospecting the precise location of background samples is of little use to anyone. In most cases a compass and tape survey with the use of prominent features on maps and/or aerial photographs to tie in lines will result in a well laid out grid. It must be stressed that a good base line is most important and it can be dispensed with only in areas with exceptionally good topographic maps. For example, in Ireland a great deal of sampling has been done over farmland which is divided up into small fields which are all plotted extremely accurately on the Ordnance Survey 6-in. (1/10 560) base maps used in the field. In such areas a reconnaissance grid can be drawn on the maps and the sample points located very accurately by simple map reading. Only in the larger fields is it necessary to locate sample points by use of compass and pacing.

In the case of grid sampling the actual pattern chosen will depend upon the type of targets sought. If it is known that the mineralization in a particular area has a long strike length in relation to its width, such as a mineralized vein or stratigraphic unit, the base line should be laid out parallel to the expected strike and the samples collected along cross lines at right angles to the strike. In these cases the interval between lines can be considerably greater than the sampling interval along lines. If the strike is not known, or, if the expected targets are likely to be equidimensional, a square grid should be used. In practice a rectangular grid is often used since it is reasoned that increasing the sampling frequency along lines does not involve a proportionate increase in the time taken. For example, if it is decided to use a 200 m × 200 m grid, it will probably only take 10–20% more time to take samples at 100 m intervals along lines, resulting in a 200 m × 100 m grid. Even if the likely shape or orientation of possible targets is not known, it is argued that the rectangular grid increases the probability of samples being located over small targets. This argument should not be stretched too far, however, and the square grid should be considered as the universal grid. Sometimes a staggered square grid or rhombic grid is used since it has a slightly better probability of locating a target than true square grid. Commonly used grid sizes are: 500 m × 100 m or 200 m × 200 m for reconnaissance and 100 m × 50 m or 50 m × 50 m for detail.

Collecting samples

In most soil surveys one usually attempts to take samples from the B soil horizon, which means that the samples are generally taken at a depth of 30–50 cm. In some instances it can be shown that shallower samples from a depth of 10–20 cm give equally good results. For a few elements, such as Ag and Hg, the A soil horizon may give better results. In areas with hard, dry soils the samples are taken by digging a small hole with a mattock or hoe. If the soil is soft and damp a small trowel or hand auger can be used. The samples are placed in small, numbered Kraft paper packets and brief notes taken. These should indicate soil type (sandy loam, clay, etc.), colour and organic content. In addition features such as outcrops, tracks, streams, etc. along the lines should be noted. It is generally easiest to do this on a base map of the grid as features can then readily be related from line to line. Sample notes can also be written on the map instead of in a field notebook or on special field sheets.

Productivity may vary from as much as 50–80 samples/man-day on a small reconnaissance grid in open country where sample location is by simple map reading to as little as five or less per man-day in rugged terrain with thick vegetation where a sampling party may consist of eight or more people necessary for running the compass line, chaining distance and cutting the line.

The numbering system used will depend to some extent on the sampling pattern, but for grid sampling a co-ordinate system is probably the best. This is most conveniently done by taking a zero point on the base line and giving reference numbers to each cross line. For example, the line through the zero point is designated 'line 0', the line 500 m east of the zero point is 'line 5E', the line 250 m west of the zero point is 'line 2·5W', and so on. The samples along each line are numbered according to their distance from the base line. Thus, the sample designated 'line 7·5E/2·5N' or '7·5E/2·5N' comes from the point 250 m north of the base line along the cross line 750 m east of the zero point. In addition the area needs to be specified and this can be done by designating the various sampling blocks with letters. For example, a complete sample number might then read

'AC/7·5E/2·5N'. This type of numbering becomes a little cumbersome for the laboratory and it is easiest to use a straight number, preferably with not more than five digits. This is the best system if the samples are not being taken from a regular grid, as, for example, in ridge and spur or traverse sampling. This is also easily done if perfectly square or rectangular blocks are being laid out by adopting a convention such as starting the numbering sequence in the SW corner and numbering up and down alternate lines. Such a system has been used for a major regional sampling programme in Ireland where very accurate 6-in (1/10 560) base maps are available. A north–south square grid was laid out over each sheet and pre-numbered by starting in the SW corner and numbering up and down alternate lines. If sample points fell in the middle of obvious obstructions such as ponds or groups of buildings, they were left unnumbered, but otherwise, all points were designated by a number and it was left to the field assistants to report the reasons why a particular sample might not have been taken. Each sheet was assigned a two-letter code and transparencies were produced from the pre-numbered sheets so that dye-line prints could be run off for use as field sheets. This system proved extremely satisfactory since hand numbering machines, specially ordered with the letter codes, could be used to number all sample packets, field sheets and lab sheets. An average of 2000 soil samples was collected per week for just over one year. Some workers use a six, seven or eight digit number which is coded to indicate project, area and sample number. For instance, the number 2040325 might mean: project 2 (**2**040325), area 04 (2**04**0325) and sample 0325 (204**0325**). This type of system is better for computer work where alpha-numeric systems should be avoided.

In areas with good climates and plenty of sunshine the samples can be dried in the open at the field camp, but, if the climate is wet or damp, it will be necessary to dry the samples on a makeshift drier at the camp or in a special drying cabinet at the laboratory. It is generally preferable to dry and sieve samples in the field unless the sampling operations are being conducted within easy reach of the laboratory. When the samples are dry, they can be 'crushed' for sieving by placing the full sample packet on a wooden block and hitting it hard several times with a wooden mallet. Unless undue force is used, this will not split the normal Kraft paper sample packet. In tropical areas where latosol soils may predominate many workers advocate crushing the samples lightly with a mortar and pestle before sieving to ensure a breakdown of the aggregated iron oxides. Tests have shown, however, that this is not necessary unless the soils are markedly pisoolitic. Stainless steel or nylon bolting cloth sieves should be used and in most cases a few shakes of the sieve will result in sufficient fines (a few grams). Before sieving another sample, it is important to ensure that the sieve is clean: a 3·5 cm or 5 cm paint brush can be used for doing this. The sieved samples are placed in small, unsealed, paper envelopes which are slipped into small polythene bags to ensure that there is no leakage and cross contamination in transporting to the laboratory. It should be possible for one person to sieve 60–100 samples/h.

The −80 mesh (0·2 mm) size fraction is most commonly used for geochemical samples, but finer or coarser size fractions may be used in certain cases. For most soil and stream sediment samples the size fraction chosen represents the natural fines of the sample, i.e. the sample is not ground before sieving. This is done for three main reasons: (i) it is generally found that the elements of interest are concentrated in the finer fractions, (ii) fine material is required for adequate digestion prior to analysis and (iii) time and cost are saved by not having to grind the samples. A coarser natural size fraction (−10 + 30 mesh for example) may be found to contain higher concentrations of the element sought, if discrete minerals of the element of interest are present or if pisoolitic latosol soils are involved. In the first case it might be advantageous to consider traditional panning methods in preference to geochemical sampling (see pp. 52–57) and in the second case crushing the samples before sieving with a mortar and pestle as mentioned earlier may be necessary. Coarser size fractions have often been found to be of more value in tin and tungsten exploration (both amenable to traditional prospecting methods) and for many base metals in some lateritic soils.

Orientation surveys

Before starting a survey in a new, untested area, it is advisable to carry out a brief orientation survey to determine the optimum size fraction for analysis, the best sampling depth and, if possible, the geochemical response to known mineralization. Although the importance of orientation surveys has probably been overstressed by academic geochemists, it is essential to be aware of the pitfalls of going ahead with a conventional geochemical prospecting programme using the −80 mesh size fraction in an unknown environment. Such an approach, for example, would have met with little success in the area of tin mineralization described on page 54. Another aspect to consider at the orientation survey stage is the possible effect of contamination due to human activity.

Figure 4.23 shows an orientation survey traverse over an alkaline complex in eastern Uganda and it can be seen that, in the case of Cu, sampling at the shallow depth of 15 cm would have been quite adequate, but in the case of Be the 45 cm depth clearly produces

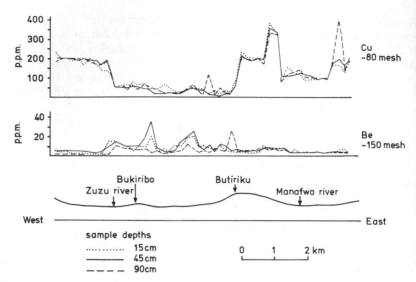

FIG. 4.23. Orientation survey traverse across the Butiriku carbonatite
complex, southeast Uganda showing results for Cu and Be.

better results. This was also shown by a number of other elements so
a 45 cm sampling depth was used for the main survey. It was also
shown that analysis of the −150 mesh fraction of the fine-grained soils
gave better contrast than the −80 mesh fraction for a number of
elements. In addition it was established that fears of contamination
from super-phosphate fertilizers (P was one of the elements of
interest), used on small coffee plantations in the area, were un-
founded.

Plotting of data
Soil surveys are normally presented as contoured maps and the
contours are referred to as *isopleths* (lines of equal concentration or
abundance). Ranges between contours can be shown by colours or
grey shading. Each sampling point with its sample value should be
shown, but sample numbers should not be included as they cause
confusion. It is permissible, however, to indicate a sequence of
numbers by plotting sample numbers at the ends of lines. Irregular
patterns of sampling, such as ridge and spur or traversing, can be
presented as dot maps (see page 162) or by using different coloured
spots at each sample site. Some workers present geochemical maps
without showing sample sites or values. If this is done, it is important
to indicate on the map the sampling pattern used, such as, 'sampled

on a 100 m × 100 m grid'. The sample depth, analytical method and size fraction analyzed should also be given.

Follow-up work

The first step in the follow-up to a reconnaissance survey will be to sample the area of interest on a closer-spaced grid. For example, an anomaly located on a 500 m × 200 m reconnaissance grid might be resampled on a 250 m × 100 m grid or an even closer-spaced interval, but it is rarely worth sampling on a grid smaller than 25 m × 25 m, unless the expected targets are very small veins or pegmatites. Detailed, close-spaced sampling of the near surface soils generally achieves very little and it is important to obtain information in depth as early as possible. The various methods of doing this are all described in Chapter 5. If indications after this stage are promising, the anomaly may be followed up with a geophysical survey before a decision to drill is finally taken.

Contamination

Contamination is a problem in old mining areas and may make the application of geochemical prospecting methods difficult, but not impossible if precautions are taken. In parts of Zambia it has been possible to ascribe some copper anomalies to old smelting sites in the vicinity of rich copper oxide deposits which were worked on a small scale by the indigenous peoples before the arrival of Europeans at the turn of the present century. In such areas one has to be extremely careful in interpreting anomalies as some may be due to natural sources. Fumes from smelters can cause widespread contamination, but is usually confined to the top 10 or 15 cm of the soil and sampling at a depth of 40 or 50 cm will generally avoid any problems. Low order lead anomalies in soils are common near busy major roads or road intersections owing to contamination from lead-based anti-knocking agents used in petrol. Fine dust from old tailings dumps can be carried considerable distances by wind, but, like smelter contamination, it usually only affects the soil to very shallow depths. In stream sediment and lake sediment sampling contamination from old tailings entering drainage systems can be a big problem and dispersion trains many kilometres long have been observed downstream from old tailings dumps. Contamination from other sources has often been overrated and one should be very cautious about writing off an anomaly as being due to contamination just because some old copper wire or galvanized sheeting was found at the site. For example, a Zn soil anomaly >1000 ppm, 50 m wide, 200 m long and persisting to a depth of 1 m contains 15 tonnes of Zn which is rather more than could have come from a few old galvanized buckets or roofing sheets!

4.8   VEGETATION AND WATER SURVEYS

Instead of sampling rocks, soils or stream sediments, both plant material and river, lake and ground waters can be used as sampling media in geochemical prospecting. The sampling of vegetation is known as biogeochemical prospecting and that of natural waters as hydrogeochemical prospecting. Of the two, hydrogeochemical methods have proved more valuable and are more widely used.

Vegetation surveys
It is common knowledge to most people that the type of soil and underlying rocks have a close bearing on the type of vegetation that will grow most readily. Some plants prefer acid soils and will not grow well on limestone soils. These plants, such as the shrub rhododendron, are known as *calcifuges*. There are other plants which prefer distinctly alkaline soils and will not grow well on acid soils. Some can tolerate wet, water-logged conditions and others cannot. There are some plants that can tolerate high levels of elements such as Ni and Cr in the calcium-poor soils over ultrabasic rocks and hence grow in preference to others that cannot tolerate this environment. Such associations between assemblages of plants and the underlying geology are used as an aid to geological mapping and are extremely important in photogeological interpretation.

Plants take up nutrients from the soil through their roots and, if one compares the concentrations of various elements in plant tissues compared to concentrations in the soil in which the plant is growing, the elements can be divided into three broad groups. The first group comprises the biogenic elements and includes H, C, O, N, P and S. These are the elements which make up the bulk of the plant tissue and generally occur in the plant at concentrations well above those in the soil. The second group comprises the essential trace elements which are necessary for healthy growth. These include B, Mg, K, Ca, Mn, Fe, Cu and Zn and commonly occur in the plant at roughly the same concentrations as those in the soil. The third group are the non-essential and toxic elements. These include Pb, Sr, Hg, Be, U, Cr, Ni, Ag, Sn, and Se and normally occur in the plant in concentrations well below those in the soil. These are broad generalizations as not only do different species of plants vary widely in their uptake of elements, but the availability of elements in a soluble form is also an important factor. In addition some of the toxic elements such as Ni and Se may be essential in very small amounts to some plants and some of the essential trace elements may be toxic if present in strong concentrations.

Since most elements that are of interest in mineral exploration are moderately to strongly toxic to plant life, vegetation in soils overlying mineralization is often strongly affected. In areas where there are high concentrations of elements such as Pb, Cu, Hg and Ni in soils in a form available to plants, the vegetation is often severely stunted or restricted to a few hardy individuals. The result is often a conspicuous clearing such as the 'copper clearings' in Zambia mentioned in Chapter 3. There are many other examples and Bolviken and Lag (1977) give an interesting account of small 'metal clearings' in Norway. In addition to plants that can tolerate concentrations of toxic elements there are others that seem to require minimum levels of certain elements toxic to other plants before they will grow. Such plants are known as *indicator plants* and there are a number of examples. The best known are probably the copper flower in Zambia and the selenium flora in the western United States. Experiments with the small mauve–white copper flower (*Beccium homblei*) in Zambia (Horizon Magazine, 1959) have shown that the seeds will not germinate in solutions containing less than 50 ppm Cu and several hundred ppm are required for most vigorous growth, but more than 600 ppm Cu is lethal. In soils where only part of the copper is readily available to the plants, the presence of the copper flower indicates several hundred to several thousand ppm Cu. In the western United States the selenium flora is a useful indicator of uranium mineralization owing to the fact that Se usually accompanies U in the sedimentary deposits of the Colorado Plateau. The most useful are various species of poison vetch (*Astragalus*). Some examples of indicator plants that have been used in various parts of the world are given in Table 4.12.

In addition to the use of indicator plants in prospecting, morphological changes may also indicate the presence of toxic elements. Stunted growth has already been mentioned, but probably the most important is yellowing of the leaves (*chlorosis*) which is caused by concentrations of elements such as Cu, Zn, Mn and Ni which interfere with iron metabolism of the plant. In some instances mutations may occur, e.g. changes in the shape and coloration of the petals of a poppy have been observed over a copper–molybdenum deposit in Armenia (Malyuga, 1964).

The recognition of distinctive plant species (indicator plants) or of changes in plants (e.g. stunting and chlorosis) as an aid in prospecting is not a geochemical method. It is known as *geobotanical prospecting*, but it is discussed here as it is a useful introduction to biogeochemical methods which involve the analysis of plant material. Research into the uses of biogeochemistry in prospecting has been carried out since the 1930's and 1940's, particularly in the Soviet Union and North America, and it has been clearly established that analysis of

TABLE 4.12

EXAMPLES OF INDICATOR PLANTS (FROM BROOKS, 1972 AND MALYUGA, 1964).

Element	Indicator plant	Region
Copper	Copper flower (*Beccium homblei*)	Zambia
	California poppy (*Eschscholtzia mexicana*)	Arizona, USA
Nickel	Alyssum (*Alyssum murale*)	Georgia, USSR
Lead	Beardgrass (*Erianthus giganteus*)	Tennessee, USA
Selenium (associated with uranium)	Woody aster (*Aster venusta*) Poison vetch (*Astragalus* sp.)	Western USA
Zinc	Calamine violet (*Viola calaminaria*)	Belgium and Germany
Boron	Sea lavender (*Limonium suffruticosum*)	Caspian lowlands, USSR
Gold	Horsetail (*Equisetum arvense*)	Czechoslovakia

plant material can be very effective in delineating anomalous concentrations of elements related to mineralization. Plant material collected for a geochemical survey is ashed prior to analysis. This removes most of the biogenic elements which make up the bulk of the plant tissue and the elements of interest are concentrated in the residue. The ash is normally 1–3% of the dry weight, so that the elements of interest are concentrated up to 100 times their levels in the original plant tissue. Some elements may reach very high concentrations in ash of plants growing over mineralization, e.g. up to 1·75% Mo has been found in the ash of fireweed (*Epilobium angustifolium*) growing on the Endako molybdenum deposit in British Columbia (Warren, 1972) and up to 10 000 ppm arsenic has been found in the twig ash of Douglas fir (*Pseudotsuga menziesii*) growing over arseniferous metal deposits (Warren *et al.*, 1964). It has been shown that metal accumulations in shrubs and trees are greatest in the first- and second-year growth of twigs and shoots. Grassy plants, almost without exception, accumulate Ni, Cu, Co, Mo, Zn, Cr, Pb, B and U in stems and leaves. In addition the capability of plants to accumulate elements increases with their age, presumably a result of the expanding root system which is able to tap a larger and larger volume of soil. It has also been shown that there are seasonal variations.

An advantage of a biogeochemical survey over a soil survey is that anomalies in plant ash may be more readily detected owing to the concentration of metals in the ash. This is generally outweighed by the fact that more work is involved in plant sampling than in soil sampling and the main advantage of biogeochemical surveys is that they provide a means of sampling at some depth below the surface by virtue of the plant's root system. For this reason deep-rooted plants are preferred for biogeochemical surveys. The effective depths not only show a variation with species, but also show a broad variation with climatic conditions. For instance in the tundra effective depths are of the order of 1–2 m, in coniferous forest of temperate regions effective depths may be 2–5 m and in dry steppe or prairie regions effective depths can be 10–20 m. Greater depths of detection of 30 m or more have been claimed in some parts of the world. For example, Chaffee and Hessin (1971) located copper mineralization under alluvium up to 75 m deep by analyzing ash from the stems of the deep rooted ironwood (*Olneya tesota* A. Gray) in southern Arizona. The Cu anomalies disclosed by the plant ash were 240 m from the nearest detectable soil anomaly.

To undertake a biogeochemical survey it is necessary to collect a minimum of 300 g of material from each plant. As a general rule it has been found that young twigs give the best results, but this can vary with different species. It is also generally accepted practice that the survey should be restricted to a single species. This is often a disadvantage with biogeochemical surveys since the distribution of plant species within a proposed survey area is often very variable. Russian workers, however, maintain that it is not necessary to restrict sampling to a single species to obtain meaningful results (Malyuga, 1964). The chosen species should be sampled exclusively as far as possible, but alternative species can be sampled if it is necessary to complete coverage of the survey area. It is also important to collect the samples from sites as near as possible to a regular grid or pattern, but this is often another disadvantage to vegetation surveys as the location of sampling points may be irregular. After the samples are collected in numbered bags, the material is left to dry to a flammable state. At this stage they can be sent off to the laboratory for ashing and analysis or they can be charred in the air in a dish on a camp gas or primus stove and the ash placed in sample packets and sent off to the laboratory. This practice is preferred as it makes handling in the laboratory much simpler. Prior to analysis the samples are finally ashed in porcelain crucibles at 450–500°C, though this temperature is too high for elements such as Sb, Hg, Se and Te and methods of wet 'ashing' should be used for these elements.

More and more research into vegetation surveys is being conducted all over the world and, although it has been shown that they have undoubted value in certain specialized cases, their application in applied exploration is definitely limited. There are numerous examples of workers demonstrating that biogeochemical surveys can be successfully applied in areas where soil surveys work equally well (e.g. Hornbrook, 1969). Very often this is simply a consequence of plants dying and decaying and dropping leaves which decay, thus releasing any metals accumulated by them back to the soil.

Water surveys
Analysis of water from rivers and lakes and groundwaters from wells and springs can be an important method in geochemical prospecting. Groundwaters may be in contact with the enclosing rocks for considerable periods of time allowing substances to dissolve and a chemical equilibrium to be established which is closely related to the chemistry of the containing aquifer. It is therefore obvious that groundwaters may vary considerably from place to place in their content of dissolved solids. For example, deep brines from oilfields with halite deposits may contain considerably more dissolved solids than seawater and groundwaters from basement rocks may be virtually pure by comparison. Groundwaters used in mineral exploration, however, are generally from shallow sources and most metal concentrations are in the ppb range.

River and lake waters are in general largely derived from surface run-off, but groundwaters may contribute significant quantities through springs and 'base-flow'. Compared with entrapped groundwaters, river and lake waters show much greater variation in content of dissolved solids as they are subject to large and sudden variations in run-off in addition to changes in pH, Eh and chemical environment that can take place over short distances. Hoag and Webber (1976) give a useful review of the sources of anomalous waters.

Samples are collected in the field in absolutely clean polythene bottles (250–500 ml) using the procedure of washing the bottles out two or three times with the water being collected. To ensure that the bottles are free of contamination, they should be cleaned with a strong metal-free acid before being taken into the field. It is also normal practice to acidify the sample upon collection with two or three drops of metal-free concentrated nitric acid to prevent precipitation of any metals present. If pH and Eh measurements or determinations of substances which would be affected by the acid are required, it will be necessary to collect a duplicate sample unless these measurements are made on the spot. In addition it may be necessary to filter the samples if they contain suspended solids, but this is more conveniently done in the laboratory prior to analysis.

When collecting lake waters, some workers advocate taking samples at depth from below the thermocline (point of temperature inversion in stratified lake waters), but other workers consider this unnecessary and claim that surface waters are quite adequate.

Hydrogeochemical surveys have been used all over the world in exploring for a wide range of elements using surface waters from streams and lakes and groundwaters from wells and springs. In the case of streams, sediments have usually proved to be the better sampling medium and are to be preferred in most cases. In the case of lake waters, it has been found that waters are as good as lake sediments and may be considered preferable. In a study of an area of the Canadian Shield, Cameron (1977) concluded that lake waters were a better sampling medium than lake sediments. Since the lake waters tend to be homogeneous across the surface, the waters are more easily collected and analyzed than the lake-bottom sediments and the indicator elements in the waters always increase in concentration towards a source, unlike the same elements in lake sediments which show a big variation depending on precipitation controls. In the case of groundwaters, hydrogeochemical methods offer the possibility of obtaining sample information from sources at depth. An example of such a survey is given by De Geoffroy et al (1967 and 1968), who used groundwater successfully for locating zinc mineralization in Wisconsin.

Hydrogeochemical surveys will probably always remain of minor importance in geochemical prospecting owing to the fact that the metal content of waters may not only show big seasonal variations, but may also display wide variations over shorter intervals of time and interpretation can be difficult as it is not always easy to determine sources. In addition productivity in sample collection is generally low in comparison with other methods. In spite of these drawbacks, however, hydrogeochemical surveys may offer a useful alternative in many areas. This is particularly true in uranium exploration where the high mobility of uranium and its daughter product radon and the relative ease with which low levels of both elements can be detected make water surveys an important method.

4.9 ANALYTICAL METHODS

A high degree of accuracy is not a prime consideration for analytical methods used in geochemical exploration. Rather it is essential that they be rapid, inexpensive and relatively simple and to this end a certain amount of precision can be sacrificed. A wide range of rapid analytical techniques have been developed and adapted for geochemical prospecting, the principal methods being chromatography, colorimetry, emission spectroscopy, X-ray fluorescence and atomic absorption. Other methods that are used in specialized cases include neutron activation radiometry and potentiometry.

Paper chromatography

This method has been widely used in the past mainly for Cu, Pb, Zn, Ni, Co, Mo, Bi, Nb, Ta, Se and U, but it suffers from poor precision at low metal concentrations and is very sensitive to variations in atmospheric humidity. For these reasons it is rarely used today as both colorimetry and atomic absorption, particularly the latter, have much better precision.

The technique for Cu, Ni, Co and Zn is briefly as follows: 0·5 g of sample (−80 mesh) is placed in a pyrex test tube and mixed with 1 g of potassium bisulphate. This mixture is gently fused for 10 min and after cooling, 2 ml of dilute HCl and HNO$_3$ are added and the test tube is gently heated on a water bath for 10 min. After the sample has settled, 0·01 ml of the solution is placed at the end of a strip of Whatman's CR-1 chromatography paper with a capillary pipette. This special slotted paper allows ten chromatograms to be run simultaneously. After drying, the paper strip is hung in a tank with the end to which the sample was applied dipped in a shallow layer of solvent made up of 15 ml HCl, 10 ml H$_2$O and 75 ml methyl-ethyl ketone. When the solvent front is almost at the top of the strip, the paper is removed, dried briefly in air, immersed in ammonia vapour for 2 min and sprayed with 0·1% rubeannic acid solution in ethyl alcohol and water. The strips are then compared against a series of standard chromatograms. Copper is greenish-grey near the top of the strip below a brown iron stain, cobalt is yellowish-brown just below copper and nickel is bright blue at the base of the strip. To determine zinc the strip is immersed in ammonia vapour again and sprayed with 1% dithizone solution. The zinc band now appears bright pink between copper and iron.

A technique for determining niobium is similar except that hydrofluoric acid is used for sample digestion and with the methyl ethyl ketone solvent and 2% tannic acid solution is used in place of rubeannic acid or dithizone. This gives an orange chromatogram.

Full descriptions of the techniques for determining various elements are given by Hunt *et al* (1955) and Ritchie (1964).

Colorimetry

This technique is based upon the principle that the absorbance of radiation and hence colour intensity of a coloured solution is proportional to the concentration of the solute (Beer's law). For analytical purposes there are a wide range of specific organic reagents which form characteristic coloured complexes with different metallic ions.

n essence analysis is carried out by adding the complexing reagent or that particular element to the sample solution and then comparing the colour intensity with a series of standard solutions. In practice it s usually more complex than this as it may be necessary to adjust the pH of the solution and to remove or suppress possible interfering elements. Very accurate determinations are possible with colorimetry by making precise measurements of the absorbance using photometers or spectrophotometers. In geochemical work, however, it is usually only necessary to make visual comparisons.

Colorimetric analysis has been adapted for a number of rapid cold extraction field tests for heavy metals, principally Cu, Pb and Zn. These tests are not used very much today, but they are still useful if one requires some quick interim results, particularly in a remote area where there may be considerable delay in receiving results from a laboratory. One of these tests was devised by Harold Bloom (1955) and is known as the 'Bloom test'. A brief description follows: Approximately 100 mg of sample (−80 mesh) is placed in a 25 ml glass-stoppered cylinder. (The sample is not weighed, but is measured out with a small plastic scoop that holds approximately 100 mg of sample.) Five millilitres of an ammonium citrate solution and 1 ml of 0·003% dithizone solution in pure xylene are added, the tube is stoppered and shaken vigorously for 5 s and allowed to settle for about 30 s until the organic and aqueous phases have separated. The colour in the xylene is observed and an index recorded according to colour (green = 0, green–blue = $\frac{1}{2}$, blue = 1, blue–purple = $1\frac{1}{2}$). If the colour is purple to red, it is titrated with 1 ml increments of dithizone solution until the blue–purple colour is obtained, the number of mls of dithizone solution required being recorded as an index. The ammonium citrate solution is made up by dissolving 25 g ammonium citrate and 4 g hydroxylamine hydrochloride in 300 ml of water. Ammonium hydroxide is added until the pH is 8·5 and the final solution is made up to 500 ml with water. A 0·01% solution of dithizone in carbon tetrachloride is used to purify the ammonium citrate solution of heavy metals by repeated shakings and separations with small amounts of dithizone solution until the colour of the carbon tetrachloride layer is green. The dithizone solution deteriorates quite rapidly and should be freshly prepared every few days. If the working solution has a yellowish hue, it indicates that the reagent has deteriorated and fresh solution should be made up. Qualitative tests for Pb, Cu and Zn are as follows: (i) *Lead*; add 3 drops of 5% KCN solution, 1 ml dithizone solution and 1 ml ammonium citrate solution. Shake vigorously. If the original colour persists, lead is predominantly present. (ii) *Copper*; to 1 g of sample add 1 ml dithizone solution and 5 ml ammonium citrate solution.

Shake vigorously. If the xylene layer is brown and turns to wine purple on addition of more dithizone, copper is predominantl present. (iii) *Zinc*; if neither lead nor copper is identified, th original colour can be ascribed to zinc. The Bloom test does not giv values in ppm as different concentrations of Cu, Pb and Zn giv different colour responses with dithizone. The sensitivity is ap proximately 50 ppm Total Heavy Metal (THM) and, as a rough guide indices greater than 3 or 4 can probably be regarded as anomalou with indices of 20 or over equivalent to several thousand ppm THM

A wide range of procedures for various elements have been des cribed by various workers for the colorimetric determination of tota metal content following a hot extraction. A list of elements with thei respective colorimetric reagents is given in Table 4.13. Full descrip tions of the procedures for 25 elements are given by Stanton (1966)

TABLE 4.13

SOME ELEMENTS AND THEIR COLORIMETRIC REAGENTS. RANGES IN PPM ARE THE LOWER RANGES. IN MOST CASES TOP RANGE CAN BE INCREASED 5-FOLD OR MORE BY TAKING SMALLER ALIQUOTS OF SAMPLE SOLUTION

Element	Reagent	Colour	Range
Antimony	Brilliant green	blue	0·4–220
Beryllium	Beryllon II	blue	1–40
Bismuth	Sodium diethyldithio-carbamate	yellow	20–160
Cobalt	Tri-n-butylamine	blue	1–250
Copper	Biquinolin	red	2–2 000
Gold	Brilliant green	blue	0·05–4
Lead	Dithizone	mixed colour*	10–2 000
Molybdenum	Dithiol	yellow–green	1–100
Nickel	α-furildioxime	yellow	1–750
Tin	Gallein	mixed colour†	0·5–100
Titanium	Tiron	yellow	100–200 000
Tungsten	Dithiol	blue–green	4–400
Phosphorus	Ammonium metavanadate + ammonium molybdate	yellow	20–2 000
Zinc	Dithizone	mixed colour*	10–2 000
Chromium	Diphenylcarbazide (from chromate)	red violet	5–250
	(from chromate)	yellow	100–3 000
Manganese	(from permanganate)	purple	50–10 000

green–blue to blue to purple to pink
†*grey to purple to pink*

For interest, procedures for copper (Almond, 1955) and molybdenum (North, 1956) are given below as examples of the technique:

Copper. Thoroughly mix 100 mg of sample with 500 mg of potassium bisulphate, place in a pyrex test tube and fuse for several minutes over a flame (a primus stove will do). Remove from the heat and rotate the tube so that the melt solidifies in a thin layer on the walls of the tube. Cool, add 3 ml of 6N HCl and heat on a water bath until the melt breaks up. Remove from the water bath and dilute to 10 ml with distilled water. Transfer a 2·5 ml aliquot to a screw-cap culture tube, add approximately 50 mg hydroxylamine hydrochloride and allow to stand, but shake occasionally to dissolve the hydroxylamine hydrochloride. Add 10 ml of buffer solution and 2 ml biquinolin solution. Shake the tube for 30 s. Allow to stand until phases separate and compare with a series of prepared standard copper solutions. The buffer solution is made up by dissolving 400 g sodium acetate and 100 g potassium tartrate in 1 litre of water. This must be free of copper and a sample of the solution can be tested for copper with hydroxylamine hydrochloride and biquinolin solution. The biquinolin solution is made up by dissolving 0·2 g of 2,2′-biquinolin in 900 ml isoamyl alcohol and then making up to 1 litre with isoamyl alcohol. This should be colourless. If it is yellowish, the reagent is not pure. This technique is sensitive down to 2 ppm in the original sample.

Molybdenum. Mix 250 mg of sample with 1·25 g of finely ground fusion mixture consisting of 5 parts of sodium carbonate, 4 parts of sodium chloride and 1 part of potassium nitrate. Place in a nickel crucible and fuse over a flame. Allow to cool and add 3 ml hot water. Leave overnight. Heat gently to free melt and transfer to a graduated test tube and make up to 5 ml mark. Heat on a water bath for 10 min. Allow to cool and settle. Transfer 2 ml aliquot to a cylindrical separating funnel (16 mm diameter), add 2 ml hydroxylamine hydrochloride solution cautiously and allow to cool. Add 0·6 ml dithiol solution and shake funnel over a period of 15 min. Drain off aqueous phase. Add 4 ml concentrated HCl and shake for 30 sec. Set aside and allow phases to separate and compare against a series of prepared standards. The hydroxylamine hydrochloride solution is made up by dissolving 2·5 g hydroxylamine hydrochloride in 10 ml water and then adding 90 ml concentrated HCl. The dithiol solution is made up by melting 1 g of toluene-3,4-dithiol at 30–40°C and dissolving in 100 ml of isoamyl alcohol. This technique is sensitive down to 1 ppm molybdenum in the original sample.

With most of the colorimetric procedures it is possible to achieve between 30 and 100 determinations per man-day at precisions of ±20–30% at the 95% confidence level.

Emission spectroscopy

Emission spectroscopy has played and continues to play an extremely important role in the determination of trace elements in pure geo-chemistry. Although the importance of this method has been reduced by the use of newer techniques such as X-ray fluorescence and atomic absorption, the development of advanced direct-reading spec-trographs has meant that emission spectroscopy remains an important analytical method. In geochemical prospecting emission spectroscopy was used more widely in the past than it is today, but the method still has considerable appeal owing to the fact that a sample can be analyzed for a large number of elements at one time. Since direct-reading spectrographs are very costly, emission spectrographs utiliz-ing photographic detection are commonly used in geochemical exploration.

In emission spectroscopy a sample is excited by an electric arc struck between two electrodes causing elements in the sample to emit radiation of characteristic wavelengths. In the spectrograph this radi-ation is dispersed by a prism or diffraction grating into the various spectral lines which make it possible to identify the elements present in the sample. In addition the intensity of the radiation of some particular wavelength depends upon the quantity of the corresponding element in the sample and quantitative measurements can be made. Highly volatile elements cannot be analyzed without special pro-cedures as they are burnt off in the arc before they can be recorded.

In a typical spectrograph utilizing photographic detection the general procedure adopted in most geochemical laboratories is as follows: 100 mg or more of sample is mixed 1:1 with pure carbon powder and packed into a recess at the end of a graphite anode. A small amount of buffer such as Li_2CO_3 or K_2SO_4 may be added to reduce matrix effects and for accurate work it is also necessary to add a small known amount of a 'specpure' oxide such as GeO_2 or Lu_2O_3 to provide what is known as an *internal standard*. The packed electrode is then preheated in an oven at several hundred degrees Centigrade to eliminate any combined water. A graphite cathode and the packed anode are clamped into place and the arc gap (usually about 2 mm) is adjusted according to the particular instrument being used. The sample is then arced for 30 s or more and the spectral lines are recorded on a photographic film. On most instruments 12 or more samples are recorded side by side on a single photographic plate and it is normal practice to include at least one standard or control sample on each complete plate. The intensities of the various spectral lines are then compared against a series of known standards run on the same instrument under identical conditions. These standards are very carefully made up in steps of increasing concentrations from 'spec-

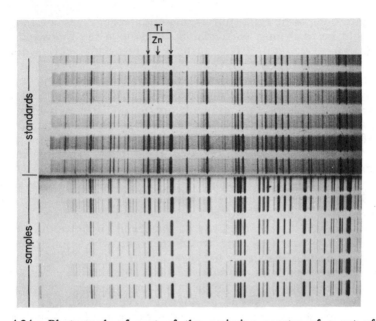

FIG. 4.24. Photograph of part of the emission spectra of a set of six standards and six unknown samples as displayed on a comparator screen. The comparator projects greatly enlarged images of developed films from a spectrograph on to a translucent screen and allows a set of standard spectra to be placed alongside spectra of unknown samples for comparison. In the example shown the 3345 Å Zn line, sandwiched between strong Ti lines, is being read. Note how the intensity of the Zn line increases in the set of standards from top to bottom as the known Zn content increases. The top sample spectra and a standard with a Zn line of similar intensity have been juxtaposed.

pure' chemicals of the various elements being analyzed. The plates can be compared visually with the standards on a comparator (Fig. 4.24) or a microphotometer can be used to measure the spectral line intensities accurately. The internal standard is mainly used to reduce errors caused by variations in the burn, but it also controls variations in the photographic process. For example, if each sample contains an internal standard of exactly 400 ppm Ge, the same spectral line for Ge should have the same intensity on all plates. If any plates show too great a variation in the Ge spectral line, samples on those plates are generally run again. If the concentration of an element is too high, the intensity of the spectral line may be too strong to measure accurately. To accommodate this most instruments have an intensity control in several logarithmic steps (e.g. 1, $\frac{1}{4}$, $\frac{1}{16}$) which enables a wide range of concentrations to be read. For example, different detection ranges of Co might be read on a particular instrument in the following steps: 5–200, 200–1000, 1000–10 000. Instead of using log sectors different lines can be read to cover ranges of concentration.

The alkalis and alkaline earths produce fairly simple spectra with relatively few wavelengths, but many elements such as the transition elements, rare earths and uranium are particularly rich in spectral lines with a vast number of different wavelengths. Some spectral lines of different elements are extremely close together and it requires a high quality instrument to resolve them. For example, niobium produces a line at 4058·938 Å and manganese a line at 4058·93 Å. This illustrates how manganese might be read for niobium or vice versa if these respective lines were being used for either element. There are a large number of lines to choose from, but the spectral line(s) selected should produce a good intensity at the levels of detection required and should not be too close to another line which might result in spectral line interference. A list of elements with respective spectral lines that are commonly read is given in Table 4.14 together with the lower detection limits that can be expected in visual work.

Some research institutions, geological surveys and large companies are able to justify the much greater expense of a direct-reading spectrograph. Essentially these instruments consist of grating spectrographs in which the film-holder has been replaced by an opaque barrier with a number of narrow slits at the spectral line positions of a number of different elements. Behind each slit is a photomultiplier tube which monitors the intensity of that particular wavelength. These instruments can be coupled to automatic recorders or even computers which can make necessary corrections and give a print-out of results. Analyses for 15 or more elements can be carried out in less than a minute on some instruments.

There are numerous sources of error in emission spectroscopy and poor standards of work have resulted in much disenchantment with the method. This is unfortunate since emission spectroscopy has much to commend it. Sources of error include fluctuations in electric power, uneven arcing, poor packing of electrodes, matrix effects in the sample, variations in film development and operator bias in reading plates. With the use of an internal standard and routine control samples, it should be possible to achieve precisions of ±? 45% at the 95% confidence level in visual work. Higher degrees of precision are possible by using microphotometers or with some direct-reading spectrographs. Between 25 and 30 samples can be run per man-day for 15 elements in visual work and this output can be more than doubled with direct-reading spectrographs. In routine base metal exploration emission spectroscopy cannot compete with atomic ab-

TABLE 4.14
LIST OF ELEMENTS WITH SPECTRAL LINES THAT
ARE COMMONLY USED IN SPECTROGRAPHIC
ANALYSIS

Element	Spectral line (Å)	Lower detection limit (ppm)
Arsenic	2 349·8	5
Beryllium	2 348·6	1
Bismuth	4 722·2	1
	3 067·7	5
Chromium	2 835·6	5
Cadmium	5 085·8	1
	4 678·5	
Cobalt	3 453·5	5
Copper	3 274	2
	2 961·2	200
Lanthanum	4 921·8	50
Lead	2 833·1	10
Manganese	2 593·7	100
Molybdenum	3 170·3	2
Nickel	3 414·8	5
	3 050·8	200
Niobium	3 163·4	50
Silver	3 382·9	0·2
Tin	3 262·3	10
Vanadium	3 185·4	2
Zinc	3 345·0	50
Zirconium	3 391·9	50

sorption for precision or cost, but for more specialized surveys where it might be necessary to analyze for 10 or more elements, emission spectroscopy is very attractive. One big advantage of the method is that plates and films can be stored as permanent records and can always be read at a later date for elements that were not requested in the original survey. It should be noted that spectrographic determinations are generally not directly comparable to colorimetric or AAS determinations (see below) owing to a number of factors, but the overall geochemical patterns should be very nearly the same.

Atomic absorption spectroscopy (AAS)
This is probably the most widely used method for routine geochemical analysis as it is quick, relatively cheap and has excellent precision

over a wide range of values. The technique is based upon the principle that ground state atoms absorb radiation at various wavelengths that are characteristic for each element. Although the phenomenon of atomic absorption was discovered almost a hundred years ago when absorption lines were first observed in the sun's spectrum owing to absorption of radiation by atoms in the sun's upper atmosphere, it was only recently that the principle was put to practical use. This work was pioneered by an Australian, A. Walsh, in the 1950's and today a wide range of commercial atomic absorption spectrophotometers are available.

Figure 4.25 shows the layout of atomic absorption equipment in diagrammatic form. The hollow cathode lamp, which emits radiation characteristic of the element being analyzed, passes a beam through

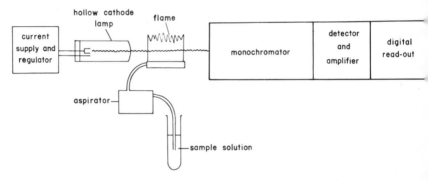

FIG. 4.25.   Basic layout of atomic absorption equipment.

an air/acetylene flame (nitrous oxide/acetylene and other mixtures are sometimes used). The sample solution is passed as a fine mist by an aspirator into the flame and pyrolysis of the solute releases ground state atoms which absorb radiation from the lamp in proportion to their concentration. The monochromator is adjusted to select a wavelength for the element being analyzed (e.g. 2320 Å for Ni) and the detector and amplifier pass the signal to a meter or digital readout. This is calibrated by spraying a number of standard solutions of the element being analyzed. The absorbance in the flame is only directly proportional to concentration over a narrow range at low concentrations and varies according to wavelength and element. Figure 4.26 shows the characteristic shape of the absorbance curve. Analysis should be carried out over the straight line portion of the curve and, if concentrations are above the critical value, the sample solution should be diluted. Failure to do this will always result in readings below the actual concentration.

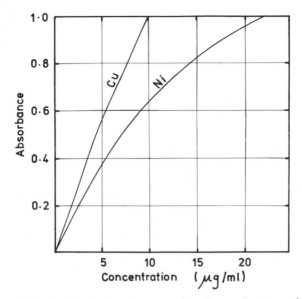

FIG. 4.26. Typical absorbance curves in atomic absorption.

Atomic absorption is sensitive to very low metal concentrations and is usually restricted to the range of 0·02–10 μg/ml in the sample solution, though it varies considerably depending upon the element and absorption line used. In the typical geochemical laboratory 200 mg of sample is digested in acid and diluted to 10 ml. This means that the method covers the range 1–500 ppm in the original sample for most elements with little difficulty. In addition it has been found that sensitivity can be greatly increased by taking up the sample in an organic solvent. This is the technique used for gold which is taken up in methylisobutyl ketone (MIBK). As little as 10 ppb in the original sample can be detected. If concentrations are above the top limit for a particular element and line being used, it is normal practice to dilute the sample by a factor of ten, doing it more than once if necessary.

The hollow cathode lamps used as the line sources consist of evacuated glass tubes partially filled with an inert gas (usually neon). The cup-shaped cathode (hence the term 'hollow cathode') is made of the specified element and, when a voltage of 100–200 V is applied across the anode and cathode, a glow discharge appears after a short warm-up period with most of the emission coming from within the hollow cathode. The radiation consists of discrete lines of the element concerned plus those of the inert gas. Multi-element lamps are produced by making cathodes of an alloy of the elements required (e.g. a brass cathode emits both copper and zinc lines), but single element lamps are generally preferred for the most accurate work. The window on the lamp is usually made of quartz, which is transparent to a wide range of wavelengths, but pyrex windows are used for wavelengths greater than 3200 Å. In theory atomic absorption spectroscopy could be carried out by using a continuous spectrum source, but the emission lines are so narrow that the band passed by the normal monochromator is much wider than the absorption line, so in practice it is much simpler and cheaper to restrict the emission lines to those whose absorption is being measured. In addition to the release of ground state atoms, many atoms become excited and with many elements emit radiation of the same wavelengths as they absorb. This reduces the absorption signal and gives readings which are too low. To overcome this the signal from the lamp is modulated and the detector tuned to the same frequency. As the source from the flame is steady, the detector cannot read this signal and only reads changes in the signal coming from the lamp.

For atomic absorption to take place in the flame it is necessary that ground state atoms be released. Thus, it is important that the flame is not too hot as this may result in raising of orbital electrons to higher energy levels or even the release of the element being analyzed in an ionic state. Likewise, the flame may not be hot enough to release ground state atoms of elements such as molybdenum, beryllium, tin, aluminium, titanium and the lanthanides, which form refractory oxides in the flame. The presence of certain anions may also reduce the concentration of dissociated metal atoms in the flame, e.g. the presence of sulphate or phosphate markedly reduces the concentration of calcium atoms by forming compounds which are blown through the flame without decomposing. There may also be some interference between metals, e.g. copper interferes with gold, calcium with lead and aluminium removes calcium by forming calcium aluminate. These are all minor drawbacks, however, and techniques have been devised for overcoming them. Atomic absorption spectroscopy is a very precise and simple analytical method for most metallic elements. The use of the air/acetylene flame is quite adequate for most of them and the nitrous oxide/acetylene flame has proved satisfactory for most of the difficult refractory elements.

For atomic absorption to take place it is only necessary to release ground state atoms in the light path between source and detector and flameless techniques can be used. The easiest element for this is mercury which is readily introduced as a vapour between source and detector. Figure 4.27 shows in diagrammatic form a common technique used for mercury analysis. Ten per cent stannous chloride solution is used to reduce the mercury in the sample solution to the

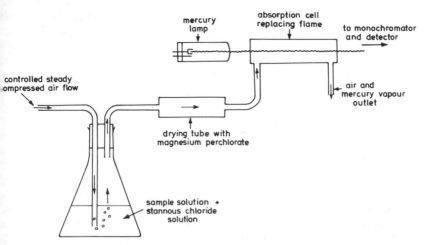

FIG. 4.27. Diagram to show the principle of a method for flameless detection of mercury in atomic absorption.

elemental state and air bubbled through the solution carries any mercury vapour present to a tube which replaces the normal burner. Mercury contents as low as 20 ppb in the original sample can be detected. Flameless techniques using electrically heated graphite furnaces in place of the burner have also been developed. The advantages are that sample solutions as small as one drop in volume can be introduced into the furnace for analysis. It is also possible to introduce sample material directly without first taking it up into solution. The use of graphite furnaces is still largely experimental and they have not been used routinely for geochemical work.

Productivity for the more common elements should work out at between 150–250 determinations per man-day in a well-run exploration laboratory. For elements such as Au and Ag productivity may be 50–100 determinations per man-day.

X-ray fluorescence (XRF)

If an element is irradiated with X-rays, some of the orbital electrons may be raised to higher energy levels. When these excited electrons fall back to lower energy levels, they emit characteristic X-rays. These X-rays can be used to yield qualitative and quantitative information about the substance being irradiated and form the basis of the technique of X-ray fluorescence (XRF) analysis. The characteristic X-ray spectra or lines are designated by the orbital shells. For instance electrons from the K-shell give rise to K-lines when they fall back from higher levels, electrons from the L-shell give rise to L-lines and so on. These lines are further subdivided by the number of energy levels through which the electrons fall. For example, K electrons falling back from the L-shell give K_α-lines and K electrons falling back from the M-shell give K_β-lines. One of the advantages of X-ray fluorescence over conventional emission spectroscopy is that there are many fewer lines in the X-ray region of the spectrum than in the optical region. For instance, iron has several hundred optical lines, but only two K-lines and six L-lines.

According to Moseley's law the reciprocal of the wavelength is proportional to the square of the atomic number. This means that the wavelengths of the lighter elements are very much longer than wavelengths of the heavier elements. The longer wavelengths or 'soft X-rays' as they are known, are difficult to detect and make elements lighter than calcium ($Z = 20$), with wavelengths longer than 2·5 Å increasingly difficult to analyze by XRF techniques. Methods have been developed to carry out measurements on soft X-rays, but, nevertheless, elements with atomic numbers less than ten are very difficult to determine.

The principal parts of an XRF analyzer are a powerful X-ray source to irradiate the sample (the primary X-rays have to have a higher energy than those it is intended to excite), an instrument to distinguish the wavelengths and a detector and recorder to enable the various intensities to be measured. The various X-rays can be distinguished by dispersion optical methods or by a non-dispersion electronic discriminator. Dispersion techniques use a crystal, the atomic spacings of which act as a diffraction grating to the short X-rays. Such X-ray spectrometers separate the X-rays in the same manner as optical monochromators. A wide variety of crystals is available for X-ray dispersion over different spectral bands. Commonly used crystals include lithium fluoride, sodium chloride, topaz and ADP (ammonium dihydrogen phosphate). The separated X-rays are detected by ionization, scintillation or semi-conductor detectors such as described on pages 307–310 for gamma-ray detection. Non-dispersion electronic discriminators operate on the same principle as the gamma-ray spectrometer described on pages 310–311.

XRF techniques are fast, non-destructive to the sample and have the added advantage in common with conventional emission spectroscopy that many elements can be analyzed at one time. The accuracy at concentrations over 1% is very good and it is still fairly good over the range 0·01–1%. At lower concentrations, however, the accuracy is generally poor owing to background 'noise' which is always present. Nevertheless, by using long counting times and applying careful background corrections good results can be obtained for low concentrations, particularly in cases where small quantities of a heavy element occur with much lighter elements.

The samples are generally prepared by grinding down to -100 or -150 mesh and placing the powders in special sample holders; sometimes presses are used to compress the powders into standard sized briquettes. Standard samples are often made up from pure chemicals of the elements being analyzed mixed with silica powder, but, since matrix effects can be quite marked, it is advisable to have standards with matrices similar to the samples being analyzed if possible.

Large automated instruments are available which can handle a large number of prepared sample holders and can be programmed to determine 25 elements or more on a pre-set schedule, make the necessary corrections and print out results. All laboratory XRF analyzers are fairly expensive, the larger ones being extremely so, and only large companies and research organizations can be expected to purchase them. Small, portable and relatively inexpensive XRF analyzers are available, however, and are extremely useful for analytical work at concentrations greater than a few tenths of a percent. These instruments are described briefly below.

Portable XRF analyzers

Certain radioisotopes give off radiation in the X-ray spectrum and, although these sources are 10^7 times weaker than conventional X-ray tubes, they can be used to excite X-ray fluorescence. Radioisotopes such as cadmium-109 or americium-241 can be used directly or beta emitters can be used to excite X-rays in other elements by the effect known as *bremsstrahlung*. One such combination source is promethium-147 and aluminium. The instrumentation that has been devised for portable isotope fluorescence (PIF) analyzers, as they are sometimes known, is relatively simple and depends on using balanced filter pairs for cutting out unwanted X-rays. All substances absorb X-rays, the amount of absorption depending on the material, its thickness and the wavelength of the X-rays. It has been observed that X-ray absorption coefficients plotted against wavelength are discontinuous functions with a sudden drop-off in absorption at a critical wavelength which depends on the substance. These discontinuities are known as absorption edges and by selecting two substances with absorption edges lying just on either side of an X-ray line it is wished to measure, the difference in response obtained with each of the filters in turn will be due to X-rays in the 'pass band' (Fig. 4.28). Figure 4.29 shows the basic principle of the PIF analyzer. A wide range of balanced filters are available and elements that can be analyzed quite adequately include Cu, Pb, Zn, Ni, Co, Sn, Ti, Ag and Mo. In addition to using different filters, it may also be necessary to change the radioisotope source when analyzing different elements. This is because radioisotope sources only give off radiation at one, two or a few wavelengths and fluorescence can only be excited if the irradia-

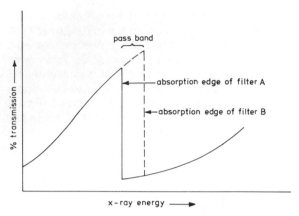

FIG. 4.28. Diagram to show how two balanced filters with closely spaced absorption edges are used to set a pass band for the measurement of X-rays of specific wavelengths.

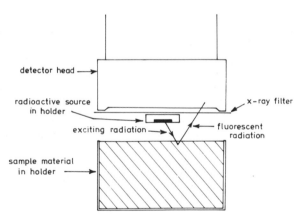

FIG. 4.29. Diagram showing basic principles of portable isotope fluorescence analyzer.

tion is at a lower wavelength than that which it is intended to excite. For example iron-55 gives off radiation at $2\cdot103$ Å (MnK_α) while americium-241 gives off radiations at $0\cdot889$ Å (NpL_α), $0\cdot698$ Å (NpL_β), $0\cdot597$ Å (NpL_γ), $0\cdot208$ Å (γ) and $0\cdot470$ Å (γ).

Calibration graphs are plotted for the various elements by making up standards or, better still, carrying out chemical assays on a selected range of samples. Best results are obtained with finely ground samples and by using standards or calibration samples with matrices similar to the samples being analyzed. The sample holders on PIF analyzers are designed so that *in situ* measurements can be made on drill core or rock outcrops by holding the source against the rock face. Such measurements, however, are generally only semi-quantitative owing to matrix effects, variability in grain size and inhomogeneity of an uncrushed sample.

Specific ion electrode

An extension to the glass electrode used for determining H^+ ion activity or pH is the specific ion or ion-selective electrode which has been developed to measure the activity of a number of ions besides H^+. Specific ion electrodes are now available for the detection of Cl^-, Br^-, I^-, S^{--}, CN^-, Ag^+, F^-, Cu^+, Ca^{++}, NO_3^-, K^+, Pb^{++} and Na^+, but in geochemistry they have found most use for the detection of fluorine. The measurements are made on an expanded-scale pH–Eh meter with the electrode placed in the test solution. There may be a number of interfering ions for which corrections have to be made, but in the case of fluorine OH^- is the only slightly interfering ion.

The determination of fluorine in a geochemical sample, as described by Finklin (1970), is as follows:

1. 250 mg of sample is mixed with 1 g of 2:1 Na_2CO_3/KNO_3 flux and fused at 500°C.
2. When cool, the fused mixture is dissolved in 20 ml water.
3. After 1 h the solution and residue are transferred to a 100 ml beaker, 10 ml of 1M citric acid is added and the solution topped up to 100 ml with water resulting in a final pH of 5·5–6·5.
4. The F^- ion activity is read with the specific ion electrode which has been calibrated against standards prepared in a similar manner to the samples.

The specific ion electrode covers the range 0·01–1900 ppm in the solution or 4–760 000 ppm in the original sample by the above dilution.

Other methods

A number of other methods have been used in geochemistry, but most are at an experimental stage and have not been developed for routine application. A notable exception is the technique of *fluorimetry* which is widely used for the determination of uranium in waters, soils and rocks. The sample is oxidized by evaporation with nitric acid and then fused with sodium fluoride. The cooled and solidified fluoride bead is then examined directly in a specially designed *fluorimeter* which measures the amount of fluorescence emitted when the bead is irradiated by ultra-violet light. The sensitivity is of the order of 5 ppb U for a solid sample.

A new technique of an old method that has much potential and is already being used commercially on a small scale is *plasma emission spectrometry*. Instead of using an electric arc to excite the optical emission lines, an inductively coupled plasma is used. The plasma is generated from argon gas in a special plasma torch consisting of a fused quartz tube by using an RF generator operating at 27·12 MHz with a power output up to 4 kW. Temperatures of 5000 K plus are attained within the plasma which forms a 'doughnut' shape within the tube. Sample solutions can be aspirated into the torch in the same manner as conventional AAS equipment. When the sample passes through the middle of the plasma 'doughnut', the atoms of the various elements within the sample are strongly excited and emit their characteristic spectral lines. These can then be separated by a conventional monochromator and detector of the type used with AAS equipment and the concentration of a particular element determined. Used in this manner only one element can be analyzed at a time for each setting of the monochromator, but the method has some advantages over ordinary AAS. The sensitivity is extremely good and a big concentration range can be covered without the necessity of making dilutions. Scott and Kokok (1975) and Scott *et al* (1974) give good descriptions of the technique. Direct-reading instruments for determination of many elements at a time have also been developed, but are too costly for ordinary geochemical laboratories.

Another technique that has been used in geochemical work by research and government organizations is *neutron activation*, but it is too costly for widespread commercial applications. The technique is based on the phenomenon that atoms bombarded by neutrons are changed to radioactive isotopes which can then be detected by their decay times or gamma-ray spectra. Samples to be tested are placed in a nuclear reactor, which may have a neutron flux as high as 10^{14} neutrons/cm^2/s, for a period of several minutes and are then measured for their radioactivity, often over a period of many hours. Standards can be made up and subjected to the same procedure. Since few people have access to nuclear reactors, the method has very limited application, but a number of neutron sources such as beryllium/antimony-124 or californium-252 are capable of activating a number of elements of interest even though the neutron fluxes produced by them are 10^9–10^{10} times less than the flux of a nuclear reactor. Field instruments employing these methods have been designed and built and include the Metalog® system of Scintrex Ltd in Canada designed for borehole logging (see Chapter 7), but they are complex and costly pieces of equipment and cannot be used in routine exploration work. Plant and Coleman (1973) describe neutron activation as applied to gold samples using a nuclear reactor and

Philbin and Senftle (1971) describe a field technique for uranium using a Cf-252 source.

Another interesting technique that has been used in a very special case is an instrument known as the *beryllometer* which has been designed specifically for the detection of beryllium (Bowie *et al*, 1960). It depends on the fact that Be is a neutron source when bombarded by gamma-rays or fast-moving positive particles. A strong radioactive source such as antimony-124 is used and the neutrons given off by Be in the sample are detected by a boron trifluoride counter. *In situ* measurements can be made in the field and values of 10 ppm and less can be detected. With its necessary shielding the instrument weighs about 40 kg and is carried and operated by two men. It has been employed very successfully in beryllium prospecting and in the evaluation of beryllium ores in various parts of the world.

Summary

There is a wide choice of analytical methods which can be used in geochemical exploration and the method selected will depend not only on the element analyzed but also on the concentration range expected. In addition cost is an important factor. Small exploration companies can set up colorimetric or atomic absorption laboratories at relatively low cost, but only large commercial laboratories and research organizations can afford X-ray fluorescence or emission spectrographic equipment. Table 4.15 lists the analytical methods most commonly used in exploration geochemistry today together with a list of selected elements. The most important method for a particular element is indicated by the use of heavy type. In some cases, however, it becomes a matter of opinion which method may be the most important. For example, colorimetry has long been regarded as the most satisfactory method for Mo, but more and more laboratories are producing satisfactory results for Mo using AAS with a nitrous oxide/acetylene flame.

For a proper understanding of geochemical analyses it is important to be aware of the different degrees of sample 'attack' and extraction of the various metals. Many geologists do not have a full appreciation of this and it is extremely important when comparing analytical results from one method to another and from one lab to another. For example, a soil sample may contain a total of 1000 ppm Ni, but 300 ppm may be contained in the lattices of silicate minerals, 200 ppm may be present as sulphides and the remainder may be loosely held by iron oxides, clays and organic matter. Both emission spectroscopy and X-ray fluorescence methods will indicate total Ni present, but colorimetry and atomic absorption, which depend on the Ni being brought into solution for analysis, may record very little of the

TABLE 4.15

ANALYTICAL TECHNIQUES COMMONLY USED IN EXPLORATION GEOCHEMISTRY. ELE-
MENTS IN BOLD TYPE INDICATE PREFERRED METHOD

Technique	Elements analyzed	Approximate 1978 cost of equipment	Typical 1978 commercial cost of analysis
AAS	**Ag, Au, Ba, Bi, Cd,** Co, Cr, **Cu, Hg, Li, Mn, Ni, Pb, Sb, Zn,** Mo, Sn, Ti	$10 000	$1·50 for first element, $0·75 for each additional element. Ag, Au, Hg $3·00–4·00 each
Colorimetry	**Mo, P, Sn, Ti, W,** Be, Bi, Co, Cr, Cu, Mn, Ni, Pb, Zn	$500–1 000	$1·50 each $3·00 each for Mo and W
Emission spectroscopy	**Be, La, V, Y,** Ag, As, Ba, Bi, Cd, Co, Cr, Cu, Li, Mn, Mo, Nb, Ni, Pb, Sn, Zn, Zr	$15 000+ (photographic detection)	$6·00 for first element, $4·00 for each additional element
XRF	**Nb, Zr,** Ba, Co, Cr, Cu, La, Mo, Ni, Pb, Sb, Sn, Ti, U, V, Zn	$30 000+	$8·00 per element
Specific-ion electrode	F	$1 500	$3·00 each
Fluorimetry	U	$2 500	$4·00 each

strongly held silicate Ni. For atomic absorption it is usual practice to digest the samples directly in a boiling acid without prior fusion except for special cases such as the analysis of Sn. The most commonly used acids are concentrated $HClO_4$–HNO_3, $HClO_4$, HCl–HNO_3, HNO_3 and HCl, which give a high degree of extraction of metals from soils and stream sediments. Even in the case of silicate rock-forming minerals these concentrated acids give more than 70% extraction except in the case of some pyroxenes and amphiboles for which extractions may be <40% (Foster, 1973). This is not a disadvantage for most geochemical exploration work, however, since metals contained in the lattices of silicate minerals are not of particular importance. For this reason 25% HNO_3 is very commonly used as it has been found to give consistent results for most soil and stream sediment samples. In the case of colorimetric analysis it is common practice to fuse the samples before taking them up in hydrochloric acid. Potassium bisulphate is the most commonly used fusion flux, but it is not effective in attacking all minerals (Harden and Tooms, 1964) and extraction generally varies from 60 to 100% depending on mineralogy, though in the case of pyroxenes and amphiboles extraction may be less than 20%. As already mentioned, this is not a disadvantage for geochemical prospecting. Other fusion fluxes are sometimes used in special cases (Table 4.16). Partial extractions are used in atomic absorption and colorimetry to determine loosely held metal, generally referred to as *cold-extractable metal*, which can be useful to assist interpretation or for rapid field tests in the case of colorimetry. Dolezal *et al* (1968) give a good account of extraction techniques in inorganic analysis.

Another important aspect of geochemical analysis is *precision*. As mentioned in the introduction, high accuracy and precision are not of major importance in exploration, but, nevertheless, it is important to control precision within certain permissible limits. Precision is defined as the ability to reproduce an analytical result for the same sample material and it does not imply accuracy. For example, the actual Cu content of a soil sample might be 500 ppm and an analytical method that consistently produced results ranging from 300 to 320 ppm would be very precise but not accurate. On the other hand, a method that produced results ranging from 450 to 550 ppm would be less precise but more accurate. The true content of a sample in exploration geochemistry is not of major importance, but it is important that the method used is consistent. The normal procedure in a geochemical laboratory is to calibrate the method being used with a series of artificial standards of known metal content. The precision, on the other hand, is determined by analyzing standard samples of the material being analyzed (soils, stream sediments, plant ash, etc.) or by

TABLE 4.16
VARIOUS EXTRACTIONS USED FOR SIMPLE DIGESTION IN
GEOCHEMICAL ANALYSES

Type of extraction	Reagent and remarks
Hot acids	nitric–perchloric ⎱ potentially explosive perchloric ⎰ aqua regia conc.nitric conc.hydrochloric 25% nitric
Fusion fluxes	potassium bisulphate—most commonly used, but extraction variable. sodium carbonate– ⎰ very good extraction, potassium nitrate ⎱ but alkali fusions sodium hydroxide ⎰ are not commonly used as temperatures are high and crucibles have to be used. ammonium iodide—used in tin analysis to attack cassiterite
Partial (cold acids and buffers)	cold, dilute hydrochloric ammonium citrate (buffered to various pH's) EDTA ascorbic acid–hydrogen peroxide (2:5 30% H_2O_2, 1% ascorbic acid)—specific for sulphides

running some samples in duplicate. In neither case is the actual content of the sample known, but the precision, i.e. the ability to reproduce a result, can be calculated from the mean value and standard deviation. For example, if ten determinations of a standard soil during a day's run in the laboratory produced the following results for Cu of 250, 260, 275, 235, 245, 265, 260, 235, 280 and 265 ppm, the mean value is 257 ppm and the standard deviation 15·5 ppm. The precision is normally expressed in terms of a percentage at the 95% confidence level, i.e. at two standard deviations on either side of the mean. Thus, the precision for the above example is:

$$\pm \frac{2 \times 15 \cdot 5}{257} = \pm 12\% \text{ at the 95\% confidence level}$$

This is the precision at a concentration of about 250 ppm Cu and variations in the precision may be found using standard samples with different Cu contents. If samples are being run in duplicate instead of using standard samples, the overall precision is given directly by:

$$\sqrt{\frac{\Sigma d^2}{2n}} \quad \text{(Thompson and Howarth, 1973)}$$

where d is the difference between duplicates and n is the number of pairs. As an example, let us assume that six duplicate analyses for Cu gave the following results:

145	155
260	280
85	105
250	260
165	175
205	225

Using the expression above, the precision is $\pm11\%$ at the 95% confidence level.

REFERENCES AND BIBLIOGRAPHY

Adler, I. (1966). *X-ray Emission Spectrography in Geology*, Elsevier Publishing Company, Amsterdam, 258 pp.

Ahren, L. H. (1950). *Spectrochemical Analysis*, Addison-Wesley Press Inc, Cambridge, Mass., 269 pp.

Ahren, L. H. (1965). *Distribution of the Elements in Our Planet*, McGraw-Hill Book Co., New York, 110 pp.

Allan, R. J. (1971). Lake sediment: a medium for regional geochemical exploration of the Canadian Shield, *C.I.M. Bull.*, **64**, 43–59.

Allan, R. J., Cameron, E. M. and Durham, C. C. (1973). Reconnaissance geochemistry using lake sediments on a 36 000 square mile area of the northwestern Canadian Shield, *Geol. Surv. Canada Paper 72-50.*

Almond, H. (1955). Rapid field and laboratory method for the determination of copper in soil and rocks, *U.S.G.S. Bull.* 1036A, 1–8.

Armour-Brown, A. and Nichol, I. (1970). Regional geochemical reconnaissance and the location of metallogenic provinces, *Econ. Geol.*, **65**, 312–330.

Baas Becking, L. G. M. and Moore, D. (1961). Biogenic sulphides, *Econ. Geol.*, **56**, 259–272.

Barsukov, V. L. (1957). The geochemistry of tin, *Geochemistry*, **1**, 41–51.

Basham, I. R. and Easterbrook, G. D. (1977). Alpha-particle autoradiography of geological specimens by use of cellulose nitrate detectors, *Trans. Instn. Min. Metall.*, Lond., **86**, B96–98.

Bates, R. G. (1964). *Determination of pH—Theory and Practice*, John Wiley and Sons, New York, 435 pp.

Beck, L. S. and Gingrich, J. E. (1976). Track etch orientation survey in the Cluff Lake area, northern Saskatchewan, *C.I.M. Bull.*, **69**, 105–109.

Beus, A. A. and Sitnin, A. A. (1972). Geochemical specialization of magmatic complexes as criteria for the exploration of hidden deposits, 24th *Intern. Geol. Congr.*, Montreal, **6**, 101–105.

Bloom, H. (1955). A field method for the determination of ammonium citrate soluble heavy metals in soils and alluvium, *Econ. Geol.*, **50**, 533–541.

Bloomfield, K., Reedman, J. H. and Tether, J. G. G. (1971). Geochemical exploration of carbonatite complexes in eastern Uganda, *Geochem. Explor.*, C.I.M. Spec. **11**, 85–102.

Bolviken, B. and Lag, J. (1977). Natural heavy-metal poisoning of soils and vegetation: an exploration tool in glaciated terrain, *Trans. Instn. Min. Metall.*, Lond., **86**, B173–180.

Bolviken, B. and Sinding-Larsen, R. (1973). Total error and other criteria in the interpretation of stream sediment data, *Geochem. Explor. 1972*, I.M.M., Lond., 285–295.

Bowie, S. H. U., Bisby, H., Burke, K. C. and Hale, F. H. (1960). Electronic instruments for detecting and assaying beryllium ores, *Trans. Instn. Min. Metall.*, Lond., **69**, 345–359.

Bowie, S. H. U., Darnley, A. G. and Rhodes, J. R. (1965). Portable radioisotope X-ray fluorescence analyser, *Trans. Instn. Min. Metall.*, Lond., **74**, 361–379.

Boyle, R. W. and Lynch, J. J. (1968). Speculations on the source of zinc, cadmium, lead, copper and sulphur in Mississippi Valley and similar types of lead–zinc deposits, *Econ. Geol.*, **63**, 421–422.

Brabec, D. and White, W. H. (1971). Distribution of copper and zinc in rocks of the Guichon Creek batholith, British Columbia, *Geochem. Explor.*, C.I.M. Spec. **11**, 291–297.

Brade-Birks, S. G. (1966). *Good Soil*, English Universities Press Ltd., London, 304 pp.

Britton, H. T. (1955). *Hydrogen Ions*, Chapman and Hall, London, 489 pp.

Brooks, R. R. (1972). *Geobotany and Biogeochemistry in Mineral Exploration*, Harper and Row, New York, 290 pp.

Brown, B. W. and Hilchey, G. R. (1974). Sampling and analysis of geochemical materials for gold, *Geochem. Explor. 1974*, A.E.G. Spec. Pub. No. 2, Elsevier Scientific Publishing Co., Amsterdam, 683–690.

Buckman, H. O. and Brady, N. C. (1969). *The Nature and Properties of Soils*, Macmillan, New York.

Cameron, E. M. (1977). Geochemical dispersion in lake waters and sediments from massive sulphide mineralization, Agricola Lake area, Northwest Territories, *J. Geochem. Explor.*, **6**, 327–348.

Cameron, E. M., Siddeley, G. and Durham, C. C. (1971). Distribution of ore elements in rocks for evaluating ore potential: nickel, copper, cobalt and sulphur in ultramafic rocks of the Canadian Shield, *Geochem. Explor.*, C.I.M. Spec. **11**, 298–314.

Cannon, H. L. (1955). Geochemical relations of zinc-bearing peat to the Lockport Dolomite, Orleans County, New York, *U.S.G.S. Bull.*, 1000-D, 119–185.

Cannon, H. L. (1960). Botanical prospecting for ore deposits, *Science*, **132**(3427), 591–598.

Carpenter, R. H., Pope, T. A. and Smith, R. L. (1975). Fe–Mn oxide coatings in stream sediment geochemical surveys, *J. Geochem. Explor.*, **4**, 349–363.

Carroll, D. (1970). *Rock Weathering*, Plenum Press, New York, 203 pp.

Chaffee, M. A. and Hessin, T. D. (1971). An evaluation of geochemical sampling in the search for concealed 'porphyry' copper–molybdenum deposits on sediments in Southern Arizona, *Geochem. Explor.*, C.I.M. Spec. **11**, 401–409.

Chakrabarti, A. K. and Solomon, P. J. (1971). A geochemical case history of the Rajburi antimony prospect, Thailand, (abs), *Geochem. Explor.*, C.I.M. Spec. **11**, 121.

Chapman, R. P. (1976). Limitations of correlation and regression analysis in geochemical exploration, *Trans. Instn. Min. Metall.*, Lond., **85**, B279–283.

Chowdhury, A. N. and Bose, B. B. (1971). Role of 'humus matter' in the formation of geochemical anomalies, *Geochem. Explor.*, C.I.M. Spec. **11**, 410–413.

Clarke, W. B. and Kugler, G. (1973). Dissolved helium in groundwater: a possible method for uranium and thorium prospecting, *Econ. Geol.*, **68**, 243–251.

Coope, J. A. (1973). Geochemical prospecting for porphyry copper-type mineralization—a review, *J. Geochem. Explor.*, **2**, 81–102.

Cox, R. (1974). Geochemical soil surveys in exploration for nickel–copper sulphides at Pioneer, near Norseman, Western Australia, *Geochem. Explor. 1974*, A.E.G. Spec. Pub. No. 2, Elsevier Scientific Publishing Co., Amsterdam, 437–460.

Craven, C. A. U. (1954). Statistical estimation of the accuracy of assaying, *Trans. Instn. Min. Metall.*, Lond., **63**, 551–563.

Cruft, E. F. (1964). Trace element determinations in soils and sediments by an internal standard spectrographic technique, *Econ. Geol.*, **59**, 458–464.

Dall'Aglio, M. (1971). Comparison between hydrogeochemical and stream sediment methods in prospecting for mercury, *Geochem. Explor.*, C.I.M. Spec. **11**, 126–131.

Davenport, P. H. and Nichol, I. (1973). Bedrock geochemistry as a guide to areas of base metal potential in volcano-sedimentary belts of the Canadian shield, *Geochem. Explor. 1972*, I.M.M., Lond., 45–57.

Day, F. H. (1963). *The Chemical Elements in Nature*, George G. Harrap and Co., London, 372 pp.

Debnam, A. H. and Webb, J. S. (1960). Some geochemical anomalies in soil and stream sediment related to beryl pegmatites in Rhodesia and Uganda, *Trans. Instn. Min. Metall.*, Lond., **69**, 329–344.

De Geoffroy, J., Wu, S. M. and Heins, R. W. (1967). Geochemical coverage by spring sampling method in the southwest Wisconsin zinc area, *Econ. Geol.*, **62**, 679–697.

De Geoffroy, J., Wu, S. M. and Heins, R. W. (1968). Selection of drilling targets from geochemical data in southwest Wisconsin zinc area, *Econ. Geol.*, **63**, 787–795.

Dolezal, J., Povondra, P. and Sulcek, Z. (1968). *Decomposition Techniques in Inorganic Analysis* (transl. from Czech), Iliffe Books Ltd, London, 224 pp.

d'Orey, F. L. C. (1975). Contribution of termite mounds to locating hidden copper deposits, *Trans. Instn. Min. Metall.*, Lond., **84**, B150–151.

Dubov, R. I. (1973). A statistical approach to the classification of geochemical anomalies, *Geochem. Explor. 1972*, I.M.M., Lond., 275–284.

Dyck, W. (1976). The use of helium in mineral exploration, *J. Geochem. Explor.*, **5**, 3–20.

Ewing, Galen W. (1969). *Instrumental Methods of Chemical Analysis*, McGraw-Hill Book Co., New York, 627 pp.

Farrell, B. L. (1974). Fluorine, a direct indicator of fluorite mineralization in local and regional soil geochemical surveys, *J. Geochem. Explor.*, **3**, 227–244.

Finklin, W. H. (1970). A rapid method for the determination of fluoride in rocks using an ion-selective electrode, *U.S. Geol. Surv. Prof. Paper 700-C*, C186–188.

Fleischer, M. and Robinson, W. O. (1963). Some problems of the geochemistry of fluorine, *Roy. Soc. of Canada, Spec. Pub. No. 6*, 58–75.

Flinter, B. H. (1971). Tin in acid granitoids: the search for a geochemical scheme of mineral exploration, *Geochem. Explor.*, C.I.M. Spec. **11**, 323–330.

Forbes, E. A., Posner, A. M. and Quirk, J. P. (1976). The specific adsorption of divalent Cd, Co, Cu, Pb, and Zn on goethite, *J. Soil Sci.*, **27**, 154–166.

Foster, J. R. (1973). The efficiency of various digestion procedures on the extraction of metals from rocks and rock-forming minerals, *C.I.M. Bull.*, **66**(736), 85–92.

Fraser, D. C. (1961). Organic sequestration of copper, *Econ. Geol.*, **56**, 1063–1078.

Friedrich, G. H. and Hawkes, H. E. (1966). Mercury as an ore guide in the Pachuca-Real del Monte district, Hidalgo, Mexico, *Econ. Geol.*, **61**, 744–753.

Friedrich, G. H. and Pluger, W. L. (1971). Geochemical prospecting for barite and fluorite deposits, *Geochem. Explor.*, C.I.M. Spec. **11**, 151–156.

Garrels, R. M. (1960). *Mineral Equilibria at Low Temperature and Pressure*, Harper and Row, New York, 254 pp.

Garrels, R. M. and Christ, C. L. (1965). *Minerals, Solutions and Equilibria*, Harper and Row, New York, 450 pp.

Garrett, R. G. (1971). The dispersion of copper and zinc in glacial overburden at the Louvem deposit, Val d'Or, Quebec, *Geochem. Explor.*, C.I.M. Spec. **11**, 157–158.

Garrett, R. G. (1973). Regional geochemical study of Cretaceous acidic rocks in the northern Canadian Cordillera as a tool for broad mineral exploration, *Geochem. Explor. 1972*, I.M.M., Lond., 203–219.

Ginzburg, I. I. (1960). *Principles of Geochemical Prospecting* (translation), Pergamon Press, London, 311 pp.

Goldschmidt, V. M. (1954). *Geochemistry*, Clarendon Press, Oxford, 730 pp.

Govett, G. J. S. (1960). Geochemical prospecting for copper in Northern Rhodesia, *Rep. 21st Intern. Geol. Congr.*, Part 2, Geological results of applied geochemistry and geophysics, 44–56.

Govett, G. J. S. and Goodfellow, W. D. (1975). Use of rock geochemistry in detecting blind sulphide deposits: a discussion, *Trans. Instn. Min. Metall.*, Lond., **84**, B134–140.

Green, J. (1959). Geochemical table of the elements for 1959, *Geol. Soc. America Bull.*, **70**, 1127–1184.

Grim, R. E. (1953). *Clay Mineralogy*, McGraw-Hill Book Co., New York, 396 pp.

Hansuld, J. A. (1966). Eh and pH in geochemical exploration, *C.I.M. Bull.*, **59**, 315–322.

Harden, G. and Tooms, J. S. (1964). Efficiency of the potassium bisulphate fusion in geochemical analysis, *Trans. Instn. Min. Metall.*, Lond., **73**, 129–141.

Harman, H. H. (1960). *Modern Factor Analysis*, University of Chicago Press, 469 pp.

Hawkes, H. E. (1954). Geochemical prospecting investigations in the Nyeba lead–zinc district, Nigeria, *U.S.G.S. Bull.* 1000-B, 51–103.

Hawkes, H. E. (1957). Principles of geochemical prospecting, *U.S.G.S. Bull.* 1000-F, 225–355.

Hawkes, H. E. (1976). The downstream dilution of stream sediment anomalies, *J. Geochem. Explor.*, **5**, 345–358.

Hawkes, H. E. and Webb, J. S. (1962). *Geochemistry in Mineral Exploration*, Harper and Row, New York, 415 pp.

Hesp, W. R. (1971). Correlations between the tin content of granitic rocks and their chemical and mineralogical composition, *Geochem. Explor.*, C.I.M. Spec. **11**, 341–353.

Hoag, R. B. and Webber, G. R. (1976). Hydrogeochemical exploration and sources of anomalous waters, *J. Geochem. Explor.*, **5**, 39–57.

Hoffman, S. J. (1977). Talus fine sampling as a regional geochemical exploration technique in mountainous regions, *J. Geochem. Explor.*, **6**, 349–360.

Hoffman, S. J. and Fletcher, W. K. (1976). Reconnaissance geochemistry of Nechaka plateau, British Columbia, using lake sediments, *J. Geochem. Explor.*, **5**, 101–114.

Horizon Magazine, (1959). A flower that led to a copper discovery, R.S.T. Company Magazine, 35–39.

Hornbrook, E. H. W. (1969). Biogeochemical prospecting for molybdenum in west-central British Columbia, *Geol. Surv. Canada Paper* 68-56.

Howarth, R. J. and Lowenstein, P. L. (1971). Sampling variability of stream sediments in broad-scale regional geochemical reconnaissance, *Trans. Instn. Min. Metall.*, Lond., **80**, B363–372.

Hunt, C. B. (1972). *The Geology of Soils*, W. H. Freeman & Co., San Franscisco, 344 pp.

Hunt, E. C., North, A. A. and Wells, R. A. (1955). Application of paper chromatographic methods of analysis to geochemical prospecting, *The Analyst*, **80**, 172–194.

Hyvarinen, L., Kauranne, K. and Yletyinen, V. (1973). Modern boulder tracing in prospecting. In *Prospecting in Areas of Glacial Terrain*, I.M.M., Lond., 87–95.

James, C. H. (1957). *The geochemical dispersion of arsenic and antimony related to gold mineralization in Southern Rhodesia*, Tech. Comm., No. 12, Applied Geochemistry Research Group, Imperial College, London.

Jedwab, J. (1955). Granites à deux micas de Guehenno et de La Villeder (Morbihan-France), *Bull. de La Loc. Belge de Geol.*, **64**, 526–534.

Jonasson, I. R. and Sangster, D. F. (1974). Variation in the mercury content of sphalerite from some Canadian sulphide deposits, *Geochem. Explor. 1974*, A.E.G. Spec. Pub. No. 2, Elsevier Scientific Publishing Co., Amsterdam.

Kauranne, L. K. (1959). Pedogeochemical prospecting in glaciated terrain: Finland, *Comm. Geol., Bull.* No. 184, 10 pp.

Krauskopf, K. B. (1955). Sedimentary deposits of rare metals, *Econ. Geol.*, 50th Ann. Vol., 411–463.

Krumbein, W. C. and Graybill, F. A. (1965). *An Introduction to Statistical Models in Geology*, McGraw-Hill Book Co., New York, 475 pp.

Lankin, H. W., Curtin, G. C. and Hubert, A. E. (1971). Geochemistry of gold in the weathering cycle (abs), *Geochem. Explor.*, C.I.M. Spec. 11, 196.

Lawrence, G. (1974). The use of Rb/Sr ratios as a guide to mineralization in the Galway granite, Ireland, *Geochem. Explor. 1974*, A.E.G. Spec. Pub. No. 2, Elsevier Scientific Publishing Co., Amsterdam.

Learned, R. E. and Boissen, R. (1973). Gold—a useful pathfinder element in the search for porphyry copper deposits in Puerto Rico, *Geochem. Explor. 1972*, I.M.M., Lond., 93–103.

Lepeltier, C. (1969). A simplified statistical treatment of geochemical data by graphical representation, *Econ. Geol.*, **64**, 538–550.

Levinson, A. A. (1974). *Introduction to Exploration Geochemistry*, Applied Publishing Ltd, Calgary, 612 pp.

Loughnan, F. C. (1969). *Chemical Weathering of the Silicate Minerals*, American Elsevier Publishing Co. Inc., New York, 154 pp.

Lovering, T. S. (1927). Organic precipitation of metallic copper, *U.S.G.S. Bull.* 795-C, 45–52.

Lukashev, K. I. (1970). *Lithology and Geochemistry of the Weathering Crust* (translated from Russian), Keter Press, Jerusalem, 367 pp.

McCarthy, J. H. (1972). Mercury vapor and other volatile components in the air as guides to ore deposits, *J. Geochem. Explor.*, **1**, 143–162.

McNerney, J. J. and Buseck, P. R. (1973). Geochemical exploration using mercury vapor, *Econ. Geol.*, **68**, 1313–1320.

Malyuga, D. P. (1964). *Biogeochemical Methods of Prospecting*, (translated from Russian), Consultants Bureau, New York, 205 pp.

Maranzana, F. (1972). Application of talus sampling to geochemical exploration in arid areas: Los Pelambres hydrothermal alteration area, Chile, *Trans. Instn. Min. Metall.*, Lond., **81**, B26–33.

Mason, B. (1958). *Principles of Geochemistry*, John Wiley & Sons Inc., New York, 310 pp.

Meyer, W. T. and Peters, R. G. (1973). Evaluation of sulphur as a guide to buried sulphide deposits in the Notre Dame Bay area, Newfoundland. In *Prospecting in Areas of Glacial Terrain*, I.M.M., Lond., 55–66.

Minatidis, D. G. and Slatt, R. M. (1976). Uranium and copper exploration by nearshore lake sediment geochemistry, Kaipokok region of Labrador, *J. Geochem. Explor.*, **5**, 135–144.

Morse, R. H. (1971). Comparison of geochemical prospecting methods using radium with those using radon and uranium, *Geochem. Explor.*, C.I.M. Spec. 11, 215–230.

Nichol, I. (1973). The role of computerized data systems in geochemical exploration, *C.I.M. Bull.*, **66**, 59–68.

Nichol, I. and Henderson-Hamilton, J. C. (1965). A rapid quantitative spectrographic method for the analysis of rocks, soils and stream sediments, *Trans. Instn. Min. Metall.*, Lond., **74**, 955–961.

Nichol, I., Garrett, R. G. and Webb, J. S. (1969). The role of some statistical and mathematical methods in the interpretation of regional geochemical data, *Econ. Geol.*, **64**, 204–220.

Nilsson, G. (1973). Nickel prospecting and the discovery of the Mjovattnet mineralization, northern Sweden: a case history of the use of combined techniques in drift-covered glaciated terrain. In *Prospecting in Areas of Glacial Terrain*, I.M.M., Lond., 97–109.

North, A. A. (1956). Geochemical field methods for the determination of tungsten and molybdenum in soils, *The Analyst*, **81**, 660–668.

Nowlan, G. A. (1976). Concretionary manganese-ion oxides in streams and their usefulness as a sample medium for geochemical prospecting, *J. Geochem. Explor.*, **6**, 193–210.

Ovchinnikov, L. N. and Grigoryan, S. V. (1971). Primary haloes in prospecting for sulphide deposits, *Geochem. Explor.*, C.I.M. Spec. **11**, 375–380.

Oyarzum, J. M. (1974). Rubidium and strontium as guides to copper mineralization emplaced in some Chilean andesitic rocks, *Geochem. Explor. 1974*, A.E.G. Spec. Pub. No. 2, Elsevier Scientific Publishing Co., Amsterdam, 333–338.

Pantazis, Th. M. and Govett, G. J. S. (1973). Interpretation of a detailed rock geochemical survey around Mathiati Mine, Cyprus, *J. Geochem. Explor.*, **2**, 25–36.

Parslow, G. R. (1974). Determination of background and threshold in exploration geochemistry, *J. Geochem. Explor.*, **3**, 319–336.

Philbin, P. and Senftle, F. E. (1971). Field activation analysis of uranium ore using Cf-252 neutron source, *Trans. Soc. Min. Eng.*, **250**, 102–106.

Philpott, D. E. (1974). Shangani—a geochemical discovery of a nickel copper sulphide deposit, *Geochem. Explor. 1974*, A.E.G. Spec. Pub. No. 2, Elsevier Scientific Publishing Co., Amsterdam, 503–510.

Plant, J. (1971). Orientation studies on stream sediment sampling for a regional geochemical survey in northern Scotland, *Trans. Instn. Min. Metall.*, Lond., **80**, B324–345.

Plant, J. and Coleman, R. F. (1973). Application of neutron activation analysis to the evaluation of placer gold concentrations, *Geochem. Explor. 1972*, I.M.M., Lond., 373–381.

Rankama, K. and Sahama, Th. G. (1950). *Geochemistry*, University of Chicago Press, Chicago, 912 pp.

Reedman, A. J. (1973). Prospection and evaluation of beryl pegmatites in southwest Uganda, *Overseas Geol. Miner. Resour.* No. 41, 86–100.

Reedman, J. H. (1974). Residual soil geochemistry in the discovery and evaluation of the Butiriku carbonatite, southeast Uganda, *Trans. Instn. Min. Metall.*, Lond., **83**, B1–12.

Ritchie, A. S. (1964). *Chromatography in Geology*, Elsevier Scientific Publishing Co., Amsterdam, 185 pp.

Robbins, J. C. (1973). Zeeman spectrometer for measurement of atmospheric mercury vapour, *Geochem. Explor. 1972*, I.M.M., Lond., 315–323.

Robinson, G. W. (1951). *Soils—Their Origin, Constitution and Classification*,

Thomas Murby and Co., London, 573 pp.

Sainsbury, C. L. (1957). A geochemical exploration for antimony in southeastern Alaska, *U.S.G.S. Bull.* 1024-H, 163–178.

Sandell, E. B. (1950). *Colorimetric Determination of Traces of Metals*, Interscience Publishers Inc., New York, 673 pp.

Scott, R. H., Fassel, V. A., Kniseley, R. N. and Nixon, D. E. (1974). Inductively coupled plasma-optical emission analytical spectrometry, *Analytical Chemistry*, **46**(1), 75–80.

Scott, R. H. and Kokok, M. L. (1975). Application of inductively coupled plasmas to the analysis of geochemical samples, *Analytic Chimica Acta*, **70**, 271–279.

Sears, W. P. (1971). Mercury in base metal and gold ores of the province of Quebec, *Geochem. Explor.*, C.I.M., Spec. **11**, 384–390.

Shaw, D. M. (1954). Trace elements in pelitic rocks, *Geol. Soc. Am. Bull.* **65**, 1151–1166.

Stanton, R. E. (1966). *Rapid Methods of Trace Analysis for Geochemical Application*, Edward Arnold, London, 96 pp.

Stevens, D. N., Rouse, G. E. and de Voto, R. H. (1971). Radon-222 in soil gas: three uranium exploration case histories in the western United States, *Geochem. Explor.*, C.I.M., Spec. **11**, 258–264.

Sutton, W. R. and Soonawala, N. M. (1975). A soil radium method for uranium prospecting, *C.I.M. Bull.*, **68**(757), 51–56.

Swaine, D. J. (1955). *Trace element content of soils*, Commonwealth Agricultural Bur., Farnham Royal, Bucks., Tech. Comm. No. 48, 157 pp.

Tauson, L. V. and Kozlov, V. D. (1973). Distribution functions and ratios of trace element concentrations as estimators of the ore bearing potential of granites, *Geochem. Explor. 1972*, I.M.M., London, 37–44.

Taylor, S. R. (1964). Abundance of chemical elements in the continental crust: a new table, *Geochem. Cosmochim. Acta*, **28**, 1273–1284.

Taylor, S. R. (1966). The application of trace element data to problems in petrology, *Physics and Chem. of the Earth*, **6**, 135–213.

Thompson, M. and Howarth, R. J. (1973). Rapid estimation and control of precision by duplicate determination, *The Analyst*, **78**, 153–160.

Tooms, J. S. and Webb, J. S. (1961). Geochemical prospecting investigations in the Northern Rhodesian Copperbelt, *Econ. Geol.*, **56**, 815–846.

Travis, G. A., Keays, R. R. and Davison, R. M. (1976). Palladium and iridium in the evaluation of nickel gossans in Western Australia. *Econ. Geol.*, **71**, 1229–1243.

Turekian, K. K. and Wedepohl, K. H. (1961). Distribution of the elements in some major units of the earth's crust, *Geol. Soc. Am. Bull.* **72**, 641–664.

Vinogradov, A. P. (1956). Regularity of distribution of chemical elements in the earth's crust, *Geochemistry*, No. 1, 1–43.

Vinogradov, A. P. (1959). *The Geochemistry of Rare and Dispersed Chemical Elements in Soils* (translated from Russian), Consultants Bureau, New York, Chapman and Hall, London, 209 pp.

Vlasov, K. A. (ed.) (1966). *Geochemistry and Mineralogy of Rare Elements and Genetic Types of their Deposits*, vol. 1 (translation), Israel Program for Scientific Translations, Jerusalem, 688 pp.

Warren, H. V. (1972). Biogeochemistry in Canada, *Endeavour*, **31**, 46–49.
Warren, H. V., Delavault, R. E. and Barakso, J. (1964). The role of arsenic as a pathfinder in biogeochemical prospecting, *Econ. Geol.*, **59**, 1381–1385.
Wedepohl, K. H. (1969). *Handbook of Geochemistry*, Springer-Verlag, Berlin.
Wedepohl, K. H. (1970). *Geochemistry* (translation), Holt, Rinehart and Winston Inc., New York.
West, W. F. (1970). Termite prospecting, *Chamber of Mines Journal*, Rhodesia, October, 32–35.

CHAPTER 5

Deep Sampling Methods

One of the first steps taken in following up a geochemical soil anomaly is to try to locate the bedrock source of the anomalous values. In areas where the soil cover is thin, it is often possible to do this by examination and sampling of outcrops, but in areas with a thick soil profile, the location of the anomalous source may involve a long and detailed deep sampling programme. In addition to following up surface anomalies, deep sampling methods may be employed in evaluating alluvial or eluvial deposits, or in reconnaissance geochemical surveys of areas covered by superficial deposits such as peat, alluvium and glacial till, or of areas with heavy surface leaching which may have removed elements of interest from the near-surface soil horizons.

5.1 PITTING AND TRENCHING

The simplest method of deep sampling is the digging of pits and trenches. In countries where labour is cheap this is generally carried out by hand, but in other parts of the world mechanical excavators, such as backhoes, are used. Pits dug for geochemical purposes can be sampled by filling sample packets with soil collected at intervals of 1 m or less, or narrow channel samples can be taken over lengths of 1–2 m by cutting a groove with a geological pick. If pits have been dug for evaluation purposes, larger samples can be taken by cutting channels 10–15 cm wide and 5 cm deep over lengths of 1–3 m.

In tropical regions where thick lateritic soil profiles are developed it is possible to dig pits to depths of 30 m or more by hand. Two labourers are used and the only tools required are a short-handled shovel, a pick or sharpened, heavy, straight crowbar, and a bucket tied to a rope for removing soil. It is important to keep the pit vertical and circular with a diameter of approximately 1 m. The man working at the bottom of the pit should always wear a hard-hat as even a small object dropped down a deep pit could result in a fatal blow on the head. A windlass is sometimes used for pulling up the bucket, but it is not really necessary as the bucket can be pulled up easily by hand. Egress and ingress are effected by footholds cut into the wall on opposite sides of the pit. A rope is sometimes used to lower a man down the pit on a bosun's chair for logging and sampling, but it is not necessary if the pit is kept narrow enough (1 m or less) and good footholds are cut into the walls. The work is quite safe provided that the soil is firm and dry. If it becomes wet or dry and crumbly, work should be stopped as the pit could collapse causing a fatal accident. The rate of progress varies widely as it depends upon the hardness of the ground. If the going is fairly good, it should be possible to complete a pit to 10 m in 10 man-days or less. Pits 20 or 30 m deep may take 40 man-days or more to complete. Very often the accumulation of carbon dioxide at the bottom of the pit makes it very difficult for the labourers to work efficiently and may even prove fatal. To overcome this, hand or motor driven pumps may be used to supply fresh air to the bottom of the pit. In most cases this is not necessary, however, and depths of 30 m have been attained without air pumps. Sometimes pits are joined by crosscuts which make it possible to sample bedrock over a critical zone at depth. Digging of crosscuts can be very much more dangerous than digging of pits and it should only be carried out if the ground is particularly firm and dry. Figure 5.1 shows an example of a pitting project carried out to evaluate phosphate reserves in soil over a carbonatite complex in Uganda and Fig. 5.2 shows a typical hand-dug prospecting pit.

Pits are best dug by mechanical excavators in areas where labour is expensive and/or where the ground is fairly soft and wet or contains many boulders which would make digging by hand extremely slow. Compared with hand-dug pits, depths are limited as ordinary backhoes are capable of digging only to 3 or 4 m and the larger mechanical excavators can attain depths of no more than 6 or 7 m. Mechanical diggers are very quick and 3 m pits can be dug, logged, sampled and filled in within half an hour. Figure 5.3 shows a profile from a pitting programme carried out with a backhoe over a zinc anomaly in central Ireland. In total 17 pits with an average depth of 3 m were dug, logged, sampled and filled in over a space of two days. The overburden is a wet boulder clay with very poor stability and it is doubtful if all of the pits could have been dug by hand as there was only enough time to dig, log and sample some of the pits before the sides started to collapse. This example shows how a short, simple, inexpensive pitting programme defined the source area of the anomaly which was ascribed to sub-economic traces of sphalerite and galena.

Pitting has distinct advantages over other deep sampling methods.

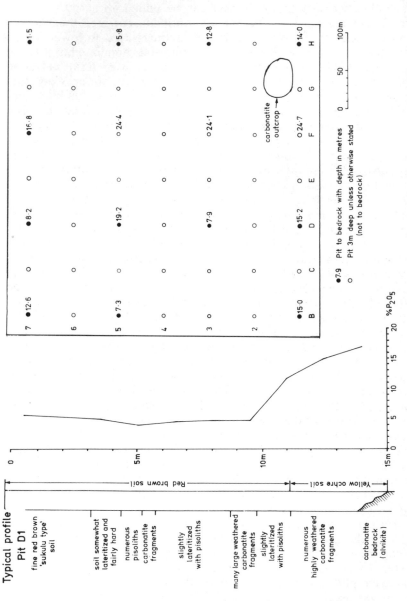

FIG. 5.1. Part of the area of an eluvial apatite deposit evaluated by hand-dug pits in the West Valley Sukulu

FIG. 5.2. Typical hand-dug prospecting pit. Depths of 30 m can be attained with such pits through weathered tropical overburden. The large spoil heap behind the worker indicates that this pit is down about 15 m.

Undisturbed samples can be collected from precise locations, large samples can be taken if necessary, detailed logging can be undertaken and comprehensive geological information can be obtained if the pits reach bedrock. The main disadvantages are that pits can only be dug to any appreciable depth in firm, dry ground, deep pits take a lot of time to dig and can be expensive (though in many parts of the world pitting is cheaper than any other method of deep sampling), and the digging of pits may create unnecessary disturbance if exploration is being undertaken over expensive farmland. For safety reasons all pits

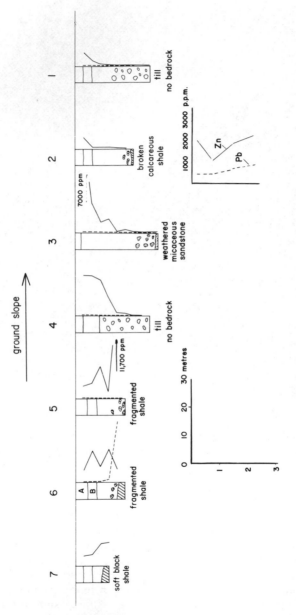

FIG. 5.3.   Pitting profile across a zinc anomaly in central Ireland.

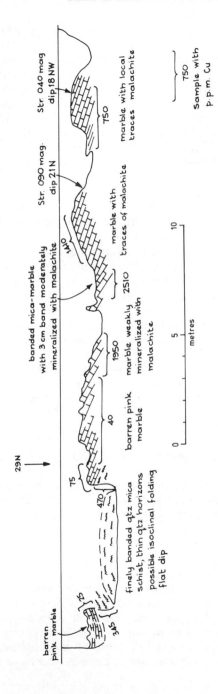

FIG. 5.4. Section of a trench dug as part of the follow-up work on a copper anomaly in central Zambia.

and trenches should be filled in when the sampling programme is completed.

Although pitting cannot normally be undertaken below the water table, Applin (1972) describes a method used to dig pits to depths of 8 m for sampling wet, diamondiferous gravel deposits in West Africa. A series of casing rings made from mild steel sheet segments which are bolted together and which vary in diameter from 1·47 to 3·35 m are set in the ground as digging progresses, continually reducing in size; a total of seven casing rings is required to reach a depth of 8 m. During the pitting operation dewatering is carried out by a diesel-driven pump.

Unlike some of the other deep sampling methods, pitting is rarely employed in reconnaissance geochemical surveys, though it has been widely used in some parts of the world for geological reconnaissance in areas of poor exposure.

Trenching is really an extension of pitting and is used to take samples over a long length, e.g. across a mineralized zone. Trenches can be dug by hand or with mechanical diggers, but the digging of trenches is confined to relatively shallow depths. It is rarely practical to dig trenches more than 5 m deep. For detailed evaluation work very large trenches are sometimes put in with bulldozers, but trenches of this size are probably more correctly classified as small opencuts. The sampling of trenches is accomplished by taking channel and/or panel samples. The old mining term *costean*, meaning to search for lodes by sinking pits or trenching, is still sometimes used synonymously for pit and trench both as a verb and noun. Figure 5.4 shows a typical trench section.

5.2   AUGER DRILLING

In areas where thick soil profiles are developed, such as are commonly encountered in the tropics, power augers are extremely useful for deep sampling, particularly where ground conditions make pitting difficult or impossible. Power augers vary in size from fairly small units, such as those commonly used for digging fence post holes, to large, powerful, truck-mounted rigs capable of reaching depths of 60 m or slightly more. The smaller machines can rarely attain depths in excess of 10 m and in practice are generally restricted to much shallower holes of 5–6 m. The augers for deep drilling are generally from 50 to 75 mm in diameter and come in lengths or flights of 1–1·5 m. These are designed to be joined together to form a smooth continuous spiral or auger to the full depth of the hole being drilled. The rigs are powered by petrol or diesel engines and have feed heads for rotating the auger flights and feeding them up or down. With large machine augers up to 140 m can be drilled in a 12 h shift, though 80–100 m is a more normal average. Drilling costs are usually about half those of percussion drilling. The main disadvantages of auger drilling are that sample contamination may be a problem and drilling may be impossible if the ground contains numerous boulders. Much disenchantment with auger drilling has resulted from attempts to use them in areas with thick boulder clay deposits. In addition the truck-mounted rigs cannot be driven on very soft or boggy ground. Figure 5.5 shows a large machine auger in action.

The procedure for drilling and sampling is as follows. The bit and first flight are run into the ground to its full length. The operator then holds the flight feed stationary while continuing to rotate the auger. This causes soil to be spiralled up to the surface where it can be shovelled to one side. Another flight is added and the process is repeated until the required depth is reached or until it becomes impossible for the auger to penetrate deeper. The material brought to the surface at the end of each run should be put carefully to one side

FIG. 5.5. A large machine auger set up to drill an inclined hole. This is fairly unusual in mineral exploration as in most applications auger drills are used for vertical holes.

in order of collection so that no confusion can arise as to the depth of origin of each sample. Geochemical sample packets can be filled from each sample pile and the remainder bagged and stored in case more sample material might be required at a later date. Some machine augers are not powerful enough to spiral samples to the surface and in such cases the sampling procedure is to run the auger flights into the ground to the desired depth, pull them up and scrape off the sample material lodged in the auger spirals. In hard or heavy overburden it is necessary to drive the auger down in stages and pull it up at intervals. It will be apparent that such machine augers are not nearly as satisfactory as the more powerful ones with higher rotating speeds and the ability to spiral sample material to the surface.

Samples should be recorded on logging sheets which can be made up for the purpose. These should include details of the property, hole number, date, grid location, depth of water table, final depth and descriptions of samples together with sample depths and numbers. It is also useful to include columns for analytical results beside the sample descriptions. For detailed logging it is often helpful to treat and mount samples on cards or boards as described for percussion drill samples in Chapter 7. Results can be shown in profile and plan. When plotting profiles, it is usually necessary to exaggerate the vertical scale unless the holes are very close together. A vertical to horizontal scale exaggeration of 10:1 is convenient. In plan, it is usual to plot either the bottom values or maximum values, though it is usually more informative to plot the latter. The hole depths should also be shown and in the case of plotting maximum values the sample depths should be shown if the maximum values do not occur at the bottom of the hole.

Auger drilling should only be used in detailed evaluation work with the utmost caution. If the material being drilled shows any tendency to flow, which is often the case, the deeper samples will be contaminated by material at higher levels being drawn into the auger spirals. In cases where the samples are of a more qualitative nature such as the profiling of a soil anomaly or deep reconnaissance geochemical sampling, auger drilling can be very effective. Machine augers have been widely used with considerable success in Zambia, both for reconnaissance sampling and testing of surface soil anomalies. The small but extremely rich Kalengwa copper mine in western Zambia was discovered as a result of a geochemical survey, but auger drilling played a fundamental role in locating the mineralization (Ellis and McGregor, 1967). Figure 5.6 shows the results of an auger drilling programme in central Zambia which successfully delineated the bedrock source of a surface anomaly.

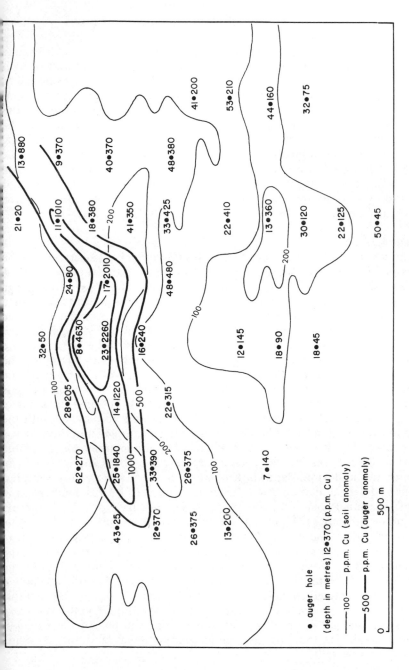

FIG. 5.6. Soil and auger copper anomalies from the Lusale area, central Zambia.

5.3 HAND-HELD PERCUSSION DRILLS

Overburden sampling at depth with hand-held percussion drills is a relatively new technique, but already a considerable amount of success can be claimed for it. The equipment was originally developed to obtain deep soil samples on engineering sites for soil mechanics tests and has been adapted for geochemical sampling. Essentially it consists of a series of hollow one metre 25 mm o.d. rods which can be screwed together by couplings, a sampler, a hand rock drill for driving the rods and a rod puller for extracting the rods from the ground. The drills, powered by two-stroke petrol engines, are of Swedish manufacture and include both the Atlas Copco Cobra and the Pionjar BR80. The rods and ancillary equipment are manufactured by the Borros Company of Sweden.

Samplers and sampling

A number of different sampling heads are available. These include the piston sampler and the window sampler, both manufactured by Borros, and the 'Holman type' sampler developed by R. C. Holman in Ireland. The piston sampler consists of a hollow steel tube with a retractable plunger or piston. To take a sample the sampler is driven into the ground to the required depth and then the rod string is rotated 30 revolutions clockwise. This causes the piston to be retracted into the sampler leaving a hollow open tube at the bottom of the hole. The sampler is then driven downwards by the rock drill a further 30 cm thus filling the open tube with sample material. The entire rod string is withdrawn from the ground by a recovery jack to retrieve the sample. The sampler is disassembled, washed clean and reassembled with the piston fully extended to block the end of the sampler and the process repeated. The window sampler or gravel spoon sampler has a sample intake opening in the side. To take a sample the sampler is driven down to the desired depth in the same manner as the piston sampler. The rods are pulled up about 25 mm, given half a turn clockwise and pushed down again to the original level; this opens the sample window or shutter. The rods are then given one full turn clockwise driven down about 3 cm and rotated again, repeating the entire process about ten times. This causes a small blade at the side of the sample intake to scrape material from the sides of the hole into the opening. On completion of the sampling process, the rods are pulled up 25 mm and given half a turn anticlockwise to close the sample window. The sampler is withdrawn from the hole by the recovery jack and the sample removed by unscrewing the end of the sampler. The sampler is then washed clean, reassembled and the sampling process repeated.

Both the piston sampler and window sampler have the disadvantage that they are larger in diameter than the rods and are therefore often difficult to recover. They are also expensive and have to be disassembled to retrieve the sample and to be cleaned before another sample can be taken. This is often a time-consuming procedure which greatly reduces productivity in the field. All these disadvantages are overcome with the extremely simple but highly effective 'Holman type' sampler (Fig. 5.7). With this type of sampler the sample material continually passes through it as it is driven down so that taking a sample is simply a matter of driving the sampler down to the required depth. The rods are then pulled from the hole to recover the sample which comes from the final depth reached. It might be thought that when the sampler is full it might become blocked and prevent the movement of sample material through it. Extensive tests, however, have shown that this does not happen. The sample is easily extracted by using a sample extractor which consists of a short vertical steel rod of slightly smaller diameter than the internal diameter of the sampler welded to a square steel plate. The bottom of the full sampler is pushed or hammered down the steel rod forcing the sample material out of the side aperture.

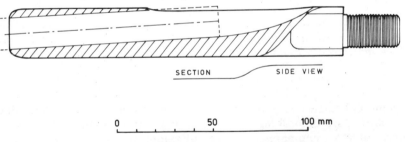

SECTION   SIDE VIEW

0   50   100 mm

FIG. 5.7. 'Holman type' sampler head.

Owing to the length of time involved in removing the sample and cleaning and reassembling piston and window samplers, it is general practice to take only one sample per hole just above bedrock. To do this the depth of bedrock is first determined by driving a probe down until it can be driven no further and vibration of the machine indicates to a trained operator that bedrock has been reached. The rods are then withdrawn and a second hole is drilled with the sampler 1 m or less away. When the sampler is 30–50 cm off bedrock, the sample is taken. With the 'Holman type' sampler this procedure is not necessary since the turn around time is so quick. Profiling with a

TABLE 5.1
COMPARISON OF SAMPLER TYPES

	Piston sampler	Window sampler	'Holman type' sampler
Cost	expensive	expensive	cheap
Weight	6·2 kg	5 kg	0·6 kg approx.
Outer diameter	39 mm	44 mm	26 mm
Overall length	665 mm	625 mm	175 mm
Ease of pulling	often difficult	often difficult	generally easy
Sample volume	125 cm^3	170 cm^3	about 25 cm^3
Time to remove sample and clean	20–40 min	15–30 min	1–5 min
Very wet sample material	often no sample*	sample generally recovered	often no sample

*A shutter is available for holding in wet samples, but it is made of thin brass sheet and should not normally be used for geochemical sampling for obvious reasons

sample collected every 1–2 m can take less time in a 20 m hole than the procedure of probing and taking one sample with a piston sampler. Table 5.1 compares the different samplers.

Pulling rods
The rods and sampler are withdrawn from the hole by means of a simple rod puller consisting of a mechanical jack and ball clamp. To remove the rods the ball clamp is slid down over the rods protruding from the ground, the jack handle is pushed down causing the ball clamp to tighten around the rods and pull them up, each stroke of the jack handle lifting the rods about 6 cm. On soft ground it is necessary to place the jack on heavy bearing timbers to prevent it sinking into the ground. If the rods become tightly jammed in the hole, two small hydraulic jacks with a lift of several tonnes are useful as a standby.

Field procedures

The equipment is most efficiently used with two-man crews, though for working on very wet and soft, boggy ground three men are useful to assist in carrying the extra bearing timbers required for jacking. Drilling progress is obviously fastest in areas where a motor vehicle for carrying personnel and equipment can be driven from site to site, but one big advantage of this sampling method is the easy transportability of equipment which permits sampling over wet and soft ground inaccessible to normal field vehicles. The drill unit weighs about 25 kg and is easily carried with a special back pack. The rods weigh 4 kg each and a rucksack frame can be adapted to hold 10 rods. This leaves the hands free to carry other ancillary equipment such as the jack, wrenches, sample packets, etc. Figure 5.8 shows the equipment in use.

FIG. 5.8. Overburden sampling with a hand-held percussion drill. Left: driving the rods and sampler down. Right: jacking to retrieve the rods and sampler head.

When driving the rods into the ground it is important that the drill unit is used in the 'breaking mode', i.e. the rods are driven down purely by the percussive action and are not rotated. The sampler or probe will push itself around or even break small cobbles or boulders, but cannot penetrate large boulders or thick hardpan duricrust. Owing to the flexibility of the rod string, however, the sampler can often be pushed down past boulders by bending around the obstructions. On encountering bedrock or a large boulder penetration will cease. The difference between bedrock and boulders is indicated by machine vibration which can usually be recognized by an experienced operator. On striking bedrock the sampler will often pick up fragments of rock which are of great value in determining geology concealed by overburden. If an obstruction is encountered at a shallow depth, the hole is abandoned and another hole is drilled 2 or 3 m away. Usually three or four attempts are made before abandoning a particular site. If the ground contains numerous boulders, the method may become impractical.

Depths up to 50 or 60 m have been attained under favourable circumstances, but the method is best suited to fairly shallow sampling up to depths of 15 m. A two-man crew can normally drill and sample an average of 20 m/8-h shift including breakdowns and holdups. The total can vary considerably, however, depending upon depth of holes, distance between holes and difficulty in retrieving rods. A typical record and log sheet is shown in Fig. 5.9. The column for lost time is for any time spent on repairs or in retrieving lost or stuck equipment. The method has now been used for a number of years in Ireland with considerable success, both for reconnaissance surveys and detailed follow-up. Figure 5.10 gives an example of a lead–zinc soil anomaly in Ireland compared with the anomaly just above bedrock defined by overburden drilling.

PIONJAR OVERBURDEN DRILLING

Project Betaghstown   Location Kildare   Survey Type Profiling 400' x 400'
Drill Operators L.G. & S.S.

Hole No.	Date	Drilling Time hrs	Drilling Time mins	Lost Time hrs	Lost Time mins	Metres Drilled probing	Metres Drilled sampling	Sample No.	Sample Depth	Bedrock Depth	Summary Log
44	9-7-73	1	05				1·00m	P351	1·00m	5·80m	brown soil
							1·00	P352	2·00m		" "
							1·00	P353	3·00m		brown soil & grey pebbles
							1·00	P354	4·00m		silt and brown soil
						0·80	1·00	P355	5·00m		med. grey clay & gravel
45	9-7-73	1	20				1·00	P356	1·00m	4·32m?	brown soil
							1·00	P357	2·00m		" "
							0·90	P358	2·90m		grey rock frags & brown soil
						0·32	1·00	P359	4·00m		grey rock frags & gravel
46	10-7-73	1	15				1·00	P360	1·00m	N. D.	brown soil & pebbles
							1·00	P361	2·00m		" " "
Totals		3	40	—	—	1·12	10·90				

FIG. 5.9. Example of an overburden drill log sheet.

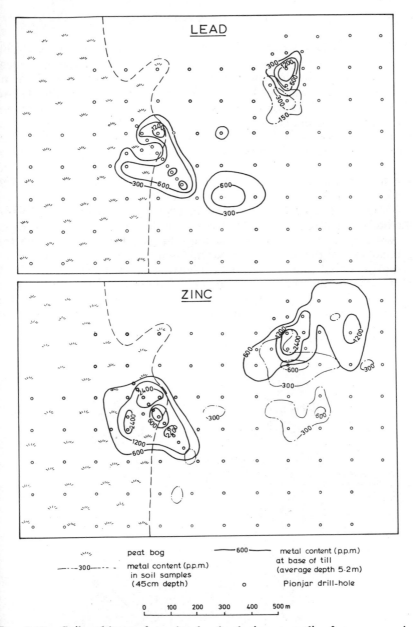

Fig. 5.10.   Soil and base of overburden lead–zinc anomalies from an area in central Ireland.

5.4   WAGON DRILLING

Percussion drills mounted on trailers are commonly known as wagon drills and are widely used in initial follow-up work on geochemical soil anomalies. The method has the advantage over other methods of deep sampling in that rock can be drilled as easily as overburden so that tests for a shallow bedrock source can be made at the same time that the overburden is sampled. The major disadvantages are that samples cannot be taken below the water table or in broken or cavernous ground. The problem of sampling below the water table can be overcome if the equipment is modified so that water is used for drilling, but this facility is not available on most wagon drills. The equipment is described under percussion drilling in Chapter 7.

Sample material is blown up the hole by the exhaust air and several methods of collecting samples have been devised. The most straight-forward method is simply to scoop up the powder as it settles on the ground around the hole collar, but this is not really satisfactory as the sample material is showered over a fairly wide area and the finer dust in particular may be lost. A more satisfactory method of sample collection is to run a short length of casing or pipe into the hole with a T-piece attached to the top and projecting to the side above the hole

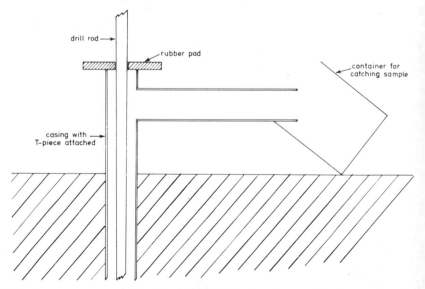

FIG. 5.11.   Diagram showing a method of collecting samples from a wagon drill.

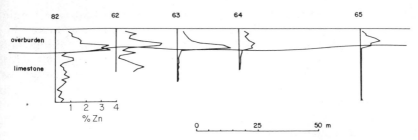

FIG. 5.12. Section of wagon drill holes across a zinc soil anomaly in central Zambia.

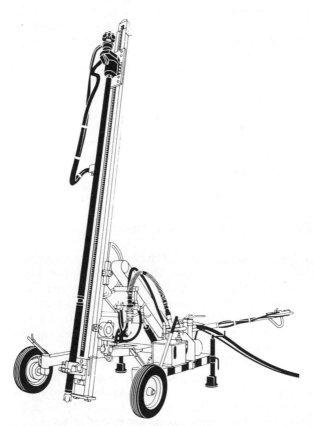

FIG. 5.13. A typical wagon drill.

collar. A thick rubber pad made from old conveyor belting with a hole in its centre, so that the drill rods can pass through it, is held over the top of the casing while drilling is in progress. The drill cuttings are then deflected through the T-piece and into a suitable container. This arrangement for sample collection is shown in Fig. 5.11. In addition to these improvised methods, the drill manufacturers produce special dust collecting cyclones which can be used for sampling. The drill chippings and dust, which emerge from the hole, enter a sleeve placed over the hole collar and pass along a hose into the cyclone where the chips and dust settle from the air stream and are collected in plastic bags.

For the purposes of geochemical sampling the sample material from the hole can be washed through a screen of aperture size 10–14 mesh. The undersize particles are retained for geochemical analysis and the cleaned oversize fragments are used for rock identification and logging. The actual treatment of the sample will depend upon the area and mineralization sought, but for most geochemical work analysis of the -80 mesh fraction will suffice.

Depths up to 50 or 60 m are possible with most wagon drills, but for general work average hole depths are more likely to be 30 or 40 m. The method is fast and it is possible to complete two 30-m holes/shift. Wagon drills are relatively light and can be manoeuvred into awkward places, but overall mobility is limited by the ancillary compressor which is bulky and heavy and has to be positioned fairly close to the drill. An example of a geochemical soil anomaly tested by wagon drilling in central Zambia is shown in Fig. 5.12 and a wagon drill is shown in Fig. 5.13.

5.5   BANKA DRILLING

The Banka drill is a hand-operated drill of Dutch manufacture designed originally for sampling alluvial and eluvial tin deposits in the former Dutch East Indies. It has been used for general overburden sampling in other parts of the world, but is now virtually obsolete as it is a slow and labour-intensive method. It is described here as frequent reference to Banka drilling is made in old reports and it is still occasionally used in Africa and the Far East.

Drilling is accomplished by four men turning an auger bit or special cutting shoe by means of four horizontal levers set at right angles and fastened to the top of the rods. The normal procedure is to use an auger bit for starting the hole and then to transfer to a bailer with a special cutting shoe screwed onto its lower end. Samples are removed periodically with the bailer as drilling progresses. If the ground is hard

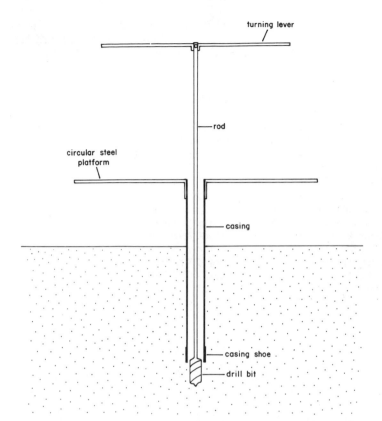

FIG. 5.14. Diagram showing basic parts of a Banka drill.

and dry, water is used to soften it and facilitate bailing. Caving ground can be cased off by inserting casing with a casing shoe so that it can be drilled down just behind the advancing drill bit. When the casing is inserted into the ground as far as it will go, a circular steel platform is screwed on the top end of the casing protruding 1 m or so above ground level. This allows the four men turning the rods to continue drilling by walking round and round the platform. The casing itself can be drilled down by fitting poles into iron rings on the platform and using additional men to turn it. A crew of eight is thus required for the full operation. For pulling tools and casing a chain sling and pulling stand is used. To free casing that is tightly jammed a special screw jack that fits over the protruding casing is available. Numerous other accessories such as fishing tools, special shoes and bits, wrenches and a hand vacuum pump for sampling very wet sands and gravels are available. Depths of 15–20 m are normally attainable and somewhat deeper holes can be drilled if conditions are favourable. Figure 5.14 shows the basic parts of a Banka drill.

REFERENCES AND BIBLIOGRAPHY

Applin, K. E. S. (1972). Sampling of alluvial diamond deposits in West Africa, *Trans. Instn. Min. Metall.*, Lond., **81**, A62–77.

Ellis, M. W. and McGregor, J. A. (1967). The Kalengwa copper deposit in northwestern Zambia, *Econ. Geol.*, **62**, 781–797.

Gleeson, C. F. and Cormier, R. (1971). Evaluation by geochemistry of geophysical anomalies and geological targets using overburden sampling at depth, *Geochem. Explor.*, C.I.M. Spec. **11**, 159–165.

van Tassell, R. E. (1968). Exploration by overburden drilling at Keno Hill Mines Limited, *Proceedings Intern. Geochem. Explor. Symposium, Quarterly of the Colorado School of Mines*, **64**, 457–478.

Wennervirta, H. (1973). Sampling of the bedrock–till interface in geochemical exploration. In *Prospecting in Areas of Glacial Terrain*, Instn. Min. Metall., Lond., 67–71.

CHAPTER 6

Geophysical Prospecting

6.1 GRAVITY SURVEYING

Gravity surveying is not as widely used in mineral exploration as some of the other geophysical methods since it is more suitable for outlining large regional structures and features and as such has more application in petroleum exploration. Nevertheless, in certain instances gravity surveying may prove extremely useful in mineral exploration, though it is rarely used as a primary exploration tool, being more suited to the advanced stages of an exploration programme when the necessary geological control has been established. Gravity anomalies are caused by density differences in the underlying rocks and, although most ore bodies have a high density contrast with the surrounding country rocks, they are generally too small and/or deep to produce appreciable gravity anomalies. In addition near surface features such as sudden changes in overburden thickness or zones of cavernous weathering generally have gravity effects in excess of those caused by an ore body. Although a particular ore body may not produce a detectable gravity anomaly, the mineralization may be controlled by a geological structure which does produce a definite gravity effect and in such cases gravity surveying can be instrumental in outlining prime exploration target areas or zones.

The force of gravity at the surface of the earth is derived from Newton's law of gravitation and is given by the formula

$$g = \frac{GM}{r^2}$$

where M is the mass of the earth, G is the universal gravitational constant ($6\cdot67 \times 10^{-8}$ cgs units) and r is the radius of the earth. The unit of gravity measurement is the milligal (mgal), which is $0\cdot001$ of a gal (an acceleration of 1 cm/s^2). Sometimes gravity values are expressed in terms of the gravity unit (gu) which is equivalent to $0\cdot1$ mgal. The value of g is not constant at different parts of the surface of the earth since r is not a constant (the polar radius is some 21·5 km shorter than the mean equatorial radius) and the centrifugal force due to the rotation of the earth also has an effect which varies from zero at the poles to a maximum at the equator. These effects are taken into account by the international gravity formula

$$g = 978·049 \ (1 + 0·0052884 \sin^2 \phi - 0·0000059 \sin^2 2\phi)$$

which gives the sea-level values of gravity at latitude ϕ. The measured value of g corrected to sea-level is generally found to be greater or smaller than the theoretical value calculated from the international gravity formula. This difference is due to in-homogeneities in the crust and upper mantle and it is the small anomalies due to near-surface inhomogeneities in the crust that are of interest in mineral exploration.

The Bouguer anomaly at any point on the earth's surface is determined from the formula

$$g = g_0 - g_B + g_f + T - g_T$$

where g_0 is the observed value of gravity, g_B is the Bouguer correction, g_f is the free-air correction, T is a terrain correction and g_T is the theoretical value of gravity at that point. Figure 6.1 shows how each of these corrections is derived.

Consider a gravity station at point A, h m above sea-level, with an observed gravity value g_0. The gravity value at A will be less than the theoretical value g_T at A' as it is h m further away from the centre of the earth. To correct the value of gravity at A to sea-level at A' a free-air correction of 0·3086 mgal/m must be added to g_0. In addition, to reduce the gravity value at A to its sea-level value at A' the gravitational attraction of the slab of ground h m thick between A and A' must be subtracted from g_0. This is known as the Bouguer correction and is given by $2\pi Gh\sigma$ where σ is the density of the slab. The density is usually taken as 2·67, but other values may be used. For a density of 2·67 the free-air and Bouguer corrections can be

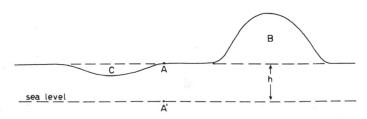

FIG. 6.1. Section to show derivation of free-air, Bouguer and terrain corrections.

combined as a single positive correction of $0 \cdot 1968$ mgal/m. Further corrections need to be applied for the effects of topography. Consider the hill at B and valley at C. The hill has an upward and thus negative attraction on the gravity station at A and its effects need to be added to g_0. The valley at C is an absence of mass, but, as it has already been included in the horizontal slab used for the negative Bouguer correction, the Bouguer correction applied to the gravity station at A is too large by an amount equivalent to the gravitational attraction of the valley filled to the elevation of the station at A with material of density σ. Thus the terrain correction is always positive whether the topographical irregularities are higher or lower than the station.

Terrain effects are usually determined by placing a template divided into segmental, concentric compartments over a topographical contoured map of the area being surveyed centred on each gravity station to be corrected in turn. The average elevation both above and below the station is estimated for each compartment and the gravity effects determined from published tables of terrain corrections (Hammer, 1939). This is a time consuming task and, although a number of computer techniques have been devised for terrain corrections, the preparation of input data still requires a considerable amount of time. Whatever methods are used, the determination of accurate terrain corrections is a laborious process and, for this reason, gravity surveys in areas of rugged terrain can be very expensive.

Field procedures

Gravity measurements are made in the field with a gravity meter or gravimeter as it is usually known. Modern gravimeters are small, portable and extremely sensitive instruments ($0 \cdot 01$ mgal) and may be classed into two main types: stable and unstable. The stable type is essentially an extremely sensitive balance. Examples of this type are the Askania and Gulf (Hoyt) gravimeters. The unstable types are the more widely used today and work on the principle of keeping the force of gravity acting on a sensitive element in unstable equilibrium with a restoring force. Examples of this type are the Worden and LaCoste–Romberg gravimeters.

In mineral exploration surveys gravity readings are usually taken on a grid with station spacings of 10–50 m depending upon the expected size of the target. The gravimeter is simple to read, though a certain amount of operator skill is required to ensure repeatable readings. On a small grid survey it should be possible for a skilled operator to establish well over 100 stations/day. The normal procedure is to establish a base station or stations and repeat base station readings every 1–3 h. This has to be done to correct for instrumental drift which is due to elastic creep in the springs. The drift is generally linear, but sudden fluctuations in the drift (tares) do occur from time to time in some instruments. There is also a tidal effect on gravity readings, but this is accounted for in the normal drift corrections. Gravimeters do not give absolute readings, but relative values which are generally adequate for mineral exploration surveys. If true Bouguer values are required, it is necessary to tie the survey to a regional base station which has been tied to a pendulum station with an absolute gravity value.

It has been shown that the combined free-air and Bouguer correction for a density of 2·67 is 0·1968 mgal/m, which means that the elevation has to be known to 5 cm for an accuracy of 0·01 mgal. Since the Bouguer anomaly is the difference between the corrected measured gravity value and the theoretical value of gravity, it is necessary to know the latitude of a gravity station to determine the Bouguer anomaly. At the equator it is necessary to know the north-south position of a station to 400 m for an accuracy of 0·01 mgal, but at high latitudes it is necessary to know the north-south position to 10 m for a similar accuracy. Tables of theoretical g are published and over short distances of 1–2 km north or south of a base station a linear latitude correction (0·081 sin 2ϕ mgal/100 m) can be applied to the theoretical gravity value of the base station (negative corrections for stations with a higher latitude and positive corrections for stations with a lower latitude than the base station).

Occasionally underground gravity surveys are carried out. The corrections that need to be applied differ slightly from those applied to surface measurements, but such underground surveys are of little importance in mineral exploration. For specialized regional surveys and for use in petroleum exploration underwater, shipborne and airborne gravimeters have been developed, but these are not normally used in mineral exploration surveys.

Interpretation
For any particular gravity field over a horizontal plane there is an infinite number of mass distributions which can produce that field. However, the number of mass distributions due to plausible geological structures is usually quite limited. The normal procedure for interpreting gravity surveys is to compare calculated gravity effects for a number of different mass distributions with the observed gravity field. The theoretical structure which produces a calculated gravity effect closest to the observed gravity effect is then adopted as the most likely interpretation. In addition there are a number of techniques which can be used to treat the raw Bouguer values to enhance particular gravity effects that may be of significance.

The first prerequisite in any interpretation is to obtain reliable rock densities for the area under investigation. For a good interpretation it is also necessary to have as much accurate geological information as possible. This is particularly important in mineral prospecting surveys as the significant gravity effects tend to be very small. Rock density measurements are usually made in the laboratory on hand or drill core specimens, but occasionally *in situ* measurements can be made from gravity observations in the field or in underground workings. Borehole gravimeters have also been developed for *in situ* measurements in large diameter drill-holes. Table 6.1 lists the densities or density ranges of a number of common rocks and minerals.

The mass distributions of many geological structures approximate to simple geometric shapes or combinations of such geometric shapes. The formulae for the gravitational effects of a wide range of

TABLE 6.1

DENSITIES OF SOME COMMON ROCK TYPES AND MINERALS

Material	Density (g/cm^3)
coal	1·20–1·50
unconsolidated sand (wet)	1·95–2·05
sandstone	2·10–2·70
limestone	2·40–2·71
shale	2·20–2·80
granite	2·55–2·70
gabbro	2·85–3·10
peridotite	3·10–3·30
basalt and andesite	2·70–3·10
gneiss	2·65–2·80
halite	2·2
gypsum	2·3
anhydrite	2·95
fluorite	3·0–3·2
sphalerite	3·9–4·2
chalcopyrite	4·1–4·3
barite	4·5
pyrrhotite	4·40–4·65
chromite	4·5–4·8
pyrite	4·8–5·1
haematite	4·9–5·3
bornite	4·9–5·4
pentlandite	5·0
magnetite	5·2
galena	7·4–7·6

shapes have been calculated by various workers and examples of four of these formulae are given in Fig. 6.2. For bodies with a very long strike length in comparison with their depth and width a number of graphical techniques have been developed for calculating gravity profiles in two dimensions. One of these is due to Jung (1937) and is illustrated in Fig. 6.3. The graticule is so constructed that the gravitational attraction in milligals of each compartment at point P is equal to $13 \cdot 34 \sigma s \times 10^{-6}$ where σ is the density contrast and s the reciprocal of the scale of the section. The gravity profile through a series of stations over the cross-section of a body is calculated by placing the point P on each station in turn and counting the number of compartments and fractions of compartments contained within the outline of the body at each position of the graticule. It must be remembered that this method is only valid for bodies with a uniform cross-section over a long strike length such as graben or horst blocks, buried channels, fold structures, etc.

The use of computers has revolutionized gravity interpretation and made many of the older methods obsolete. There is a large literature on the subject, but one of the simplest methods in mineral surveys is to divide the mass distribution into a large number of cubes. A very close approximation to any irregular body can be made if the cubes are sufficiently small. Figure 6.4 illustrates the procedure used. Let P be a point at the surface with coordinates x, y, z. Let P_1 be a point at the centre of a small cube with coordinates x_1, y_1, z_1. Then the vertical component of gravity at P due to the cube Q is given by the expression:

$$g = \frac{Gm}{PP_1^2} \cos \theta \qquad \text{(i)}$$

where G is the gravitational constant and m is the mass of the cube which can be assumed to act at the centre of the cube if the cube is sufficiently small. Using the coordinates of P and P_1, formula (i) becomes:

$$g = \frac{Gm(z_1 - z)}{[(x_1 - x)^2 + (y_1 - y)^2 + (z_1 - z)^2]^{3/2}} \qquad \text{(ii)}$$

if there are n cubes each with centre coordinates x_i, y_i, z_i, then the total vertical component of gravity at P becomes:

$$g = Gm \sum_{i=1}^{n} \frac{(z_i - z)}{[(x_i - x)^2 + (y_i - y)^2 + (z_i - z)^2]^{3/2}} \qquad \text{(iii)}$$

Although these calculations are extremely onerous, a big computer can sum thousands of cubes in a matter of seconds. In addition the gravity effects at any number of points, P, at the surface can be

Sphere

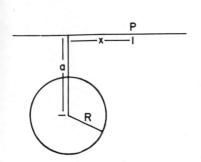

$$g_P = \frac{4}{3} \frac{\pi G R^3 \sigma a}{(x^2+a^2)^{3/2}}$$

Horizontal cylinder

$$g_P = \frac{2\pi G R^2 \sigma a}{x^2 + a^2}$$

Vertical cylinder

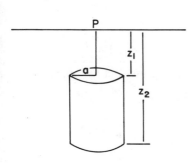

$$g_P = 2\pi G \sigma \left[z_2 - z_1 + \sqrt{(z_1^2 + a^2)} - \sqrt{(z_2^2 + a^2)} \right]$$

if $z_1 \to 0$

$$g_P = 2\pi G \sigma \left[z_2 + a - \sqrt{(z_2^2 - a^2)} \right]$$

if $z_2 \to \infty$

$$g_P = 2\pi G \sigma a$$

Infinite rectangular prism

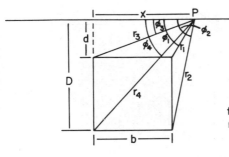

$$g_P = 2G\sigma \left[x \ln \frac{r_1 r_4}{r_2 r_3} + b \ln \frac{r_2}{r_1} + \right.$$
$$\left. + D(\phi_2 - \phi_4) - d(\phi_1 - \phi_3) \right]$$

for faulted slab or scarp $b \to \infty$,
$r_1 \to r_2$, $\phi_1 \to \phi_2 \to \pi$

$$g_P = 2G\sigma \left[x \ln \frac{r_4}{r_3} + D(\pi - \phi_4) - d(\pi - \phi_3) \right]$$

FIG. 6.2. Formulae for determining the gravitational effects of some simple solids (from Parasnis, 1962).

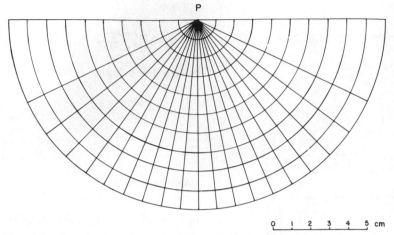

FIG. 6.3. Jung's chart for calculating gravity profiles in two dimensions.

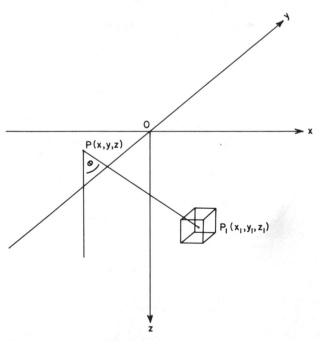

FIG. 6.4. Diagram to show the derivation of the vertical gravity attraction at P of a small cube at P_1

calculated and programs can be prepared which will contour the results. The computer can also be programmed to sum different combinations and densities of cubes until the nearest fit to the observed data is obtained.

This method is only approximate since it assumes that the mass of each cube acts at its centre. More precise calculations can be made by using formulae for the exact gravitational effects of rectangular prisms (Nagy, 1966), but the computations are much more involved. Most ore bodies are easily divided into cubes by using sections or level plans and the method described should give values to better than $\pm 10\%$ of the true theoretical value. For example, the gravitational attraction on the axis of a vertical cylinder of density contrast $1\,g/cm^3$, radius 160 m and depths to top and bottom faces of 100 m and 340 m respectively is 2·22 mgal. By dividing the cylinder into 288 equal cubes of edge length 40 m the gravitational attraction determined by the method described is 2·17 mgal.

Residual gravity maps
In mineral exploration surveys we are generally concerned with small gravitational effects and it becomes necessary to remove large-scale regional effects from the Bouguer anomaly map. The result is known as a 'residual gravity map'. There are a number of ways of doing this. One method is to plot a series of profiles from the Bouguer anomaly map across the main trend of regional pattern. Smooth curves are fitted to the data and any departures from these curves are taken as the residual gravity anomalies (Fig. 6.5). Another method is to draw smooth contours along the main trends which are usually quite obvious and any departures from these smoothed contours are taken as the residual gravity effects. This is illustrated in Fig. 6.6. Computers are now widely used in producing residual gravity maps. The basic procedure is to fit a surface defined by an orthogonal polynomial of 2nd, 3rd, 4th, 5th, 6th or higher orders to the observed Bouguer values. The best fit to the observed data is obtained by using the method of least squares to determine the coefficients of the polynomial. Departures in the original data from this calculated surface are plotted as the residual gravity anomalies. It must be stressed, however, that blind faith in the computer often results in quite erroneous interpretations. In many cases the regional trends are obvious and a simple graphical technique used by someone with a good geological knowledge of the area and a certain amount of commonsense can produce satisfactory residual gravity maps. It must always be remembered that residual gravity maps do not show features that are not present in the original data; they merely enhance the features of interest.

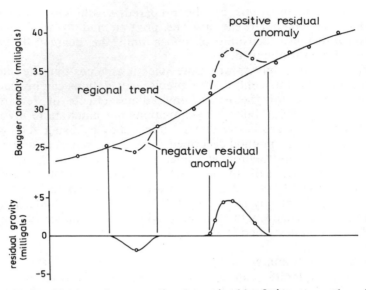

FIG. 6.5.   Residual gravity anomalies determined by fitting a smooth regional trend to a gravity profile.

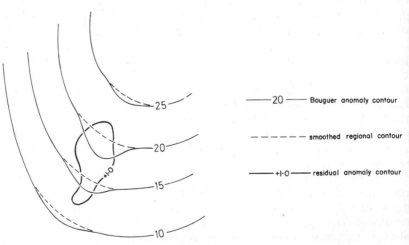

FIG. 6.6.   Residual gravity anomaly determined by fitting smoothed contours to a Bouguer anomaly map.

Second derivative gravity maps

These maps, which show the vertical rate of change of gravity with depth, are useful for resolving and sharpening anomalies of small areal extent. A number of graphical techniques for calculating second derivatives have been described, but all are somewhat subjective. Computers have now made the computation of second derivatives a much easier task. However, in most instances second derivative maps do not bring out features that are not also apparent on residual gravity maps.

Estimation of mass

Although it is not possible to determine a unique mass distribution from gravity measurements, it is possible to make an accurate estimate of the total mass. This observation, first made by Hammer (1945), is extremely useful in mineral exploration as it enables one to make an estimation of total ore reserves from a gravity survey if the approximate density contrast of the ore with the country rock is known. The total tonnage in tonnes can be calculated from the formula:

$$23 \cdot 9 \, \frac{\rho_1}{\rho_1 - \rho_2} \sum \Delta g \Delta s$$

where Δg is the anomaly in milligals, Δs is the areal extent of the anomaly in square metres, ρ_1 is the density of the ore and ρ_2 the density of the country rock. Let us consider an example of a gravity anomaly covering an area of 80 000 m^2 (Fig. 6.7) where the ore has a density of 3·0 in country rock of density 2·7. If we divide the anomaly up into eight equal square blocks (any number could be used) of 10 000 m^2 with gravity values at the centres of each square of 0·25, 0·60, 1·10, 1·10, 0·50, 0·30, 0·50 and 0·30 mgal, the total tonnage becomes:

$$23 \cdot 9 \times \frac{3 \cdot 0}{3 \cdot 0 - 2 \cdot 7} \times 10\,000$$

$$\times (0 \cdot 25 + 0 \cdot 60 + 1 \cdot 10 + 1 \cdot 10 + 0 \cdot 30 + 0 \cdot 50 + 0 \cdot 50 + 0 \cdot 30)$$

$$= 11 \cdot 1 \text{ million tonnes}$$

As mineral exploration surveys are frequently carried out on a square grid, the calculations are easily carried out in the above manner using the survey grid. Figure 6.8 gives an example of a successful gravity survey carried out at the Pine Point lead–zinc deposit, Northwest Territories, Canada (Seigel *et al*, 1968). The ore reserves of the No. 1 pyramid ore body were estimated at 7·5 million tons from the gravity survey. The actual tonnage subsequently proved by drilling was 9·2

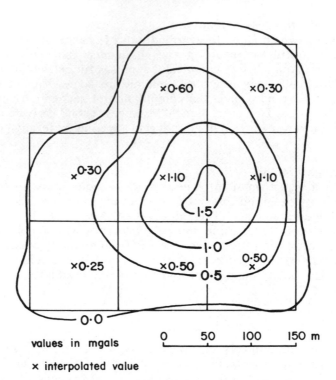

values in mgals

× interpolated value

FIG. 6.7.　Example to illustrate how the total mass of an ore body can be estimated from a gravity anomaly.

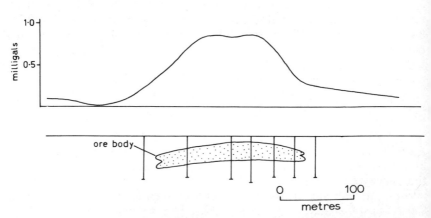

FIG. 6.8.　Gravity profile over a lead–zinc ore body at Pine Point, Northwest Territories, Canada (after Seigel *et al*, 1968).

million tons containing 12% Pb + Zn, the discrepancy between the actual and estimated tonnages being due to a slight underestimation of the actual density of the ore.

6.2 MAGNETIC SURVEYING

Magnetic surveying is the oldest geophysical method and is reputed to have been used in prospecting for iron ore in Sweden in the 17th century. By the 19th century magnetic prospecting for iron ore had become quite common using refined dip needles. However, only the strongest magnetic sources could be detected and it was not until the early part of the present century, when the first precision magnetometers were designed, that magnetic prospecting as we know it today really began.

Ever since the publication of Gilbert's *De Magnete* in 1600, it has been known that the earth may be likened to a giant magnet with magnetic north and south poles. These magnetic poles do not correspond to the geographic poles, being located approximately at 72° N, 102° W and 68° S, 146° W. A compass needle lines itself up in the earth's field with its north-seeking end (positive pole) pointing in the direction of the north magnetic pole and its south-seeking end (negative pole) pointing in the direction of the south magnetic pole. Since unlike poles attract and like poles repel, it will be appreciated from the above that the magnetic north pole is in reality a negative or south pole and the south magnetic pole a positive or north pole. The earth's magnetic field can be resolved into a horizontal component (X) and a vertical component (Y) and the angle, arctan (Y/X), known as the inclination varies from 0° at the magnetic equator to 90° at the magnetic poles. It may also be 90° at any number of other points owing to local disturbances and the inclination is an important factor in interpreting magnetic anomalies. The direction of the magnetic field generally lies east or west of true north owing to the fact that the geographic and magnetic poles do not coincide. This variation from true north is known as the declination.

The force between two poles, which may be attractive (unlike poles) or repulsive (like poles), obeys an inverse square law like gravitation and is given by the formula:

$$F = \frac{1}{\mu} \frac{PP_0}{r^2}$$

where P and P_0 are the respective pole strengths, r the distance between them and μ the permeability constant (equal to one in air). Unit poles are defined as those poles which attract or repel each other with a force of 1 dyne when separated by 1 cm in air. Magnetic poles always exist in pairs (positive and negative), but, as their separation may be considerable (e.g. a very long bar magnet), they may be considered as isolated poles. The magnetic field strength (H) at a point is defined as the force exerted on a unit pole placed at that point. The field strength due to a pole P_0 at a distance r is:

$$H = \frac{P_0}{\mu r^2}$$

The unit of H is the *oersted* which is defined as 1 dyne per unit pole. In magnetic surveying this unit is too large and variations in the field strength are measured in *gammas* (1 gamma = 10^{-5} oersteds). The total field strength of the earth varies from approximately $30\,000\gamma$ near the equator to $70\,000\gamma$ near the poles.

If magnetic material is placed in a magnetic field, H, it will have magnetic poles induced upon its surface. The intensity of induced magnetization (I) is given by:

$$I = \kappa H \cos \theta$$

where θ is the angle of the external field with the normal to the surface of the magnetic material and κ is a constant known as the *susceptibility*, which may be positive (paramagnetic materials) or negative (diamagnetic materials). The induced poles give rise to their own magnetic field and in moderately magnetic materials the net field (B) is given by:

$$B = \mu H$$

where $\mu = 1 + 4\pi\kappa$. In strongly magnetic materials it is found that B does not always fall to zero when the external field H is removed, but retains a residual magnetism (R). Although most rocks are weakly magnetic and might not be expected to show residual magnetism, it has been found that many different rocks display what is known as Natural Remanent Magnetization (NRM). Instruments have been developed for measuring the direction and intensity of NRM and it has been shown that rocks of varying geological age may have NRM directions very different from the present direction of the earth's field. The study of this phenomenon, known as *palaeomagnetism*, has shown that the magnetic poles have wandered and even been reversed in the geological past. NRM can be attributed to several different processes, but the most important are Thermo-remanent Magnetization (TRM), Chemical or Crystallization Remanent Magnetization (CRM) and Detrital or Depositional Remanent Magnetization (DRM). TRM is acquired by rocks cooling from high temperatures and is the main component of NRM in igneous rocks

CRM is acquired when a magnetic substance has chemically formed or crystallized in a magnetic field. This type of NRM is exhibited by red beds, which often display a very stable NRM and are thus useful in palaeomagnetic studies. The NRM of red beds is attributable to haematite much of which is considered to have formed during diagenesis. DRM is acquired by some sedimentary rocks and is a result of detrital magnetic grains aligning themselves in the earth's field during deposition. In addition some rocks may acquire a highly localized NRM caused by lightning strikes. The NRM of all rocks is lost if they are heated above the Curie point which is approximately 600°C for magnetite.

Since both the NRM and induced magnetization of a rock have direction as well as magnitude, the net field can be expressed by the vector equation:

$$\mathbf{J} = \kappa\mathbf{H} + \mathbf{J}_n$$

where $\mathbf{H}$ is the external field and $\mathbf{J}_n$ is the NRM. $\mathbf{J}_n$ is difficult to estimate as it varies widely from point to point and, as TRM is the most important cause of NRM, $\mathbf{J}_n$ is generally less than $\kappa\mathbf{H}$ in most sediments and metamorphic rocks and can usually be ignored. In igneous rocks, however, $\mathbf{J}_n$ is very often greater and may be very much greater than $\kappa\mathbf{H}$. In addition the intensity of magnetization acquired by a rock depends upon the grain size of the ferrimagnetics and for this reason volcanic rocks generally have much stronger NRM's than plutonic rocks. This can be shown by the ratio, $J_n/\kappa H$ (Q), which is approximately one for coarsely crystalline rocks, 10 for volcanics, 30–50 for rapidly cooled volcanics and generally less than one for sediments. For this reason NRM is of considerable importance in interpreting magnetic surveys over areas of igneous rocks.

The magnetic susceptibility (κ) of rocks can be measured *in situ* in the field by an induction balance or in the laboratory by suitable instruments. Laboratory measurements of κ using fields much greater than the earth's are often of dubious value and *in situ* measurements are always to be preferred. Table 6.2 lists the susceptibilities of a number of minerals and common rock types. In cases where the direction of the NRM and earth's field are close, an apparent susceptibility (κ_A) may be defined as:

$$\kappa_A \simeq J/H$$

Some orders of magnitude for κ_A are: iron ores 0·1 emu, basic volcanics 10^{-3}–10^{-2} emu, metamorphic rocks 10^{-4} emu and sediments 10^{-5} emu.

TABLE 6.2
MAGNETIC SUSCEPTIBILITIES OF
SOME MINERALS AND COMMON
ROCK TYPES. UNITS in 10^{-6} emu

Material	Susceptibility (κ)
magnetite	100 000–1 000 000
pyrrhotite	50 000–500 000
ilmenite	20 000–300 000
haematite	200–3 000
quartz	$-1\cdot2$
halite	$-0\cdot82$
peridotite	12 000
gabbro	100–3 000
basalt	120–4 000
limestone	2–280
sandstone	2–1 600
shale	2–1 500

Magnetometers

Schmidt variometers

Instruments of this type were the first precision magnetometers designed and consist basically of a bar magnet pivoted on an agate knife edge at a point just off the centre of gravity. This results in the torque on the magnet due to the earth's magnetic field being opposed by a gravitational torque. The angle made by the magnet in its equilibrium position depends upon the magnetic field strength. The position of the centre of gravity can also be altered by moving a small weight to adjust the sensitivity to different field strengths. In addition auxiliary magnets of known magnetic moment can be positioned along a brass rod below the instrument both in order to calibrate the instrument and to compensate for very strong field strengths which may make a direct reading impossible. These instruments, which are available in separate models for measuring the vertical and horizontal components, have to be built to a high degree of mechanical and optical precision and the finer instruments can be read to better than $\pm5\gamma$. Although they are accurate and only the size of a large theodolite, they have to be levelled precisely on a tripod and a single reading can take 10 min or more. For this reason, they are now obsolete and have been replaced by the small lightweight modern flux-gate and proton magnetometers.

Torsion magnetometers

These instruments work on a principle similar to the Schmidt variometer except that the magnet is pivoted on a torsion fibre and the magnetic intensity is measured by the amount of torque required to bring the magnet to a horizontal position. Although these instruments need to be levelled precisely on a tripod, they can be read in a matter of minutes by a skilled operator. They are accurate to 1 or 2γ and can be read in any azimuth since the moment of the horizontal component is zero when the magnet is horizontal.

Flux-gate magnetometers

These magnetometers, designed in the 1940's, were the first of the modern electronic instruments and they made it possible to undertake airborne and shipborne surveys. The detector in the instrument consists of two identical parallel coils wound in series in opposite directions around ferromagnetic elements of extremely high permeability. The coils are energized by an alternating sinusoidal current which drives the cores beyond saturation at the top and bottom of each cycle. When the coils are held parallel to the earth's field, the magnetic field in one of the cores is reinforced causing saturation to be reached slightly earlier in the cycle than would be the case in the absence of the ambient field, and the magnetic field in the other coil is reduced causing saturation to be reached slightly later in the cycle than would be the case in the absence of the ambient field. Secondary coils around each core are connected in opposition to a voltmeter and the maxima of this resultant voltage are approximately proportional to the ambient magnetic field. Small hand-held instruments, which measure the total field to ± 5 to 30γ, are manufactured by a number of different companies in various countries. The instruments are extremely simple to operate and readings can be obtained in a minute or less. The instrument has to be held level and steady so that the elements are vertical and a direct reading in gammas is obtained on a meter by pressing a button. The instruments are equipped with a range selector switch so that on-scale readings can be obtained over a wide range of field strengths. Airborne instruments consist of gimbal-mounted elements with orienting inductors set at right angles in a plane perpendicular to the element. The inductors are connected to servomotors which keep the element parallel to the earth's field. The airborne flux-gate magnetometer records a continuous profile and is accurate to 1γ. In airborne work the detector element is towed in a 'bird' behind the aircraft while the recording instruments are kept on board.

Proton magnetometers

These magnetometers, which are the most widely used today, were developed in the 1950's and are based on the phenomenon of nuclear magnetic resonance. A strong magnetic field (100 times greater than that of the earth) is applied to a bottle containing a proton-rich liquid (water or a hydrocarbon) by a coil wound round it. By virtue of their magnetic spins, the protons align themselves parallel to the applied field. When the external field is removed, the protons will return to their original magnetic moment by precessing round the direction of the earth's field with angular velocity $\omega = \gamma_p H$, where H is the field strength and γ_p a constant (the gyromagnetic ratio of the proton). This induces a small voltage in the coil, the frequency of which is the same as the frequency of precession. Suitable electronic circuitry makes it possible to measure the frequency of the induced voltage and the total field strength is simply equal to the measured frequency times a constant ($23\cdot4874\gamma/\text{Hz}$). These instruments give absolute measures of the earth's field and are accurate to 1γ. The latest instruments are small, lightweight and have extremely fast recycling times of 1 s or less. In airborne work the sensor bottle is towed in a 'bird' behind the aircraft while the recording instruments are kept on board. Unlike the flux-gate magnetometer, which records a continuous magnetic profile, the proton magnetometer gives a series of readings at discrete time intervals. This was considered a disadvantage for airborne work, but the recycling times of the latest instruments are fast enough for it not to make any practical difference.

Optical absorption magnetometers

These are the most sensitive magnetometers and have only been developed in recent years. They are based on the phenomenon known as 'optical pumping', the theory of which is extremely complex. Atoms of Rb or Ce vapour in a cell are excited by a modulated beam of light from a Rb or Ce lamp which is filtered for high energy wavelengths. Orbital electrons around the Rb or Ce atoms shift from one energy level to another. Any electrons which jump into the highest energy levels are no longer excited by the light beam from which the corresponding wavelengths have been filtered. After a certain amount of time all electrons are above the levels excited by the light beam and the vapour is said to be 'optically pumped'. This state is interrupted if the cell is swept by a magnetic field because of the Zeeman effect. In one form of optical absorption magnetometer the modulated light beam transmitted through the Ce or Rb cell is detected by a photocell. The current from this photocell is shifted in phase by 90° and used as a feedback signal to a coil which applies an alternating magnetic field to the vapour cell. This arrangement is an oscillator whose frequency is proportional to the total ambient field affecting the cell. The frequency required to maintain a constant absorption in the vapour cell gives a measure of the earth's total field strength. These instruments are extremely sensitive and measurements of $\pm 0.01\gamma$ are possible. Ground instruments similar in size to proton magnetometers are available and are accurate to 0.1γ or better. The instruments are also used in airborne work.

Field procedures

All modern magnetometers are extremely simple to operate and surveys can be conducted by semi-skilled personnel, though flux-gate instruments are more difficult to read than proton or optical absorption magnetometers as they have to be held absolutely level and steady to obtain accurate readings. The operator should ensure that he is not carrying any steel or iron objects such as large belt buckles, pocket knives or small magnets which may affect the readings slightly. Observations are taken along traverses across the geological strike with station intervals of 15–30 m commonly employed in mineral exploration surveys. A base station is selected at a point within or near the survey area where there is little disturbance of the normal magnetic field. With flux-gate instruments the magnetic readings are expressed as positive or negative differences from the base station which is taken as zero. This procedure can also be adopted with proton or optical absorption magnetometers, but, as these instruments give absolute readings, it is common practice simply to plot the actual readings. For very precise surveys it is necessary to correct for the *diurnal variation*. This generally varies between 10 and 20γ, but may be up to 50γ or more on days of magnetic activity. If large anomalies are encountered, corrections for the diurnal variation can be ignored. The diurnal variation is determined by using a base instrument as a reference or by returning to a base station every 1–2 h. With flux-gate magnetometers it is advisable to adopt the procedure of reoccupying a base station at regular intervals as this also takes into account temperature corrections which can be up to 50γ or more in a day. Proton magnetometers are not affected by normal temperature changes and optical absorption magnetometers are maintained at 35°C, their optimum operating temperature, by means of a thermostat. It is sometimes useful to take several readings at each station 3 or 4 m apart as a check against very strongly localized sources, though this precaution is only really necessary if stations are widely spaced (100 m and more). Strong variations in topographic relief can give spurious magnetic anomalies, but there are no definite rules for carrying out terrain corrections. Generally, anomalies showing strong correlation with terrain are regarded as less significant than others, particularly in airborne surveys where terrain effects can be very marked. When carrying out magnetic surveys, the operator should always be on his guard against taking readings close to iron or steel objects such as pipelines, fences, bridges and cars. For example, a car at 3 m may give an effect of approximately -700γ, but at 30 m distance the effect is negligible. A north-south wire fence may give an effect of -350γ at 1 m, but the effect disappears at a distance of 35 m.

Interpretation

Magnetic field strength obeys an inverse square law and a magnetic anomaly field can be explained by potential theory as is the case for gravitation. However, magnetic interpretation is much more difficult than gravity interpretation since two poles are involved and remanent magnetization may play an important part with a direction and intensity very different from the earth's present field. Nevertheless, many magnetic anomalies can be considered as being due to induced polarization and interpretations can be made from a knowledge of susceptibilities and approximate geometric shapes of structures. As in the case of gravity interpretation a number of workers have derived formulae for calculating the magnetic effect of various regular geometric shapes (e.g. Nettleton, 1942; Cook, 1950). Figure 6.9 gives some examples of the formulae for a few simple cases. In addition the size and shape of anomalies is also governed by the inclination of the earth's field and, in the case of bodies with a long strike length compared to their cross section, by the orientation of the bodies with respect to the field. This relationship is illustrated in Fig. 6.10 which shows the magnetic anomaly over a dipping dyke with two strike directions (N–S and E–W) in a field with two different inclinations (14° and 63°). At the magnetic equator a long N–S structure will produce no anomaly, whereas an E–W striking body produces a small negative anomaly. Bodies with short strike lengths compared to their cross-sectional area, on the other hand, always produce anomalies, though of much smaller magnitude at or near the equator than at higher latitudes (Fig. 6.11).

Although magnetic poles always exist in pairs, the effect of the more distant pole may be negligible if the separation of poles is large. Bodies elongated in the direction of the magnetizing field will produce induced poles with a wide separation, the deeper poles can be ignored and the magnetic anomaly ascribed to a so-called monopole or line of monopoles. If, on the other hand, a body is magnetized across a narrow width, it will produce two poles close together and will act as a dipole or line of dipoles. These effects are illustrated in Fig 6.12.

Sphere

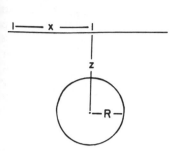

$$V = 8{\cdot}38 \times 10^5 \frac{R^3 I}{z^3} \frac{1 - \frac{x^2}{2z^2}}{\left(1 + \frac{x^2}{z^2}\right)^{5/2}}$$

Horizontal Cylinder (infinite length)

$$V = 6{\cdot}28 \times 10^5 \frac{R^2 I}{z^2} \frac{1 - \frac{x^2}{z^2}}{1 + \frac{x^2}{z^2}}$$

Horizontal Infinite Slab (Fault)
(third dimension infinite)

$$V = 2 \times 10^5 \frac{I t x}{z^2} \frac{1}{1 + \frac{x^2}{z^2}}$$

Vertical Sheet (third dimension infinite)

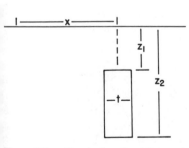

$$V = 2 \times 10^5 I t \left[\frac{1}{z_1 \left(1 + \frac{x^2}{z_1^2}\right)} - \frac{1}{z_2 \left(1 + \frac{x^2}{z_2^2}\right)} \right]$$

FIG. 6.9.   Formulae for determining the magnetic effects of some regular bodies (after Nettleton, 1942).

The rate of fall-off of an anomaly from a monopole is inversely proportional to the square of the distance from the pole and in the case of a dipole the rate of fall-off is inversely proportional to the cube of the distance from the centre of the dipole. The shape of the anomaly produced by a dipole or monopole depends upon the inclination of the inducing field. In a vertical field a monopole produces

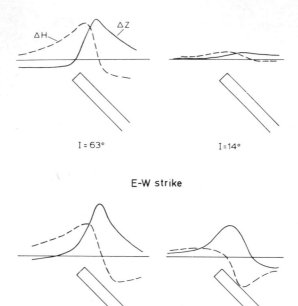

FIG. 6.10. Magnetic profiles over identical inclined dykes with different strike directions in fields with low and steep inclinations (after Haalck, 1953)

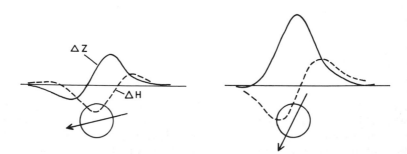

FIG. 6.11. Magnetic profiles over a sphere in fields with low (14°) and high (63°) inclinations (after Haalck, 1953).

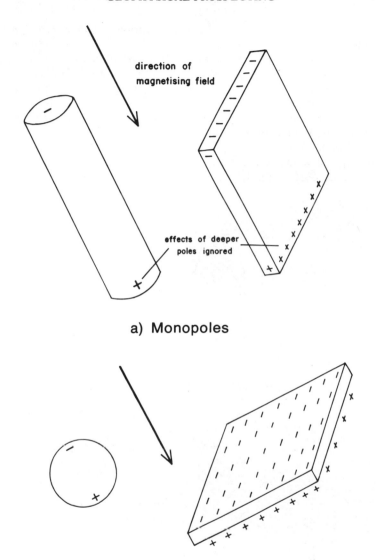

direction of
magnetising field

effects of deeper
poles ignored

a) Monopoles

b) Dipoles

FIG. 6.12.   Diagram showing origin of monopoles and dipoles in a magnetizing
field.

a single positive anomaly whereas a dipole produces a positive anomaly flanked by two negative anomalies. In a horizontal field a monopole produces positive and negative anomalies on either side of the pole and a dipole produces a negative anomaly flanked by two positive anomalies. In inclined fields both monopoles and dipoles produce a negative and positive anomaly, though the negative anomaly is very small in the case of the monopole. As the earth's field is usually inclined in most survey areas, magnetic anomalies almost always occur in positive and negative pairs. Figure 6.13 shows the characteristic shape of a total field magnetic anomaly over an inclined dipole in the northern and southern hemispheres. This shows how the

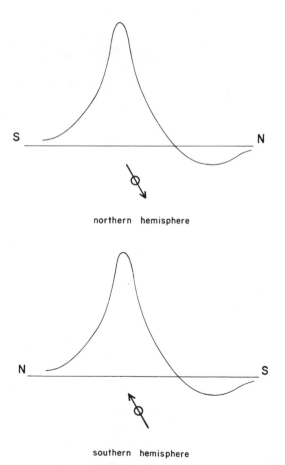

northern hemisphere

southern hemisphere

FIG. 6.13. Characteristic total field magnetic profile over an inclined dipole.

small negative anomaly lies on the north side of the main positive anomaly in the northern hemisphere while the reverse is true for the southern hemisphere.

Depth estimates can be made from magnetic anomalies due to monopoles and dipoles from the size of the anomaly. Nettleton (1940) gives the following rules for estimating depth:

1. Single pole   depth $= 1·305 \times \frac{1}{2}$ width
2. Sphere (dipole)   depth $= 2 \times \frac{1}{2}$ width
3. Horizontal cylinder   depth $= 2·05 \times \frac{1}{2}$ width
   (line of dipoles)

where $\frac{1}{2}$ width is the distance from the centre of the anomaly to the point where the magnitude is one half.

The magnetic effects and anomalies described so far are all due to induced magnetization in the earth's field, but, as mentioned earlier, it may be important to consider remanent magnetization in interpreting magnetic anomalies, particularly in areas with volcanic rocks. Remanent magnetization can give rise to very strong anomalies many times greater than those that can be explained by induced magnetization calculated from the geometry of the bodies involved and their magnetic susceptibilities. There are even cases on record of remanent magnetization causing anomalies opposite to those expected. For example, Yüngül (1956), who conducted a magnetic survey over chromite bodies in ultrabasic rocks in east central Turkey, observed strong, positive anomalies $(1000\gamma+)$ over the chromite masses, although the magnetic susceptibility of the chromite ore is 2 to 18 times smaller than that of the surrounding country rocks and negative anomalies were expected. The observed anomalies could be due only to permanent magnetization in the chromite bodies pointing downwards. Remanent magnetization can also be suspected when dipole anomalies are observed with a reverse polarity. For example, a high on the southside of a low in the southern hemisphere and a high on the northside of a low in the northern hemisphere are both the opposite of what is expected from an induced dipole and indicate a permanent dipole with polarity opposite to the present earth's field.

Aeromagnetic (AM) surveys
Magnetic surveying is the most widely used airborne geophysical method and the aeromagnetic map has come to be regarded as of fundamental importance in understanding any area as the geological map. For this reason, it is customary for Government departments to finance aeromagnetic surveys of their countries. Nevertheless, there are still large areas of the world which have not been covered by such surveys.

Airborne surveys are carried out with flux-gate, proton or optical absorption magnetometers, though proton magnetometers are probably the most widely used today. The survey is usually flown at a ground clearance of 60–100 m for mineral exploration, but higher altitudes may be used for regional surveys looking for deep basement structures. One of the problems in any airborne survey is accurate position location, which is usually accomplished with radio navigation equipment such as the Decca system. In addition a film strip with fiducial marks tied to the magnetic recording is taken of the ground beneath the flight path to assist in fixing the flight lines to topographic features. Over water or featureless country, however, photographic control is not applicable. A series of parallel flight lines with a spacing of 400–2000 m is flown with cross lines, known as control or tie lines, flown as checks at intervals across the main survey lines. The tie lines are generally spaced up to 10 km apart. A base magnetometer on the ground in the survey area is usually used to record diurnal variations. If there are strong disturbances in the diurnal cycle due to magnetic storms, it may be necessary to suspend survey work until conditions quieten down. The magnetic input data is recorded directly on magnetic tape which is suitable for direct computer processing. Closure errors at line/tie line intersections, level corrections, random error corrections, diurnal variation corrections and co-ordinate fixes are then added to produce the final working tape which is used for interpretation and producing the aeromagnetic maps. With the latest techniques machine-drawn maps may be very similar in appearance to hand-contoured maps.

The amplitude of anomalies on an airborne map is dependent on the depth to source and altitude of the survey. Small shallow sources will produce little magnetic effect at high altitudes. Figure 6.14 shows a ground magnetic profile compared to the aeromagnetic profile flown at a ground clearance of 200 m. The airborne anomaly reaches peak values of about 60γ compared to 800γ for the ground measurements, which indicates that the sources are shallow. In addition the ground survey indicates a number of different sources which are not resolved by the airborne survey. In general, airborne surveys will not resolve separate magnetic sources if the terrain clearance is greater than the horizontal distance between the sources less their depth below ground surface.

The interpretation of aeromagnetic surveys has become highly specialized using various computer techniques. To determine depth to basement beneath a sedimentary cover the shorter wavelength anomalies from shallow sources are filtered out to enhance the longer wavelength anomalies due to deep sources. This is what is commonly done in petroleum exploration where one is attempting to map sedi-

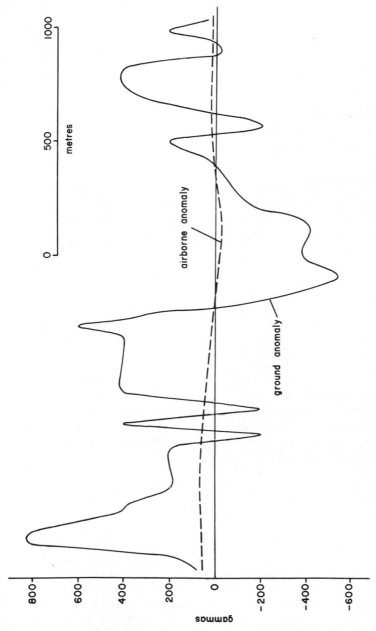

FIG. 6.14. A ground magnetic profile compared to an airborne magnetic profile flown with a terrain clearance of 200 m.

mentary basins. In mineral exploration one may wish to do the opposite, i.e. enhance shallow, localized features by the use of selective filters. Computer processing can also be used to assist in trend identification and to simplify anomaly patterns. It should be stressed, however, that not too much reliance should be placed on purely mathematical interpretations however complex and involved the computer program. Geophysical interpretation can never be automated and a good interpretation will always depend on the interpreter having a good understanding of the geology of the area and a full grasp of the application and limitations of the mathematical techniques being used.

Magnetic gradients
It is sometimes of value to measure either the vertical or horizontal magnetic gradients in the field, since magnetic gradients may assist in resolving the shallower anomalies which are often of interest in mineral exploration. By the same token, in areas with very shallow localized sources, such as strong soil anomalies, which are referred to as 'magnetic noise', the measurement of magnetic gradient is of little value. The horizontal or vertical gradients can be readily measured in the field with a proton magnetometer. For vertical gradients readings are taken at each station with the sensor held at two different heights above the ground (e.g. 1 and 3 m). The vertical gradient is then simply the difference in magnetic readings divided by the difference in heights above ground. For consistency it is important that the sensor is always held at the same heights for each station and the magnetic readings for the higher sensor position should be subtracted from the magnetic readings for the lower sensor position. The horizontal gradient can be measured in the same manner by taking two readings at each station with the sensor at the same height above the ground, but at a separation of 2 or 3 m. Figure 6.15 gives an example of the type of resolution of an anomaly that can be obtained by measuring the vertical magnetic gradient.

Gradient measurement can be usefully employed in shipborne and airborne surveys. By towing two sensor heads from an aircraft or ship a direct recording of magnetic gradient is made. Since diurnal changes and magnetic storms affect both sensors equally, these naturally occurring time variations are automatically removed from the field data. This is particularly valuable at sea where shore-based monitor instruments may be some distance from the survey vessel. In addition this type of survey is applicable to high latitudes where magnetic disturbances often render normal magnetic field data useless for exploration purposes. When measuring magnetic gradients it is important that the instruments used are accurate to $0\cdot2\gamma$ or better.

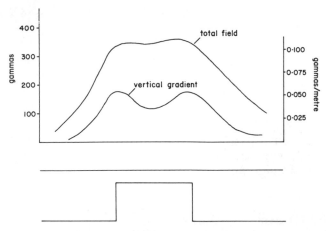

FIG. 6.15.   Example of the use of magnetic gradients in resolving anomalies
caused by buried structures.

6.3   RESISTIVITY SURVEYS

In the earth resistivity (ER) method the apparent resistivity of the
ground is determined by measuring the potential difference across
two electrodes while introducing a current into the ground through
two other electrodes (Fig. 6.16). The resistivity, ρ, of a conductor is
defined by the relation:

$$\rho = \frac{RA}{L} \tag{1}$$

where R is the resistance, A the cross-sectional area and L the
length. The unit of resistivity is the ohm-metre. The current I passing

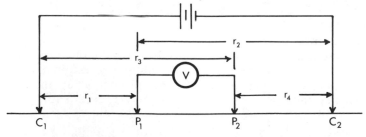

FIG. 6.16.   Arrangement of current electrodes (C_1 and C_2) and potential
electrodes (P_1 and P_2) for measuring earth resistance.

through a conductor with resistance R under an impressed voltage V is given by Ohm's law:

$$I = \frac{V}{R} \tag{2}$$

From the relations (1) and (2) we derive:

$$V = \frac{I\rho L}{A} \tag{3}$$

If a current $+I$ is passed into level ground through an electrode, the potential difference across a hemispherical shell of radius r and thickness dr around the electrode is given by:

$$dV = \frac{I\rho \, dr}{2\pi r^2} \tag{4}$$

Integrating (4) gives:

$$V = \frac{I\rho}{2\pi r} \tag{5}$$

If a current is passed into the ground through electrodes C_1 and C_2 in Fig. 6.16, the potential at P_1 is given by:

$$V = \frac{I\rho}{2\pi} \left[\frac{1}{r_1} - \frac{1}{r_2} \right]$$

Likewise the potential at P_2 is given by:

$$V = \frac{I\rho}{2\pi} \left[\frac{1}{r_3} - \frac{1}{r_4} \right]$$

Thus, the potential difference between P_1 and P_2 will be

$$V = \frac{I\rho}{2\pi} \left[\frac{1}{r_1} - \frac{1}{r_2} - \frac{1}{r_3} + \frac{1}{r_4} \right] \tag{6}$$

Solving for ρ gives:

$$\rho = \frac{2\pi V}{I} \frac{1}{\dfrac{1}{r_1} - \dfrac{1}{r_2} - \dfrac{1}{r_3} + \dfrac{1}{r_4}} \tag{7}$$

Electrode arrays
Quite a number of different electrode arrays have been used in resistivity work, but the two most commonly employed are the Wenner and the Schlumberger. In the Wenner array the distance between electrodes is equal and expression (7) reduces to:

$$\rho = 2\pi a \frac{V}{I}$$

where a is the distance between electrodes. In the Schlumberger

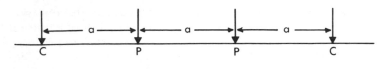

WENNER ARRAY

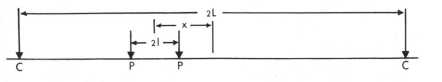

SCHLUMBERGER ARRAY

FIG. 6.17. The two most common electrode arrays used in resistivity work.

array (Fig. 6.17) the distance between the centre potential electrodes is very small compared to the distance between the current electrodes, i.e. the distance of the potential electrodes from either current electrode should be at least ten times the potential electrode separation. The apparent resistivity can be shown to be given by:

$$\rho = \frac{\pi}{2l} \frac{(L^2 - x^2)^2}{L^2 + x^2} \frac{V}{I}$$

where $2l$ is the distance between potential electrodes, $2L$ the distance between current electrodes and x the distance of the centre point of the potential electrodes from the centre point of the current electrodes. If the potential electrodes are located at the mid-point of the array, the apparent resistivity is given by:

$$\rho = \frac{\pi}{2l} \frac{L^2 V}{I}$$

Field procedures
Basic resistivity equipment is quite simple requiring a generator to supply current to the ground, an ammeter to measure the applied current and a voltmeter to measure the potential difference across the potential electrodes. In fact it is not necessary to measure the current and potential difference separately to obtain a value for the resistance and on most instruments the potential across the potential electrodes is balanced on a potentiometer incorporated in the current electrode circuit and calibrated in ohms. In d.c. equipment the current is commutated to counteract polarization at the electrodes and the leads

from the potential electrodes to the voltmeter are on the same commutator so that the voltage is only measured during current flow to avoid spurious measurement of self potentials. Some equipment uses an alternating current, the frequency of which is very low varying from a fraction of a cycle to 20 Hz. Since a.c. measuring instruments are used, interference by d.c. self potentials presents no problems. There are two main methods of measurement in resistivity surveying: electric 'drilling' or sounding and electric mapping or traversing.

Electric 'drilling'

In this method the electrode spacings are continually increased in steps for each measurement to give greater and greater depths of penetration, hence the term 'drilling' or 'sounding'. With the Wenner array the electrode separations are always kept equal, but with the Schlumberger array the potential electrode separation is kept constant while the current electrode separation is increased. For example, 2*l* may be fixed at 2 m and measurements made with 2*L* at 20, 30, 50, 70, 100, 200 m and so on. Results are presented as a graph by plotting apparent resistivity along the ordinate and *a* and *L* along the abscissa in the Wenner and Schlumberger arrays respectively.

Electric mapping

In this method the electrode spacing is kept constant and the array moved about the area being surveyed, thus mapping the lateral variations in resistivity. With the Wenner system the array is moved as a whole. This procedure can also be adopted with the Schlumberger system, but very often the current electrode separation is kept constant at several hundred metres and the potential electrodes are moved between them with a constant small separation (5–20 m, say). Results are presented as resistivity contours with the plotting point at the centre of the array in the Wenner system and at the mid-point of the potential electrodes with the Schlumberger array.

Interpretation

Although a great deal has been written on the interpretation of resistivity surveys, it still remains very complex as the theories only apply to relatively simple models such as inclined sheets, horizontal layers or buried spheres. However, much valuable information has been obtained from model experiments which allow actual resistivity measurements to be made over more complex structures. Nevertheless, owing to complexities in nature, much interpretation is still essentially qualitative or only semi-quantitative.

Figure 6.18 gives the common resistivity ranges for a number of

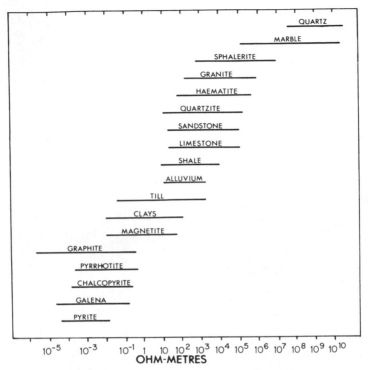

FIG. 6.18. Resistivity ranges of some common rocks and minerals.

rocks and minerals which vary quite widely from quartz which is a good insulator to sulphides which are mainly good conductors with the notable exception of sphalerite. There is also quite a wide range of resistivities for any given rock type depending on composition. For instance, argillaceous limestones may have quite low resistivities, whereas pure limestones have high resistivities. In addition, water content is a significant factor in governing resistivity *in situ*. A hard, compact, dry quartzite has a very high resistivity, whereas a saturated, porous, permeable, quartz sandstone has a low resistivity.

In cases where there is a horizontal or near-horizontal layering depth determinations can be made using the method of 'electric drilling'. Figure 6.19 shows three characteristic resistivity curves for three different cases of horizontal layering. There is no simple relation between the shapes of the curves and the resistivities of and depths to the different layers. In practice the resistivity of the top layer can be determined by taking a mean value of the measured

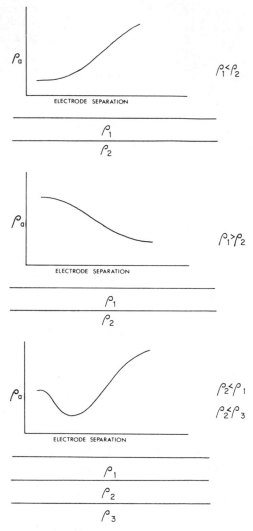

FIG. 6.19.   Characteristic resistivity curves for three generalized cases of horizontal layering.

apparent resistivities at small electrode spacings on the assumption that depth penetration is largely confined to the top layer. A large number of theoretical curves for different cases have been prepared by different workers (e.g. Mooney and Wetzel, 1956; La Compagnie Générale de Géophysique, 1955) and a general method of interpretation is to compare the actual curves with the theoretical ones. For the two-layer case a method by Tagg (1934) gives a family of curves with ρ_a/ρ_1 as a function of h/a for different values of k, a resistivity factor, where h = depth to the interface, a = electrode separation for the Wenner array and $k = (\rho_2 - \rho_1)/(\rho_2 + \rho_1)$. There are two families of curves, one for negative k values ($\rho_1 < \rho_2$) and one for positive k values ($\rho_1 > \rho_2$). Tagg's method may be extended to the three-layer case if the resistivity of the bottom layer does not affect the curve too much. Let us consider an example for the two-layer case using Tagg's method. Table 6.3 gives the results of a resistivity depth probe which indicates a two-layer case with $\rho_2 > \rho_1$. Using the small electrode spacings of 5 and 10 m an estimate for ρ_1 is determined at 87 Ωm. Then values for ρ_1/ρ_a are obtained for the other electrode spacings as shown in Table 6.3. Values of k and h/a are then read from the upper set of Tagg's curves ($\rho_2 > \rho_1$) in Fig. 6.20 for each value of ρ_1/ρ_a in Table 6.3. For example, this gives the following set of values for $\rho_1/\rho_a = 0.70$ and $a = 20$ m:

k	h/a	h
0·2	0·18	3·6
0·3	0·41	8·21
0·4	0·55	11·0
0·5	0·68	13·6
0·6	0·78	15·6
0·7	0·85	17·0
0·8	0·92	18·4
0·9	0·99	19·8
1·0	1·07	21·4

TABLE 6.3
RESULTS OF A RESISTIVITY PROBE
USING THE WENNER ARRAY

Electrode spacing (m)	ρ_a	ρ_1/ρ_a
5	85	—
10	110	—
20	125	0·70
30	155	0·56
60	175	0·49
90	190	0·46
120	210	0·41
150	230	0·38

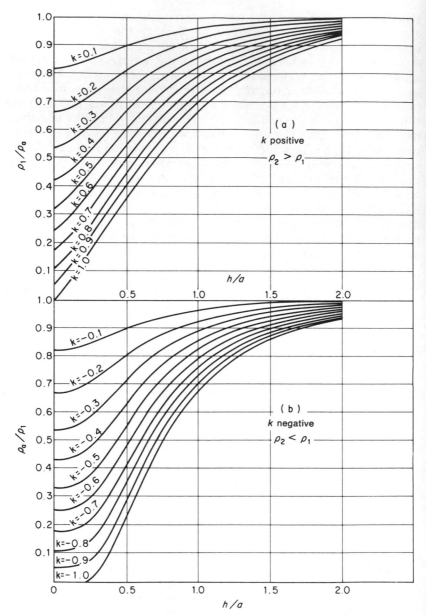

FIG. 6.20. Tagg's curves for two-layered cases (Tagg, 1934).

This is repeated for different values of ρ_1/ρ_a and a to give additional sets of values for k and h/a. The k values are then plotted against h values to give a number of curves as shown in Fig. 6.21. These lines intersect in the region of $k = 0.4$ and $h = 10$ m. Thus, the depth probe indicates an upper layer of resistivity 87 Ωm ten metres thick resting on a lower layer of resistivity 203 Ωm.

As in other geophysical methods it is rarely possible to obtain a

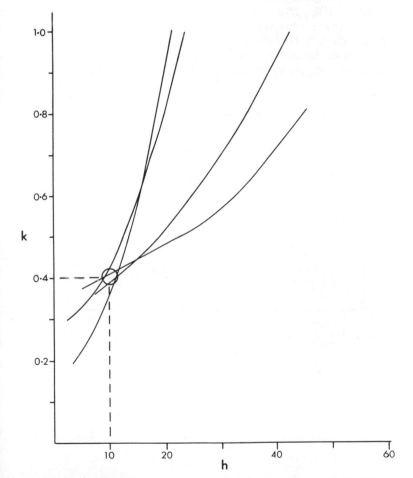

FIG. 6.21. Example to show how Tagg's curves are used to find the resistivity factor (k) and the depth to the second layer (h) from a resistivity depth probe.

unique solution to resistivity measurements. A given curve may represent widely different subsurface configurations. However, these can be reduced to one or a few possibilities if there is sufficient geological control.

Interpretation of resistivity mapping is generally more qualitative, though theoretical calculations can also be made to aid interpretation as in the case for depth probes. The shapes of the resistivity profiles often depend upon the orientation of the electrode array with respect to the structure, particularly in the case of steeply dipping conductors or insulators. This effect is illustrated in Fig. 6.22 which shows the resistivity curves obtained in a model experiment over a vertical conductor plate. When the electrode array is moved parallel to the plate, a simple resistivity 'low' is obtained, but, when the array is moved at right angles to the plate, a more complex curve is observed

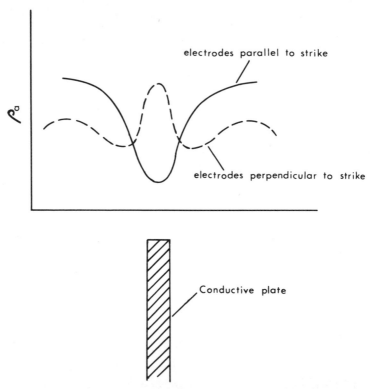

FIG. 6.22.   Examples of resistivity profiles obtained in a model experiment over a conductive plate.

with a 'high' directly over the plate flanked by two small 'lows'. This serves to show that interpretation of resistivity data is often far from straightforward. Nevertheless, lateral variations in resistivity often followed a simple pattern and meaningful qualitative interpretations are possible. In Ireland resistivity surveys have proved useful in outlining sub-outcrops of reef limestones which are often the hosts of sulphide mineralization. The reef limestones have much higher resistivities than the off-reef facies, which consist largely of argillaceous limestones, and the reef limestones are often clearly delineated by resistivity 'highs' (Fig. 6.23).

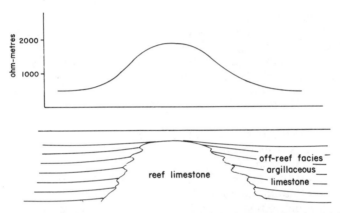

FIG. 6.23.   Reef limestones in Ireland defined by resistivity high.

Nowadays, resistivity surveys are rarely used for direct prospecting in mineral exploration, though resistivity measurements are always carried out in conjunction with IP surveys (p. 268) and play an important role in their interpretation. For indirect prospecting such as determining depth of overburden and delineating sub-outcrops of different rock types resistivity surveys can be very useful. In hydrogeology resistivity is an important method and is widely used in all parts of the world for siting water boreholes.

Airborne surveys
Recently, methods have been developed for carrying out resistivity mapping from an aircraft. The method known as E-PHASE, which has been developed by Barringer Research in Canada, is really an EM method which uses radio sources in the VLF (15–25 kHz), LF (200–400 kHz) and BCB (550–1100 kHz) bands (McNeill and Barringer, 1970; McNeill et al, 1973), but it is included in this

section as the final product of the survey is a contoured resistivity map. The method is useful for locating shallow resistive targets such as gravel deposits or for the geological mapping of layered structures.

6.4   INDUCED POLARIZATION (IP) SURVEYS

Although the phenomenon of *induced polarization* or *overvoltage* has been known and studied in the laboratory by physical chemists for a long time, it was only comparatively recently that it was put to practical use in applied geophysics. Overvoltage was discovered accidentally in the field during the course of resistivity surveys when it was observed that the voltage across the potential electrodes did not fall back to zero as soon as the current was cut off, but persisted for a short time as a fading residual voltage. Experimental exploration surveys using overvoltage were carried out in various parts of the world in the late 1940's and early 1950's, but it was not until the late 1950's that the method became more widely used and it was only during the 1960's that IP became the widely used prospecting tool it is today.

The phenomenon of overvoltage is extremely complex, but there appear to be two main effects: *electrode polarization* and *membrane polarization*. It is the first effect that is of most importance in IP surveying. When a current is passed through the ground, ionic charges tend to build up on conductor–electrolyte interfaces and a voltage is created which tends to oppose the flow of the current. When the external current is cut off a residual voltage continues to exist and slowly fades away as the ionic charges diffuse into the electrolyte and equilibrium is restored. Membrane polarization is rather more complex, but it is exhibited by non-metallic clay minerals which also give an IP effect. As overvoltage is essentially a surface effect, the IP response increases with the number of conductor–electrolyte interfaces present and is thus an excellent method for detecting highly disseminated sulphide ores. In measuring the IP effect either direct current or alternating current can be used. In the case of d.c. the IP effect is said to be measured in the *time-domain* and in the case of a.c. it is said to be measured in the *frequency-domain*.

Time-domain
There are two main methods of measuring the IP effect in the time-domain: the residual voltage can be read at a specific time interval (usually a few seconds) after cessation of the current, or the voltage decay curve can be integrated between two time limits. In the first case results are usually expressed as per cent IP effect.

$$\text{IP effect} = \frac{V_t}{V} \times 100(\%)$$

where V_t is the residual voltage in millivolts after time t and V is the voltage in volts of the applied current. In the second case the area under the voltage decay curve is expressed as millivolt–seconds. This is divided by the voltage in volts of the applied current to give millivolt–seconds/volt or simply milliseconds.

Frequency-domain
Measurements of the decaying residual voltage are not made in the frequency-domain, but rather, use is made of the fact that the apparent resistivity decreases as the frequency of a current passing through the ground is increased. This effect is caused by the over-voltage phenomenon and is analogous to passing an alternating current across a capacitor and a resistor wired in parallel. The potential difference across the potential electrodes is measured while the current is being applied to the current electrodes and the apparent resistivity is calculated in the same manner as for ordinary resistivity surveys. In practice the apparent resistivity is determined for two different frequencies (usually $0\cdot1\,\text{Hz}$ and $10\,\text{Hz}$) at each station. Results are expressed as the so-called frequency effect or in terms of the metal factor:

$$\text{frequency effect} = \frac{\rho_1 - \rho_2}{\rho_2} \times 100(\%)$$

$$\text{metal factor} = \frac{\rho_1 - \rho_2}{\rho_1 \times \rho_2} \times 2 \times 10^5$$

where ρ_1 is the apparent resistivity at frequency 1 and ρ_2 is the apparent resistivity at frequency 2.

Recently, use has been made of the observed phase shift in the frequency-domain which is analogous to time delay in the time-domain. This phase shift, which is measured in milliradians, is defined as the phase difference between the fundamental harmonic of the transmitted and received signals. An increase in the phase shift is observed over conductors and good resolution has been claimed for the method.

Field procedures
Typical IP equipment consists of a generator, a transmitter unit, a receiver and a cycling timer coupling transmitter and receiver. Fairly high voltages are usually necessary and a typical transmitter would deliver 1000 V at 2 A to the current electrodes. Such currents are extremely dangerous and great care should be exercised in carrying

out IP surveys to avoid accidents. When using a direct current in the time-domain, it is common to use a pulse of short duration followed by a short gap of current off when the decay curve is measured. The direction of the current pulses is repeatedly reversed to minimize effects due to polarization at the electrodes.

Resistivity measurements are extremely useful in assisting with the interpretation of IP results and a resistivity survey is always carried out concurrently with an IP survey. In the frequency-domain the apparent resistivities are determined routinely to calculate the frequency effect or metal factor values, but in the time-domain it is necessary to measure the potential across the potential electrodes during current on to determine the apparent resistivity. Standard features of time-domain equipment enable the potential to be measured both during current on (apparent resistivity) and during current off (IP effect).

Quite a number of different electrode arrays have been used for IP work, but the most commonly used arrays are the *three-electrode* or *pole–dipole* in the time-domain and the *dipole–dipole* in the frequency-domain (Fig. 6.24). For small scale reconnaissance work the *gradient array* is sometimes used. In this system the two current electrodes are placed outside the survey grid and remain fixed while the two potential electrodes are moved about taking the IP measurements. Metal stakes are generally used as electrodes, but in dry areas where there is a high resistivity surface layer it is common practice to use sheets of metal foil as electrodes. If the ground is very dry, it is usually necessary to water the area around the electrodes to reduce the resistivity. Distance between electrodes may be as little as 10 m or as much as 200 m, but spreads of 50–100 m are most commonly used. The effective depth of penetration is increased with increased electrode spacings. In the pole–dipole and dipole–dipole arrays to give increasing depth of penetration the surveys are usually carried out for $n = 1, 2, 3, 4$, *etc.* with a fixed at 30 or 50 m (Fig. 6.24). Maximum depth penetration for IP surveys is of the order of 200 m.

As IP equipment is relatively heavy and bulky, it is best to position the apparatus near a central point on a survey line to minimize movement of the equipment. IP is a relatively slow and costly method and is unsuitable for large scale reconnaissance work. A good crew should be able to complete 20–60 line-km/month depending upon electrode spreads and terrain.

A number of extraneous factors can affect IP surveys, two of the most important being atmospheric conditions (electrical storms) and artificial conductors. If atmospheric conditions are very bad, they can have a drastic effect on the receiver signal and it is sometimes necessary to cease field measurements for one or more days until

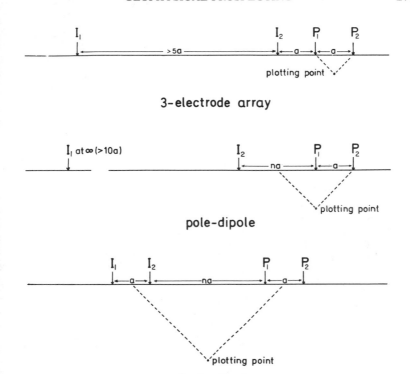

FIG. 6.24. Some electrode arrays used for IP surveys.

conditions quieten down. Artificial conductors such as power-lines, wire fences and metal pipes can also affect IP responses and their presence should always be noted during surveys. In interpreting IP work, one has to be very careful to try to distinguish formational anomalies caused by barren rocks such as graphitic shales, cultural effects caused by artificial conductors and anomalies due to mineralization. It is often very difficult or impossible to do this, but it should always be borne in mind that very large IP anomalies can be caused by barren argillaceous, carbonaceous or pyritic rocks.

Plotting of data
IP results are generally plotted along sections, though contoured plans can also be prepared. In cases where measurements are taken for different values of n it is common practice to plot contoured pseudo-sections in the frequency-domain (Fig. 6.25) and a series of profiles in the time-domain. It is also quite common to show strong

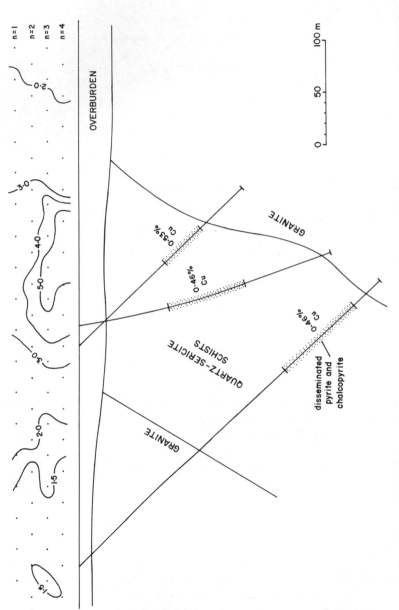

FIG. 6.25. IP pseudo-section showing frequency effect over a copper deposit in Zambia.

anomalous zones along survey lines diagrammatically in plan with a heavy black line and weakly anomalous zones with a dashed line (Fig. 6.26). A typical IP profile obtained over a disseminated sulphide ore body using time-domain equipment is shown in Fig. 6.27.

Magnetic induced polarization (MIP)

One of the major drawbacks of conventional induced polarization surveys is that they are slow and costly and are not suited to reconnaissance work. This problem is overcome to a large extent by a new method proposed by Seigel (1974). Current is supplied to the ground through two current electrodes approximately 1000 m apart using either time-domain or frequency-domain IP transmitters. Then, instead of using ground electrodes for measuring the IP effect in the normal manner, the induced magnetic field is measured, obviating the need for the detection system to be in contact with the ground. In the case of time-domain the transient magnetic field due to the polarization current flow is measured. In the case of frequency-domain, either

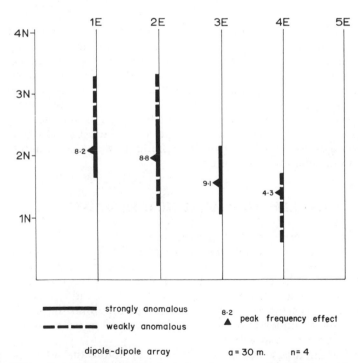

FIG. 6.26. Conventional method of showing anomalous IP zones in plan.

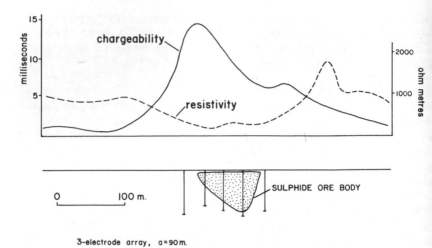

FIG. 6.27.   IP and resistivity profiles over a lead–zinc ore body at Pine Point, Northwest Territories, Canada (after Seigel *et al*, 1968).

the change of the magnetic field with frequency, or the phase shift of the magnetic field for a given frequency can be measured. A vector magnetometer of an advanced flux-gate type is used to measure the horizontal component of the magnetic field orthogonal to the line joining the current electrodes. The method essentially detects areas of anomalous polarization and should be able to locate both disseminated and massive sulphide conductors. The technique is still in a development stage, but it may well prove to be a useful exploration tool, though interference from magnetic rocks and minerals may present considerable difficulties.

6.5   ELECTROMAGNETIC (EM) SURVEYING

The basic methods of EM prospecting were largely developed in North America and Scandinavia during the 1920's and 1930's, though many refinements to the techniques have been made in recent years. The theory is quite complex and for this reason EM surveying is very much a specialist's field and is generally poorly understood by most geologists. Nevertheless, the basic principles behind the method are not difficult to understand and the field procedures for some of the EM methods are relatively simple.

The EM methods are based upon the principle that electromagnetic waves induce currents in conductors. These induced currents are themselves the source of new electromagnetic waves which can be detected by suitable instruments. Familiar applications are metal detection devices widely used in security checks and mine detectors used in wartime.

A primary magnetic field of frequency $\omega/2\pi$ acting on an electrical circuit induces an emf giving rise to a secondary magnetic field which lags $(\pi/2) + \phi$ behind the primary field, where $\phi = \tan^{-1}(\omega L/Z)$, L being the inductance and Z the resistance of the conductor. In a very good conductor $Z \to 0$, $\phi \to \pi/2$ (90° out of phase). Thus the degree of phase shift can be used as a measure of the conductance of a buried conductor. The primary (**P**), secondary (**S**) and resultant (**R**) fields can be shown on a vector diagram (Fig. 6.28) where it can be seen that the component of the secondary field in phase with **P**, known as the real or in-phase component, is equal to $-S \sin \phi$ and the component of the secondary field lagging 90° behind **P**, known as the imaginary component or quadrature, is equal to $S \cos \phi$. The real and imaginary components are usually expressed as a percentage of the primary field and the ratio, real/imaginary, is used as a measure of the conductance of a buried conductor, a ratio greater than one indicating a good conductor and a ratio less than one a poor conductor.

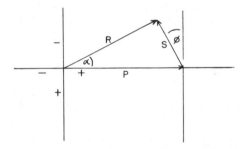

FIG. 6.28.   Vector diagram showing primary, secondary and resultant EM fields.

The resultant field of the primary and secondary fields can be resolved into vertical (X) and horizontal (Y) components which form the axes of an ellipse described by the resultant vector $2\pi\omega\sqrt{(X^2 + Y^2)}$ times per second and the field is said to be elliptically polarized. The attitude of this ellipse depends upon the size, depth, shape and nature of conductors and some early methods in EM surveying were based upon finding the plane of the ellipse of polarization, determining its azimuth and dip, and locating the major and minor axes. Over a very good conductor the ellipse degenerates to a straight line and over a very poor conductor it degenerates to a circle.

EM methods can be divided into two basic classes:

1. Fixed source—e.g. Tilt angle, Sundberg, Turam, Beeler–Watson
2. Moving source—e.g. EM gun, Slingram, Max–Min

Tilt angle
This is probably the simplest EM method and is therefore quite popular, though it has a number of disadvantages. A transmitting coil connected to an oscillator (a frequency of 1000 Hz is commonly used) is usually held in a vertical plane, though it can also be held horizontally. In the vertical plane a horizontal primary EM field is transmitted. A search coil or receiver connected to an amplifier and detector (usually earphones) is held at right angles to the transmitting coil and tilted to either side until the signal is minimum. This occurs when the search coil is in the plane of the ellipse of polarization and no current is induced. Figure 6.29 shows two possible ways of transmitting and receiving. The angle of tilt is easily measured by a clinometer on the search coil. It is usual to adopt the convention that tilting to the left is negative and tilting to the right positive. The transmitter coil is set up at a point in the area to be surveyed and a series of readings is taken with the receiver coil along traverse lines across the assumed strike of possible conductors. It is usual to

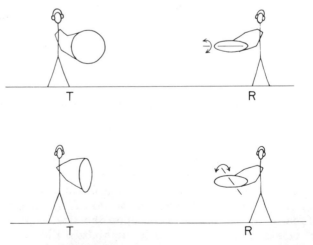

FIG. 6.29. Two possible methods of transmitting and receiving in tilt angle EM.

orientate the plane of the transmitter coil to pass through the point of observation for each reading. When the distance between the transmitter and receiver is of the order of 600 m, the transmitter is moved to a new position closer to the receiver before taking further readings. Instead of using a fixed transmitter position, one variation of the method uses two people who alternately transmit and receive in a manner similar to that shown in Fig. 6.29, keeping a constant separation between the two coils along the survey lines. In the absence of conductors the field remains horizontal and the tilt angle is zero. In passing over a conductor a cross-over is observed with the tilt angle changing from negative to positive (Fig. 6.30). The main disadvantages of the method are its poor resolving power (only shallow conductors are detected) and the difficulty of finding sharp search coil positions because of out-of-phase fields. Methods which measure amplitude and phase shift have much better resolving power and are more widely used.

Turam
This method was devised in the 1930's in Sweden by H. Hedström and the name comes from the Swedish, 'tva ram', or two frame. As the name implies, two search coils are used and are moved along the traverse lines 10–50 m apart. The ratios of the amplitudes of and the phase difference between the induced voltages in the two coils are measured on a bridge type compensator. The coils are usually held horizontally but they may be kept vertical or one may be horizontal and the other vertical. The source of the primary field is usually a long cable perpendicular to the traverse lines and grounded at both ends, but a large loop, usually rectangular, laid out on the ground may also be used. When measuring the amplitudes in the two search coils, the variation of the primary field with distance has to be taken into account. In the case of a long cable, the normal ratio of the primary field is simply the inverse ratio of the distances of the two coils from the cable. If a loop is used instead of a cable as a source, it is necessary to make additional calculations (Fig. 6.31) to find the required Turam ratios. In the case of a long cable source, if V_1 and V_2 are the induced voltages and α_1 and α_2 the phases of the vertical field at positions 1 and 2, the Turam quantities measured are $V_1 r_1 V_2 r_2$ and $\alpha_2 - \alpha_1$ where r_1 and r_2 are the respective distances of the coils from the cable. When the coils are on a different level from the source, a correction needs to be applied. In the absence of conductors the amplitude ratio is equal to one and the phase difference is zero. Over a conductor the Turam ratios attain a maximum and the phase differences attain a negative minimum. Distances to which measurements can be made from the source are limited as the voltages in the

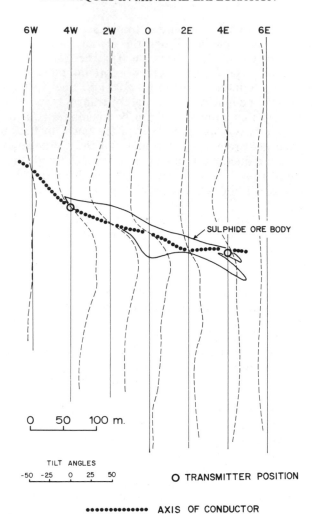

FIG. 6.30.   Tilt angle survey over the Mobrum Copper Ltd massive sulphide
deposit, Quebec (after Seigel *et al*, 1957).

search coils become very weak and are difficult to measure at large
distances from the source. Typical Turam equipment consists of; (1) a
motor generator, (2) a primary source cable, (3) two receiver coils and
(4) a ratio and phase meter. The Turam method is ideal for detecting
relatively shallow and steeply dipping conductors. Figure 6.32 shows
a typical Turam profile over a sulphide ore body.

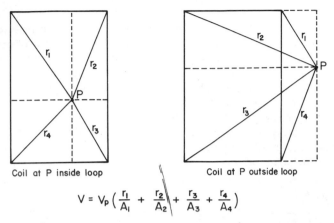

Coil at P inside loop   Coil at P outside loop

$$V = V_P \left(\frac{r_1}{A_1} + \frac{r_2}{A_2} + \frac{r_3}{A_3} + \frac{r_4}{A_4} \right)$$

A's = areas of rectangles of which respective r's are diagonals

FIG. 6.31. Formulae for determining Turam ratios with a rectangular loop.

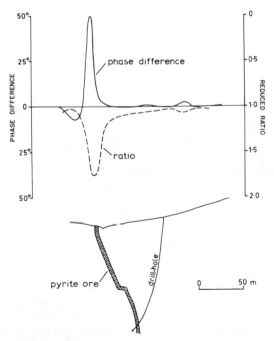

FIG. 6.32. Typical Turam anomaly over a massive sulphide ore body (after Rocha Gomes, 1958).

EM gun

This is a system of EM surveying where the source and search coils are moved for each new reading. It is probably the most popular EM method owing to the simplicity and flexibility of operation. A crew of two is required, one member carrying the transmitter and the other the receiver. The coils are kept a fixed distance apart (usually 25–100 m) by a reference cable which connects them as shown in Fig. 6.33. The primary voltage is supplied to the transmitter coil by an oscillator usually operating in the range 600–1800 Hz and a fixed reference voltage is tapped from the transmitter coil and fed to a compensator. The induced voltage in the receiver coil is decomposed

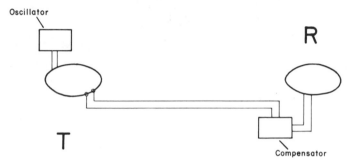

FIG. 6.33.   Basic layout of EM gun equipment.

into two components, one in phase (real) with the reference voltage and one 90° out of phase (imaginary) with it. The magnitudes of the real and imaginary components are compared to the reference voltage and each expressed as a percentage of the primary field. The coils are usually held horizontally but they may also be held vertically. When carrying out an EM gun survey, it is important to keep the coils the same distance apart for each new reading as only a small variation in the separation can lead to significant errors. In the case of horizontally held coils, the real and imaginary readings are both negative and reach a minimum over a conductor. If the conductor is very wide with respect to the coil separation, however, both real and imaginary values may be positive. The nature of the conductor is indicated by the ratio, real/imaginary, a large ratio indicating a good conductor. Over vertical conductors the real and imaginary curves are symmetrical, but over a dipping conductor they become asymmetrical, the amount of asymmetry being a rough guide to the dip of the conductor. Figure 6.34 shows a typical profile obtained with the EM gun over a sulphide ore body.

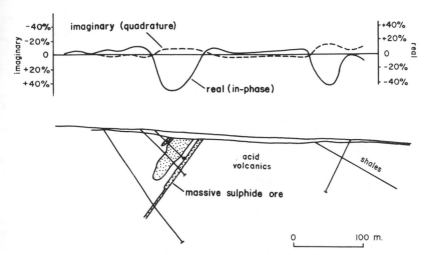

FIG. 6.34. EM gun profile over a sulphide ore body, Note the anomaly over the barren shales (after Malmqvist, 1958).

Pulse EM (PEM)

This is a time-domain EM method developed by Newmount Exploration Ltd in Canada in the 1950's. The original equipment was very bulky, but smaller, more portable equipment was produced by Crone Geophysics Ltd in 1972 under an agreement with Newmont. A 3-m transmitter loop is momentarily energized with a strong d.c. pulse. After power cutoff, a receiver coil 30 m away picks up the secondary decay signal at various time intervals ranging from 0·15 ms to 8·85 ms. PEM surveys have the advantage that they can penetrate a weathered surface and achieve a good depth penetration without large coil spacings. It also has the advantage in common with the similar INPUT method that there is no coupling between the transmitter and receiver and, since secondary signals are measured after the primary field is cut off, weak secondary fields can be detected without being obscured by overriding primary fields, which is a principal factor limiting the sensitivity of other EM methods. Figure 6.35 shows an example of a PEM profile over a sulphide conductor.

Distant source EM

In fairly recent years there has been a development of EM methods employing distant sources which are outside the control of the person operating the receiving instrument. One, known as AFMAG, makes use of natural electromagnetic radiation and the other known as VLF-EM makes use of low frequency radio waves.

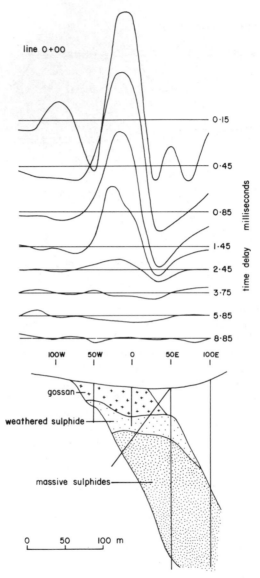

FIG. 6.35.   PEM profile over the Lasail massive sulphide deposit, Sultanate of Oman (Crone, 1975).

AFMAG

The name is derived from *audio frequency magnetic fields* (1–10 000 Hz) and makes use of natural electromagnetic fields which are largely derived from local and distant thunderstorms. These vary seasonally; in northern latitudes they are strongest from June to September and in southern latitudes from December to February. They also display a diurnal variation which is generally strongest in the early morning and late afternoon and weakest at midday. These fields have a wide range of frequencies and normally have a very small vertical component with a consequent horizontal or near horizontal plane of polarization. In addition the azimuth of polarization tends to be random. In the presence of conductors, however, the azimuth becomes more definite and the plane of polarization tilts away from the horizontal. Instrumentation allows the azimuth and tilt of the resultant field to be determined at a selected frequency. Results are plotted in plan as a series of AFMAG vectors, the direction showing the azimuth and the length of the vector being proportional to the tilt (Fig. 6.36). The presence of conductors is shown by a cross-over as is observed in the normal tilt angle EM method. It is common practice to take readings at two different frequencies (e.g. 200 and 500 Hz) at each station. The response ratio, 'low/high', is indicative of the nature of a conductor; a high ratio (greater than one) indicates a good conductor and a low ratio (less than one) a poor conductor.

The AFMAG technique has a lot of appeal because of its simplicity. Owing to high background noise and poor resolution, however, its precision is inferior to other methods and not much success can be claimed for it.

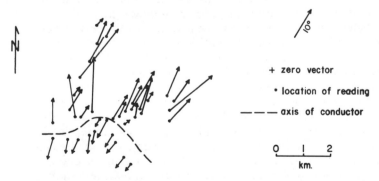

FIG. 6.36.   AFMAG vectors (after Jewell and Ward, 1963).

VLF-EM

The name of this method is derived from the fact that it makes use of very low frequency radio waves. These radio waves in the frequency range 10 000–25 000 Hz have the ability to penetrate a limited depth of water and are used to communicate with submerged nuclear submarines. A number of stations around the world (e.g. Cutler, Maine; Rugby, England; Odessa, USSR; Bilboa, Panama) are continuously broadcasting on various frequencies in this VLF range. Unfortunately the name, VLF-EM, is very misleading for, although the frequencies are very low for radio waves, they are very high for EM surveying which is usually carried out in the range of a few tens to a few thousand hertz.

VLF-EM receivers are light and extremely simple to use and for this reason the method is very popular. The quantities measured are: amplitude, azimuth, and tilt; in addition some instruments measure a quadrature component. The instruments are equipped with a tuner which selects the station, a meter and/or earphones for indicating signal strength, a clinometer for measuring tilt and a compass for indicating azimuth. The direction to the transmitting station selected for a survey should be perpendicular, or as nearly perpendicular as possible, to the strike of the features being investigated, i.e. parallel to the traverse lines. At each station the meter is held so that the axis of the detecting coil, which is wound on a ferrite core, is horizontal and then it is moved from side to side until either a maximum (coil parallel to the resultant field vector) or minimum (coil perpendicular to the resultant field vector) is obtained. The maximum orientation is used on some instruments which have meters for measuring the signal strength or amplitude. The azimuth of either the maximum or minimum signal is shown by the compass on the instrument. The detecting coil is then held in a vertical position and tilted from side to side about a horizontal axis parallel to the minimum signal azimuth until a minimum signal is obtained. The clinometer then gives the tilt or dip-angle using the convention that tilting to the left is negative and tilting to the right positive. In addition the operator should use the convention of facing in the same direction on all traverse lines before taking a reading. When plotting the survey data, the direction to the transmitting station used should be shown on the plans.

Although the method is simple and extremely flexible, it suffers from high background noise, responding to a wide range of sources such as sulphide ore bodies, faults, dykes, variations in overburden thickness, etc. In addition depth penetration is poor because of the high operating frequencies. The method is useful for detailed work in areas of good geological control to trace features such as faults, dykes and mineralized veins. In spite of its ease and flexibility of operation, it is of little value for reconnaissance work unless a high degree of geological control is available. The method should not be used as a follow-up to more conventional EM surveys, which have better resolution and depth penetration.

Airborne electromagnetic (AEM) surveys

One of the advantages of EM surveying is that no direct connections with the ground are required and it has proved possible to adapt the method to airborne work. This has meant that the method has become a widely used reconnaissance prospecting tool with numerous successes claimed for it, particularly in Canada and Sweden. A great variety of systems have been employed, but they can be divided into three broad classes:

1. Source and search coils flown at same height—mounted on a beam below a helicopter, mounted on the wing tips of a fixed wing aircraft, or by using two aircraft.
2. Source and search coils at different heights—'bird' towed behind aircraft.
3. Source on the ground and search coils in aircraft—e.g. airborne Turam (Turair).

Methods 1 and 3 can be flown as low as 30–50 m, but method 2 is generally flown at 150 m or more, which is a big disadvantage. Some of the more common systems are briefly described below:

Helicopter or wing-tip

In this system transmitter and receiver coils are mounted on the wing tips of a fixed wing aircraft or at either end of a rigid boom 10–20 m long installed underneath a helicopter. The system can be flown at a height of 30–50 m and the in-phase and out-of-phase components picked up by the receiver are continuously recorded and expressed as a fraction (usually parts per million) of the primary field. Subsurface conductors give rise to anomalies of several hundred to a thousand parts per million and the response ratio, in-phase/out-of-phase, is a rough measure of the conductivities. With this method, it is important to keep coil separation constant as only small variations give spurious in-phase signals.

Dual frequency

In this system only the out-of-phase component is monitored by the receiver at a low frequency (400 Hz) and a high frequency (2300 Hz). The response ratio, 'low/high', is indicative of the conductivity of any anomalous body. The transmitter is mounted on the aircraft while the receiver is towed in a 'bird' up to 150 m behind the plane. Flying height has to be at least 120 m and depth penetration is poor. In addition there is a lot of background noise which makes discrimination between superficial and deeper conductors difficult. The in-phase component cannot be measured in this system owing to the bumpy movements of the bird.

Rotary field
In this system, which was devised in Sweden, the transmitter and receiver both consist of two coils mounted at right angles in vertical and horizontal planes. The two transmitting coils are fed with a signal at the same amplitude and frequency but with a phase difference of 90° which results in a rotating elliptically polarized field. The induced voltages in the receiver coils are balanced against each other after shifting the phase of one by 90° so that there is a zero signal on a recording meter in the absence of conductors. Secondary fields from subsurface conductors affect the receiver coils unequally and a deflection is picked up by the recording meter. Both the in-phase and out-of-phase components are measured, usually as a percentage of the primary field induced in either receiver coil. The system can be employed by placing the transmitter in an aircraft and receiver in a bird, but, since no connection is required between transmitter and receiver, it is usual to use two aircraft flying in tandem 300 m apart with the receiver towed behind the leading aircraft on a short cable about 15 m long. Such a system can be flown fairly low (60 m) and good resolution and depth penetration are claimed.

INPUT
This system which stands for *induced pulse transient* was developed by Barringer Research in Canada and is one of the most widely used systems today. A large transmitting coil mounted on an aircraft is energized by short intermittent current pulses at the rate of 288 pulses/s ($1 \cdot 1$ ms current pulse followed by $2 \cdot 37$ ms 'silence'). A receiver coil towed in a bird records the amplitude variations of the transient decay curves in the intervals between pulses on a number of different channels. For example, Mark III equipment records on four channels at 200, 600, 1000 and 1600 μs after termination of the primary pulse. Later equipment uses even more channels. The method has good depth penetration and very good resolving power, as superficial conductors do not usually show responses beyond the first channel.

In addition to the more conventional AEM systems, airborne versions of the distant source AFMAG and VLF-EM methods are available. AEM methods are under continual research and development and modified versions of the various systems are being brought out all the time.

Depth penetration

The depth penetration of EM surveying is directly proportional to the square root of the resistivity and inversely proportional to the square root of the frequency. From this relationship, it is readily apparent that a survey carried out at 10 000 Hz will only have one tenth the depth penetration of a survey at 100 Hz. If ground resistivity is high, depth penetration may be considerable, but depth penetration will be greatly reduced if a high conductivity near-surface layer is present. For example, a survey carried out at 1000 Hz over ground with a resistivity of 2000 Ω m will be 700 m, but a ground resistivity of 100 Ω m will reduce the depth penetration to only 150 m. In addition the practical depth penetration may be a lot less than the theoretical depth penetration as it depends upon the detection limit of an EM anomaly over background 'noise'. In general, the maximum depth penetration will not be more than five times the separation between transmitter and receiver; very often it will be considerably less than this. In summary, EM methods are most suited to detecting relatively shallow massive sulphides concealed beneath a high resistivity overburden.

6.6 SELF-POTENTIAL (SP) SURVEYS

SP is one of the oldest geoelectric prospecting methods and was used quite extensively in the 1920's and 1930's. There are even cases on record of the method being tried in the 1800's. Today it is very much out of favour for a number of reasons and is rarely used, though SP surveys are sometimes carried out in conjunction with other methods. The method is extremely simple from a field operation point of view, the only equipment required being a sensitive voltmeter or potentiometer for measuring the potential difference between two electrodes in the ground.

The method depends upon the fact that a potential difference tends to develop between the top (negative) and bottom (positive) of an ore body. The reason for this is not clearly understood, but the generally accepted view is that oxidation of the upper part of the ore body is necessary for a potential difference to develop. Oxidation of sulphides in the aerated, oxidized zone produces salts in solution in contact with the ore body at a different electric potential to solutions in contact with the lower, unoxidized part of the ore body. According to the oxidation theory SP effects are dependent upon climatic conditions. In a theory proposed by Sato and Mooney (1960) oxidation is not necessary, but rather the SP effect is due to a redox potential (Eh) difference between substances in solution above the water table and those below. The sulphide ore body does not need to enter into the electrochemical action, but simply transfers electrons from reducing agents at depth to oxidizing ones at the top. If this theory is correct, SP effects should be expected to occur in a wide variety of climates. The theory also explains SP effects caused by substances such as graphite in which no oxidation occurs. Whatever explanations are correct SP effects are a near-surface phenomenon.

Field procedure
There are two basic methods used in carrying out SP surveys: one is to keep one electrode fixed and move the second electrode in progressive, equal steps along the survey line and the other method is to move both electrodes for each new measurement (Fig. 6.37). In the first method, which is the more commonly used, results are expressed in millivolts, which may be positive or negative, though negative anomalies are observed over sulphide-ore bodies. This does not mean that a negative anomaly indicates sulphides, but a positive anomaly is highly unlikely to be due to sulphides. In the second method the potential gradient is measured and results are expressed in millivolts/metre. In both cases non-polarizing, porous-pot electrodes must be used. The measurements are most commonly made using a

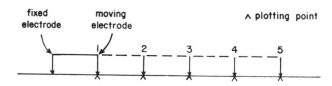

a) <u>Moving electrode method</u>. Numbers refer to successive electrode positions and plotting points.

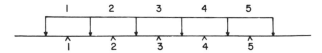

b) <u>Moving electrode array</u>. Numbers refer to positions of successive electrode pairs and plotting points.

FIG. 6.37.   Electrode arrays used in SP surveys.

sensitive voltmeter with a very high internal resistance, though it was common at one time to make SP measurements with a compensating potentiometer. SP effects are readily affected by a number of different factors and it is often very difficult to repeat survey results. It is also necessary for ore to be in contact with electrolyte for best effects and, for this reason, SP responses are generally weak in areas with a dry near-surface high resistivity layer. Atmospheric conditions and telluric currents can also adversely affect SP surveys. If it is necessary to dampen the ground around the electrodes, there may be quite a time lag before steady readings are obtained. In addition changes in ground elevation can cause spurious effects.

Interpretation of SP surveys is often difficult owing to stray interferences and fluctuating levels of response. In addition, certain rocks other than sulphide ore bodies may give strong SP effects. For example, SP anomalies of many hundreds of millivolts have been observed over graphitic shales. For these reasons, SP surveys are rarely diagnostic and must be used in conjunction with other evidence. Nevertheless, a number of shallow sulphide ore bodies have

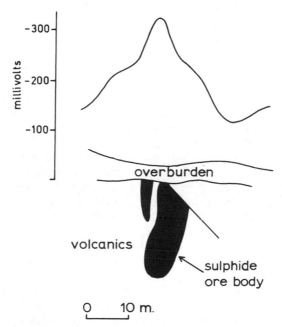

FIG. 6.38. SP profile over a massive sulphide ore body at the Temagami Mine, Ontario (after Bergey *et al*, 1957).

been discovered in various parts of the world by SP surveys. Figure 6.38 gives an example of a typical SP anomaly over a sulphide ore body.

6.7 EQUIPOTENTIAL (EP) SURVEYS

Equipotential methods were among the earliest types of electrical geophysical methods used, but, with the exception of the *mise-à-la-masse* variation, they have now become obsolete and have been replaced by IP and EM methods, which have greater depth penetration and lend themselves to more rigorous interpretation. Nevertheless, equipotential methods are briefly described here as they are of historical interest and are frequently mentioned in old reports. A number of mineral discoveries in various parts of the world can be attributed partly or wholly to EP surveys.

If a current is passed into homogeneous ground, flow lines will have a regular and symmetrical pattern. If there are any inhomogeneities, the pattern of flow lines will be disturbed. When the inhomogeneity is caused by a conducting sulphide ore body, the disturbance may be considerable and result in a pronounced distortion of the current flow line pattern. In practice, it is not possible to measure the current, but it is a simple matter to measure potential difference. Since current flow lines are at right angles to equipotential lines, the distorted current flow line pattern will be represented by a distorted pattern of the equipotential lines which is the basis of the equipotential method.

The field procedure is to use two stake electrodes or two line electrodes 600 m apart as the current electrodes and then trace the equipotential lines with two movable, copper-jacketed, pointed steel search electrodes between the fixed current electrodes. A voltage of 100–200 V is commonly used with a power output of 250–1000 W using an audio frequency of 100–500 Hz. Then, with an amplifier and headphone set connected to the search electrodes one search electrode is placed in the ground and the other moved about until no sound is heard in the headphones. When this occurs, both search electrodes are on a line of equipotential. Often it is not possible to obtain a clear null because of inductive effects which cause phase shifts. Several methods have been devised for overcoming this. An obvious one is to use a d.c. source and read the potential difference between the search electrodes on a voltmeter, but this is very cumbersome as it requires the use of non-polarizing electrodes and stronger currents. In addition self potentials can interfere. Another is to use low frequency a.c. sources of about 25 Hz together with a suitable a.c. voltmeter for reading the potential difference between

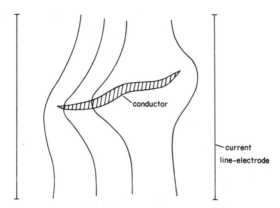

FIG. 6.39. Example of an EP survey showing distortion of equipotential lines by a conductor.

the search electrodes. The equipotential lines can then be interpolated between reading points. There is no quantitative theory for equipotential surveys and interpretation is purely qualitative. Figure 6.39 shows the type of response to be expected over shallow sulphide mineralization.

The *mise-à-la-masse* method, which is still occasionally used, depends on having a drill-hole intersection in mineralization. Then, with one current electrode placed at the depth of the mineralization in the drill hole (hence the name 'placed in the mass') and one on the surface, the equipotential lines are mapped. The method has been employed successfully in outlining probable extensions to mineralization discovered in drill holes.

6.8 MAGNETO-TELLURIC (MT) SURVEYS

This method, which was first proposed by Cagniard (1953), uses natural variations in the earth's magnetic field and the associated telluric currents, which are induced by them. Thus, the MT signal is composed of a magnetic field H and an electric field E. Cagniard (1953) has shown that the following relationship holds for a homogeneous earth:

$$H/E = \sqrt{(0.2T/\rho)} \qquad (1)$$

where H is in gammas, E is in millivolts/kilometre, T is the period in seconds and ρ the ground resistivity in ohm-metres. The ratio H/E can be computed for any number of horizontal layers, but the problem is much more complicated if the layers are not flat. The depth of penetration in kilometres is given by:

$$\frac{1}{2\pi}\sqrt{(10\rho T)} \tag{2}$$

Since frequencies of 1 Hz to 0·001 Hz or less are involved, theoretical depth penetrations of many tens and even several hundred kilometres are possible. In practice the ratio H/E is determined for a set of frequencies and interpretation consists of comparison with a set of H/E values computed for a set of frequencies and a given layered model. Alternatively, an apparent resistivity can be defined from equation (1) as

$$\rho_A = 0·2T(E/H)^2$$

Apparent resistivities can be determined for various frequencies and then plotted against the period on a graph. Theoretical plots calculated for various layered models can be compared to the field curves for interpretation. In addition to using the apparent resistivity the observed phase shift can be plotted against the period. In a homogeneous earth, the phase of E lags $\pi/4$ behind H and any departure from this indicates non-homogeneity. Computed curves of phase shift against period are also available for interpretation.

The field procedure consists of using two copper electrodes 250–500 m apart for measuring the telluric currents. Very large coils with highly permeable cores are needed to measure the magnetic fields at the low frequencies involved and very high gain, low-noise amplifiers have to be used to measure the low voltages induced in the coils. The equipment is bulky and productivity in the field is low. The method seems best suited to soundings of layered structures in sedimentary basins where the method has the ability to penetrate to considerable depths even in the presence of a high conductivity near-surface layer by the choice of an appropriately low frequency. Magneto-telluric surveys appear to be of limited value as a direct method in mineral exploration.

6.9 SEISMIC METHODS

Seismic prospecting is the most widely used of the geophysical methods in terms of cost and time and is very much a specialist's field. It is the backbone of the petroleum exploration industry and has become highly developed using many of the latest techniques in electronics and computer science. Seismic methods have had little application in mineral exploration and are mainly used as an indirect method, for example in locating fault structures, determining over-burden depths or tracing sedimentary beds for strata-bound deposits. As mineral exploration comes to depend more and more on indirect prospecting, seismic methods may play a more important role in the future.

The seismic method is based on tracing induced shock waves transmitted through the earth's crust. There are two types of waves propagated in elastic solids: *longitudinal* or *primary* (P) waves and *transverse* or *shear* (S) waves. In P waves, which travel faster than S waves, the direction of particle oscillation is the same as the direction of propagation and they are in fact ordinary sound waves. In S waves the particle oscillation is perpendicular to the direction of propagation and these waves cannot be transmitted through liquids. There are other waves known as Rayleigh and Love waves, which travel along surfaces. P waves are of most importance in applied seismology, but increased use is being made of S waves in the latest equipment. They are useful in engineering site test work, but are difficult to detect as they arrive after the P waves and have to be selected from a complicated signal. The velocity of P waves in a medium is given by:

$$v = \sqrt{\left(\frac{K + 4\eta/3}{\rho}\right)}$$

where K is the bulk modulus, η the shear modulus and ρ the density. Explosives are the most commonly employed sources of waves in seismic prospecting, but other sources include electric sparks, mechanical vibrators, dropped weights and hammer blows. Some seismic velocities are given in Table 6.4.

TABLE 6.4
SOME VELOCITIES OF P WAVES IN
METRES/SECOND

Air	330
Soil, sand	170–800
Water	1 450
Sandstone	2 000–2 800
Marl	2 000–3 000
Ice	3 670
Chalk	2 200–4 200
Shale	2 750–4 270
Limestone	1 000–4 500
Slates	3 200–5 000
Granite	4 000–5 500
Salt	4 500–7 000
Basic igneous rocks	5 500–8 000

Basic seismic equipment consists of electric detectors known as *geophones*, amplifiers and special recorders. The exact instant when the shock wave is initiated is recorded and all signals from the geophones are amplified, filtered and recorded on paper, photographic film or magnetic tape. The record, known as the *seismogram*, shows the precise arrival times of the various signals detected by the geophones. There are two basic methods in seismic prospecting, refraction and reflection.

The refraction method

This method depends upon the refraction of P waves at interfaces between solids with different seismic velocities in the same manner as light waves are refracted. An incident ray AB in Fig. 6.40(a) striking the interface between layers with velocities v_1 and v_2 with incident angle i_1 is refracted along path BC with angle i_2 where

$$v_1/v_2 = \sin i_1/\sin i_2$$

according to Snell's law. When the incident ray strikes the interface at

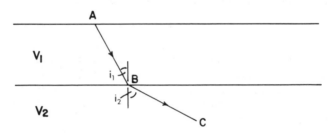

a) P wave being refracted at interface

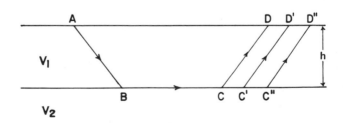

b) P wave striking interface at critical angle

FIG. 6.40.   Refraction of P waves at an interface between two horizontal layers with velocities v_1 and v_2 $(v_2 > v_1)$.

the *critical angle*, i_2 becomes 90° and the ray travels along the interface (Fig. 6.40(b)). As the ray travels along the interface, it sends out secondary waves which arrive back at the surface as shown, for example, at D, D' and D" in Fig. 6.40(b). In refraction seismic work the first arrival times after the shock instant at each geophone are plotted on a graph against their respective distances from the shock source. For geophones near the shock source, the first arrivals will be the surface wave, but, if a lower layer has a higher velocity, there is a *critical distance* at which a refracted wave from the lower layer overtakes the surface wave and reaches the geophone first. This is seen on the time–distance graph as a break in slope (Fig. 6.41) and it can be shown that the slopes represent the reciprocals of the velocities of the different layers. A number of different geophone arrangements is used depending on the targets. The simplest and most common is known as *profile shooting* and consists of spreading the geophones along a line. Other arrangements include fan shooting, arc shooting and triangle shooting.

Two-layer case
At the critical distance x_c in Fig. 6.40(b) the time taken for the ray to travel along the path AD is the same as the time to travel along ABCD. If t is the time to travel along AD and t_1, t_2 and t_3 the times to travel the distances AB, BC, and CD respectively,

$$t = t_1 + t_2 + t_3$$

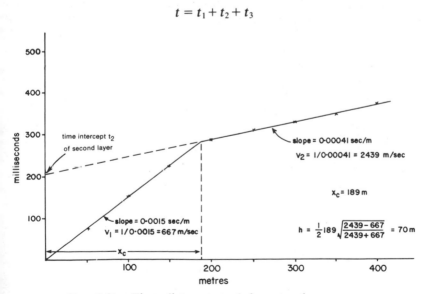

FIG. 6.41. Time–distance graph for a two-layer case.

At the critical angle $\sin i = v_1/v_2$, so

$$AB = CD = h/\cos i = h/\left(1 - \frac{v_1^2}{v_2^2}\right)^{1/2}$$

$$BC = x_c - 2h \tan i = x_c - \frac{2hv_1}{\sqrt{(v_2^2 - v_1^2)}}$$

Now,

$$t = \frac{AB}{v_i} + \frac{BC}{v_2} + \frac{CD}{v_1}$$

$$t = \frac{2h}{v_1} \frac{1}{\left(1 - \frac{v_1^2}{v_2^2}\right)^{1/2}} + \frac{x_c}{v_2} - \frac{2h}{v_2} \frac{v_1}{\sqrt{(v_2^2 - v_1^2)}} = \frac{x_c}{v_1}$$

$$\frac{x_c(v_2 - v_1)}{v_1 v_2} = \frac{2h(v_2^2 - v_1^2)}{v_1 v_2 \sqrt{(v_2^2 - v_1^2)}}$$

which reduces to

$$x_c = 2h \sqrt{\left(\frac{v_2 + v_1}{v_2 - v_1}\right)}$$

so that

$$h = \tfrac{1}{2}x_c \sqrt{\left(\frac{v_2 - v_1}{v_2 + v_1}\right)}$$

Detection of faults
For the faulted two-layer case a time–distance curve such as is shown in Fig. 6.42 is obtained when the shock source is on the upthrown side

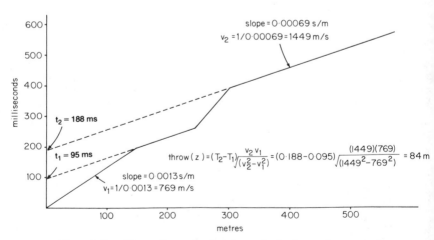

FIG. 6.42.   Time–distance plot for a faulted two-layer case.

of the block. In this case the fault throw (z) can be shown to be given by the expression

$$z = (t_1 - t_2) \frac{v_2 v_1}{\sqrt{(v_2^2 - v_1^2)}} \qquad \text{(Dobrin, 1960)}$$

Dipping beds

To determine the dips of an interface time–distance graphs are plotted for two directions of shooting, updip and downdip (Fig. 6.43). The angle of dip (α) can be shown to be given by:

$$\alpha = \tfrac{1}{2}(\sin^{-1} v_1 m_d - \sin^{-1} v_1 m_u) \qquad \text{(Dobrin, 1960)}$$

where m_d and m_u are the downdip and updip slopes of the time–distance lines respectively (Fig. 6.42). The perpendicular distance to the interface from the shot point is given by:

$$z_d = \frac{v_1 t_u}{2 \cos \alpha}$$

for downdip shooting and

$$z_u = \frac{v_1 t_d}{2 \cos \alpha}$$

for updip shooting (Dobrin, 1960).

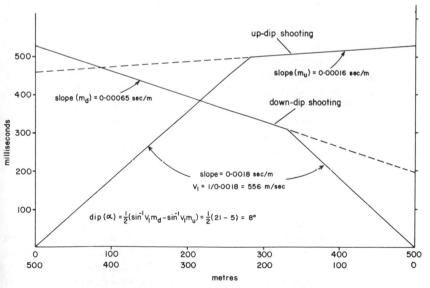

FIG. 6.43. Time–distance plots for dipping beds.

Three-layer case

For the three-layer case where $v_3 > v_2 > v_1$ the derivation of the formula is similar to the two-layer case, though it is somewhat more complicated. The distance (z_1) between the second and third layers can be shown to be

$$z_1 = \tfrac{1}{2}\left[t_3 - 2z_0 \frac{\sqrt{(v_3^2 - v_1^2)}}{v_3 v_1} \right] \frac{v_3 v_2}{\sqrt{(v_3^2 - v_2^2)}}$$

where z_0 is the depth to the second layer and t_3 the time intercept for the third layer. These cases can be extrapolated to apply to any number of layers as long as the velocity in each layer is higher than in the one just above.

The cases illustrated above are of the simplest types and in practice it is often very much more complicated than this. Refractive interfaces are rarely horizontal or smoothly inclined and interpretation can become complex. In addition velocities are often non-linear, i.e. they may show an increase with depth or very often may show a marked anisotropy with velocities parallel to bedding 10–15% greater than across the bedding. All examples discussed here are based on first arrivals, but more sophisticated equipment makes possible the detection and resolution of later arrivals to aid interpretation.

Disadvantages of the refraction method

If a lower layer has a velocity lower than a layer above, it cannot be detected by refraction. Such a layer is known as a *blind zone* and ignorance of the existence of such a layer will result in the computation of the depth to the next higher velocity layer being too high. If a particular layer is very thin and the velocity of the layer immediately below very much higher than the thin layer, the ray from the higher velocity layer may overtake the ray from the thin layer and arrive first. When this happens, the thin layer will not be detected and is known as a *hidden layer*. In cold regions of the world where the ground is frozen, refraction techniques may be nullified completely as the frozen surface layer may often have a higher velocity than the unfrozen ground below it. The refraction method is also at a disadvantage for deep exploration as the shot spreads become very long and shot energy requirements are high.

The reflection method

The reflection method has a number of advantages over the refraction method. Shot energy requirements are less for reflection than refraction, much greater subsurface detail can be obtained and depth penetration is greater. The reflection method is not affected by hidden layers and blind zones and much shorter geophone spreads are used which makes the method easier to deploy in the field. For these reasons the reflection method is used far more widely than the refraction method, particularly in petroleum exploration where depth penetration to 5000 or 6000 m might be required. In addition the detection and resolution of deep structures is almost as good as that of shallow ones which makes the reflection method unique among geophysical techniques.

As a seismic wave travels downwards, part of its energy will be reflected back up at any interface where there is a change in *acoustic impedance*, which is defined as the product of seismic velocity and density. Thus, it is not strictly necessary for the layers to have different seismic velocities to be detected by the reflection method. Figure 6.44 shows a reflection at an interface where v = the velocity of the top layer, t = the arrival time of the first reflection, h = the depth to the interface and x = the distance from shot to geophone, then

$$2\sqrt{\left(h^2 + \frac{x^2}{4}\right)} = vt$$

$$\sqrt{\left(h^2 + \frac{x^2}{4}\right)} = \frac{vt}{2}$$

$$h^2 = \frac{v^2 t^2}{4} - \frac{x^2}{4}$$

$$h = \tfrac{1}{2}\sqrt{(v^2 t^2 - x^2)} \tag{1}$$

Thus, the depth can be determined if the velocity of the top layer is known. This velocity can be determined from *velocity logs* measured

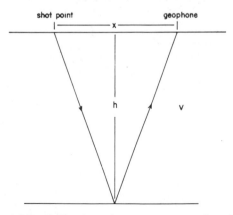

FIG. 6.44.   Reflection of a P wave at an interface.

in boreholes or it can be abstracted from the reflection data itself. It is normal practice to have a number of geophones spread out along a line from the shot point and therefore an arrival time for each geophone can be determined for the first reflection. Solving equation (1) for t^2, we get

$$t^2 = \frac{x^2}{v^2} + \frac{4h^2}{v^2} \qquad (2)$$

If t^2 is plotted against x^2, the slope of the graph will be equal to $1/v^2$. The depth can also be determined from the graph, since the time intercept is equal to $4h^2/v^2$. If several reflecting interfaces are present, a series of lines will be obtained on the graph for the different reflections. In each case the velocities determined from the slopes of the lines will be equal to the average velocities to the various reflecting horizons and not the interval velocities of the layers. The depths to the successive reflecting horizons are readily determined from the time intercepts. In the derivation of the formulae (1) and (2) the effects of refraction were ignored as they are very small, but corrections for refraction can be applied for accurate work.

In reflection seismic work the signals received at each geophone are recorded on a separate line or channel. On the old seismograms recorded mechanically or photographically on paper, up to 48 signal traces are recorded simultaneously. This produces a series of wavy lines (one for each geophone) across the seismogram and the reflected signals, which arrive after the surface wave, are recognized by a lining up of the crests and troughs. As the time delay increases with the geophone distance from the shot point, there is a step-out of the arrival times (Fig. 6.45). If the step-out is greater than normal, the dip is directed away from the shot point (Fig. 6.46) and, if the step-out is

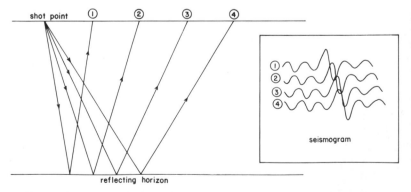

FIG. 6.45.   Step-out for reflecting rays from a horizontal interface.

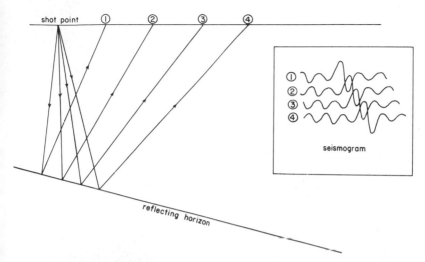

F<small>IG</small>. 6.46.   Step-out greater than normal indicates dip away from shot point.

less than normal, the dip is towards the shot point (Fig. 6.47). Arrivals
of refracted waves may also appear on the seismogram, but they can
be recognized by the much longer step-outs. Usually the maximum
shot-detector distance in reflection work is kept equal to or less than
the depth to the shallowest horizon, which ensures that all signals
received are from reflected rays. A simple graphical procedure for
determining the position of a dipping reflecting horizon is shown in
Fig. 6.48. Arcs equal to the product of the average velocity v and the
travel times t_1, t_2, t_3, etc. to geophones G_1, G_2, G_3, etc. are drawn with

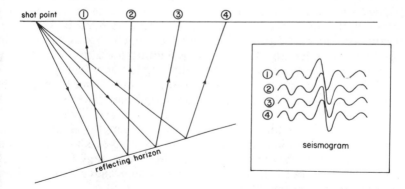

F<small>IG</small>. 6.47.   Step-out less than normal indicates dip towards shot point.

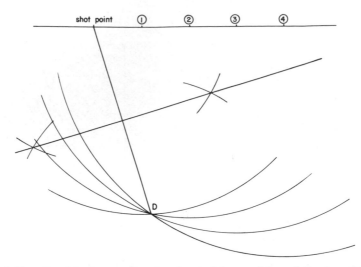

Fig. 6.48. Graphical procedure for determining position of dipping reflecting horizon.

centres at the respective geophone centres and intersect in a point D. The perpendicular bisector of a line joining the point D and the shot point is the reflecting horizon. Modern seismic recording is made on magnetic tapes and there are a number of techniques to transform the data to visual records, which are similar to the old seismograms, though the various methods of presentation used make them very much clearer. Figure 6.49 shows an example of a shallow reflection seismogram, with interpretation, from the North Sea.

If the seismic shooting has been carried out so that the reflections recorded on the last trace of the seismogram come from the same source as the first trace on the next seismogram, it is known as *continuous profiling*. The seismograms from each spread can then be placed edge to edge and the various reflecting horizons traced along the profile. One common method of shooting is known as *split-spread* (Fig. 6.50). The geophones between A and B and B and C record the reflections from a shot at B. Then the geophones between A and B are moved to between C and D to record the next shot at C. This 'leap-frogging' procedure continues for the length of the profile required. Since continuous profiling is expensive, shots are often spaced at distances greater than the geophone coverage for each shot. This leaves gaps, but it is adequate for reconnaissance work in areas where the reflection quality is good.

The field procedures adopted are often quite complicated and will

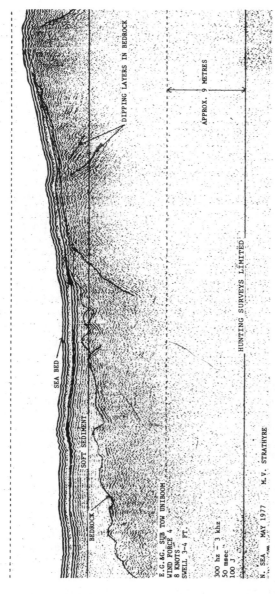

FIG. 6.49. Seismic record with interpretation from a shallow survey in the North Sea (courtesy of Hunting Surveys Ltd). Such surveys using electric sparks or electro-mechanical sources known as boomers have a limited penetration, but are being rapidly developed as a technique for determining layering and thickness of soft sediments on the sea floor. Using a repetitive source continuous profiling can be done in the same manner as a sonar survey. The surveys have particular application in civil engineering projects and in searches for sand and gravel deposits (see McQuillon and Ardus, 1977).

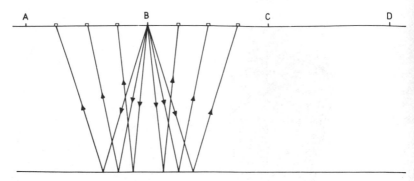

FIG. 6.50.   Split-spread shooting.

vary from area to area and with the experience of the geophysicist in charge. Geophones are often coupled together in groups so that a number of geophones are recorded as a single trace. This is done to minimize noise due to surface waves with a predominantly horizontal motion. The actual theory is quite complex, but basically the method depends on selecting the geometry of the layout so that the vertical component of the surface wave will be upwards at some geophones while it is downwards simultaneously at the other geophones. If these geophones are coupled together, the net output will be zero. When a reflected wave travelling almost vertically upwards arrives at the geophones, the motion will be in the same direction at the same time and a trace will be recorded by the coupled geophones. In addition to the problems involved in choosing geophone spreads and shooting procedures it is usually necessary to make corrections for elevation differences between the shot point and geophones and for the near-surface weathered zone. There are a number of methods for doing this (Dobrin, 1960).

Seismic records may be complicated by *multiple reflections*, which can be a problem if there is a horizon present with an unusually high acoustic impedance. These multiple-reflected rays result when a ray is reflected back and forth between two horizons several times before it reaches the surface. Since the distance travelled is much greater than the singly reflected ray from the same horizon, depths calculated from multiple reflections will appear to be much greater than they really are. The spurious data and distorted pictures caused by multiple reflections can be a real problem in some areas and can make reflection seismic work very difficult.

The account of seismic prospecting presented here has only been a very brief sketch of the basic principles. As already mentioned, seismic work has become very specialized and methods of treating and interpreting data may be very involved and complicated. For instance, filtering of the signals is often done to suppress or remove unwanted frequencies. Also, magnetically recorded signals may be played back and mixed in various combinations to produce results similar to those gained by using multiple geophone arrays. It is also normal practice to employ computers for the various calculations used in interpretation and synthetic seismograms can be computed for different reflecting horizons and compared to actual ones determined in the field.

In addition to using single explosive charges as wave sources, it has become fairly common practice to use repetitive shock sources produced mechanically, hydraulically or with controlled explosives. One such technique, developed by the Continental Oil Company, is known as *Vibroseis*®, in which a truck-mounted vibrator makes contact with the ground and generates repetitive controlled-frequency wave trains. Another technique, developed by the Sinclair Oil Company, is known as *Dinoseis*®. With this method, mixtures of propane and oxygen are exploded repeatedly in a cylinder with a bottom plate free to move and in contact with the ground.

The hammer seismograph

This is a type of seismic equipment, for shallow surveys, in which a sledge hammer is used to provide the shock wave by striking a steel plate on the ground. A trigger on the hammer head is connected to the recording instrument by a cable so that the instant of impact starts the electronic timers. In its simplest form, it is used for refraction surveys by measuring precisely in milliseconds the first arrivals at the geophone or geophones. With the latest multichannel equipment, reflection surveys can also be undertaken. Shallow penetration in the range of 30–50 m is possible and the equipment is most commonly used for determining overburden depths in engineering site investigations.

6.10 RADIOMETRIC SURVEYING

Ever since the discovery of radioactivity by Becquerel in 1896, it has been known that a number of materials spontaneously emit invisible high-energy radiation. This phenomenon is caused by the transformation of atomic nuclei into those of the other elements. Although numerous types of transformation take place in nature ranging from atomic fusion in the sun and stars to nuclear disintegration caused by cosmic ray bombardment of the atmosphere, radioactivity in rocks and minerals is due to three principal transformations: (1) emission of alpha (α) particles, (2) emission of beta (β) particles and (3) electron capture. An α-particle is equivalent to a helium nucleus which consists of two protons and two neutrons and its loss by an atomic nucleus results in the reduction of atomic number by two and atomic weight by four. A β-particle is equivalent to a nuclear electron and its loss by an atomic nucleus increases the atomic number by one. In the case of electron capture by an atomic nucleus the atomic number is reduced by one. The transformation due to emission of a β-particle or electron capture is usually accompanied by the emission of electromagnetic radiation of extremely short wavelength known as a gamma (γ) radiation. In some cases γ-radiation also accompanies the release of an α-particle. The energy of this radiation is measured in millions of electron volts (MeV). An electron volt is the kinetic energy acquired by an electron falling through a potential difference of 1 V and is equivalent to $1 \cdot 6 \times 10^{-12}$ erg. α-particles commonly have energies of several MeV, β-particles have a continuous spectrum of energy up to several MeV and γ-rays usually have energies in the region of 1 MeV.

The disintegration of a radioactive element takes place at a constant rate with the number of atoms disintegrating per unit time being proportional to the number of atoms present.

$$\frac{-dN}{dt} = \lambda N \tag{1}$$

where λ is the decay constant.
Integrating (1) gives:

$$-\lambda t = \log_e(N/N_0)$$

where N is the number of atoms at time t and N_0 the original number present at $t = 0$. The time taken for half the atoms to transform ($N = N_0/2$) is known as the *half-life* and is equal to:

$$(\log_e 2)/\lambda = 0 \cdot 693/\lambda$$

Half-lives vary with different elements and isotopes from fractions of a second to billions (10^9) of years. Examples of some half-lives are uranium-238 $4 \cdot 51 \times 10^9$ years; radium-226 1622 years; lead-210 20 years; bismuth-214 $19 \cdot 7$ min and polonium-216 $0 \cdot 16$ s. In a radioactive decay series the element undergoing transformation is known as the *parent* element and the product is known as the *daughter* element. Sometimes an element can undergo two different transformations to form two different daughter elements. An example of this is potassium-40 of which 89% transforms to calcium-40 by emission of a β-particle with decay constant λ_β and 11% transforms to argon-40 by electron capture with decay constant λ_e. The total decay constant λ is the sum of the decay constants, λ_β and λ_e, and the ratios λ_β/λ and

$_e/\lambda$ are known as the *branching ratios*. The rates of decay of certain elements such as uranium, rubidium-87 and potassium-40 are used to measure absolute ages of rocks and minerals and form the basis of the science of *geochronology*.

There are over 60 naturally occurring radioactive isotopes and another 1000 or more have been created artificially. The majority of these naturally occurring radioactive isotopes are part of the uranium-238, uranium-235 and thorium-232 decay series (Table 6.5). There are also a number of singly occurring radioactive isotopes (Table 6.6) in addition to the three decay series. From the radiometric surveying point of view the most important ones are potassium-40, the thorium series and the uranium-238 series.

The unit of radioactivity is the *curie* (Ci) which is the amount of radiation emitted by 1 g of radium and is equivalent to 3.7×10^{10} radioactive disintegrations/second. Since this unit is very large, units of *millicuries* or *microcuries* are more commonly used. Radiation intensity is measured in counts/second, counts/minute or in milliroentgens/hour (mR/h). The *roentgen* is the quantity of γ- or X-radiation which produces 2.083×10^9 ion pairs/cm^3 of air at STP. Instruments which measure radiation intensity can be calibrated with known radioactive sources. For example, 1 mg of radium enclosed in a platinum capsule 0.5 mm thick emits a radiation flux of 0.84 mR/h at a distance of 1 m.

Although α- and β-particles carry considerable energy, their penetration or range is very limited. α-particles can only travel a few centimetres through air and are stopped by a few sheets of paper. β-particles have a considerably greater range, and are stopped by a thin sheet of metal or a few centimetres of sand. The range of γ-rays is very much greater, but their penetration of matter depends on the density and falls off exponentially. As a rough guide 50% of γ-radiation is absorbed by 6 cm of rock, 10 cm of soil or 130 m of air and 90% is absorbed by 20 cm of rock, 35 cm of soil or 440 m of air. Thus, radiometric surveying essentially depends upon the detection of γ-radiation.

Detectors

There are three basic types of γ-ray detectors: (1) gas-filled tube counters such as the Geiger–Müller counter, (2) scintillation counters and (3) semi-conductor detectors.

Geiger–Müller counters counters consist of a thin metal cylindrical cathode with a wire anode along the central axis all enclosed in a glass tube. The tube is filled with argon at low pressure and a small amount of a quenching agent such as ethyl alcohol or chlorine is added to prevent 'cascading'. A potential difference just below that

TABLE 6.5
THE URANIUM AND THORIUM RADIOACTIVE DECAY SERIES

Uranium-238	Uranium-235	Thorium-232

^{238}U
α
↓
^{234}Th
β
↓
^{234}Pa
β
↓
^{234}U
α
↓
^{230}Th
α
↓
^{226}Ra
α
↓
^{222}Rn
α
↓
^{218}Po
α   β
↙   ↘
^{214}Pb ^{218}At
β   α
↘   ↙
^{214}Bi
β   α
↙ ↓
^{214}Po
α
↓
^{210}Tl
β
↓
^{210}Pb
β
↓
^{210}Bi
α   β
↙   ↘
^{206}Tl ^{210}Po
β   α
↘ ↙
^{206}Pb

^{235}U
α
↓
^{231}Th
β
↓
^{231}Pa
α
↓
^{227}Ac
α   β
↙   ↘
^{223}Fr ^{227}Th
β   α
↘ ↙
^{223}Ra
α
↓
^{219}Rn
α
↓
^{215}Po
α   β
↙ ↘
^{211}Pb ^{215}At
β   α
↘ ↙
^{211}Bi
β   α
↙ ↓
^{211}Po
α
↓
^{207}Tl
β
↓
^{207}Pb

^{232}Th
α
↓
^{228}Ra
β
↓
^{228}Ac
β
↓
^{228}Th
α
↓
^{224}Ra
α
↓
^{220}Rn
α
↓
^{216}Po
α   β
↙ ↘
^{212}Pb ^{216}At
β   α
↘ ↙
^{212}Bi
β   α
↙ ↘
^{212}Po ^{208}Tl
α   β
↘ ↙
^{208}Pb

TABLE 6.6
SOME NATURALLY OCCURRING SINGLE RADIOACTIVE ISOTOPES

Parent	Type of decay	Daughter	Half-life (years)
bismuth-209	α	thallium-205	2.7×10^{17}
carbon-14	β	nitrogen-14	5 568
lanthanum-138	β	caesium-138	7×10^{10}
lutectium-176	33% β	hafnium-176	2.4×10^{10}
	67% e capture	ytterbium-176	
potassium-40	89% β	calcium-40	1.33×10^{9}
	11% e capture	argon-40	
rubidium-87	β	strontium-87	6.15×10^{10}
tellurium-130	double β	xenon-130	$\sim 10^{21}$

required to discharge the tube (usually about 1000 V) is applied across the anode and cathode. Any of the gas atoms struck by γ-rays entering the tube are ionized causing the tube to discharge. These discharges are monitored by suitable electronic circuitry and relayed to earphones where they are heard as a series of clicks. In addition Geiger counters often have meters which indicate the radiation intensity in milliroentgens/hour or counts/second and in such cases they are commonly referred to as *ratemeters*. γ-rays are only weakly ionizing and, as only 1% of γ- rays entering the tube are counted, Geiger counters are rather inefficient γ-ray detectors. On the other hand charged particles have strong ionizing power and high energy β-particles which can penetrate the walls of the tube give a good response. However, the detection of β-particles is of little importance in radiometric surveying and, since scintillation counters are much more efficient γ-ray detectors, ratemeters are rarely used today except for rough reconnaissance work.

Crystals of certain substances such as caesium fluoride, cadmium tungstate, anthracene and sodium iodide emit small flashes of light when bombarded by γ-radiation and this phenomenon is made use of in scintillation counters. The most commonly used phosphor is sodium iodide with a minute amount of thallium added. In a typical instrument a large crystal of NaI(Tl) is positioned against a photomultiplier tube which is connected to a recording circuit and meter. The number of flashes produced by the phosphor is roughly proportional to the incident radiation and the meter is calibrated to read in milliroentgens/hour or counts/second. Small portable scintillation counters similar in appearance to ratemeters are available.

It has been found that Ge(Li) semiconductors are excellent γ-ray detectors with a very good resolution up to 10 times that of NaI(Tl) scintillometers. The main disadvantage is their small volume and low efficiency for high energy γ-rays. They have not been developed as field instruments because of the necessity for the Ge(Li) semiconductor to be cooled by liquid nitrogen. However, they are useful laboratory instruments for the detailed study of complex γ- or X-ray spectra.

γ-ray spectrometers

γ-rays from different sources have quite a wide spectrum of energy from fractions of MeV to several MeV. For instance, the main source of γ-radiation in the ^{238}U decay series is emitted by ^{214}Bi with an energy peak at 1·76 MeV and the main source of γ-radiation in the ^{232}Th decay series is emitted by ^{208}Tl with an energy peak at 2·62 MeV (Fig. 6.51). Instruments which can distinguish these different energy peaks are known as gamma-ray spectrometers and they make it possible to identify the radiation source (e.g. K, U or Th). In a

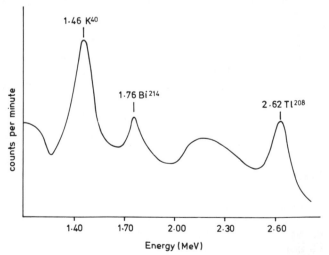

FIG. 6.51. ^{40}K, ^{214}Bi(U) and ^{208}Tl(Th) peaks showing up on part of the γ-ray spectrum of a rock sample.

scintillation counter the phosphor crystal and photomultiplier tube produce an output pulse whose height is proportional to the intensity of the light flash. The intensity of the light flash is in turn proportional to the energy expended in the phosphor by an incident γ- photon. These various pulses are monitored by a multichannel pulse height analyzer which is able to discriminate various energy bands through fairly complicated electronic circuitry. For example, a typical field γ-ray spectrometer would distinguish energy bands at 1·37–1·57, 1·66–1·86 and 2·41–2·81 MeV corresponding to K, U and Th respectively. In addition field γ-ray spectrometers also measure a total radiation energy band, usually from 0·41–2·81 MeV. The measurement is complicated by a number of factors the main one being due to *Compton scattering* in the crystal. When a γ-photon strikes an orbital electron, it may lose all its energy (photoelectric effect) or it may only expend part of its energy and glance off at an angle with reduced energy (Compton scattering). The scattered photons may lose a part or all of their energy by repeated Compton scattering or eventual photoelectric absorption. The result of this Compton scattering is that some of the γ- photons with initial high energies may be recorded at a lower energy level if and when they are eventually photoelectrically absorbed. The net effect from a practical point of view is that radiation recorded in the U energy band may be partly derived from Th, and radiation in the K energy band may be partly derived from both Th and U. For accurate determinations corrections have to be applied in the form:

$$U_c = U_u - ATh_u \qquad \text{for uranium}$$

and

$$K_c = K_u - BU_u - CTh_u \quad \text{for potassium}$$

where U_c and K_c are the corrected U and K values, U_u, Th_u and K_u are the uncorrected U, Th and K values and A, B and C are correction coefficients for the Compton scattering. In the case of Th no correction is required since Th radiation is in the highest energy band. The Compton scattering is peculiar to each detection system and known radiation sources have to be used to determine the correction coefficients for each instrument. If no correction for Compton scattering is applied, readings in the U and K channels may be too high by 30% or more. In addition to Compton scattering in the detector crystal, the energy levels of γ-rays are also reduced by Compton scattering in the air and this effect shows an increase with altitude. In uranium exploration the Compton scattering is often referred to as the *uranium stripping ratio* which is defined as the ratio of counts detected in the U window from a Th source to those detected in the Th window.

Radioactive equilibrium

A decay series is said to be in equilibrium when the rate of decay of the daughter elements is as rapid as their rate of formation from their respective parent elements. This relationship may be expressed in the form:

$$\lambda_1 N_1 = \lambda_2 N_2 = \lambda_3 N_3 = \lambda_4 N_4 = \cdots$$

where 2 is the daughter of 1, 3 is the daughter of 2, 4 is the daughter of 3 and so on. When a series is in equilibrium, the only variables are the diminishing parent and increasing stable end product. If the half-life of the parent is shorter than the daughters', a state of equilibrium cannot be attained. Only when the half-life of the parent is very much longer than the daughters', such as in the uranium decay series, is equilibrium possible. If the half-lives of the parent and daughter are similar, only a transient equilibrium will be reached. In actual fact absolute equilibrium is never attained in any series, though the variances may be very small.

Radioactive equilibrium is extremely important in the evaluation of uranium prospects from radiometric measurements. If the decay series is in equilibrium, γ-ray logging of boreholes, pits, trenches and outcrops can be used to make a fairly accurate evaluation of the prospect. If the series is not in equilibrium, however, an evaluation based on γ-ray logging will be incorrect and may err very much on the optimistic side. When the extent of the disequilibrium for a particular deposit is known and if it is constant over the deposit, a correction can be applied so that γ-ray logging can be used to evaluate the deposit. The main cause of disequilibrium is chemical fractionation of the various members of the series. For example, uranium is more readily oxidized than thorium and has greater mobility. This may result in a depletion of uranium with respect to the various daughter elements and, since γ-rays emanate from daughter elements such as ^{214}Bi and ^{226}Ra, radiometric measurements will indicate a higher uranium content than actually present. Disequilibrium is also caused by the escape of radon gas, but this is not as important a factor as chemical fractionation.

Field procedures

In addition to their use in uranium exploration, radiometric surveys may be carried out to assist with geological mapping (e.g. distinguishing granites, tracing slightly radioactive beds) or to search for deposits of other radioactive minerals such as monazite and pyrochlore.

Ground surveys are carried out by holding a scintillometer or ratemeter up to 1 m above the ground and taking a series of readings along traverse lines. The readings can be taken at fixed station intervals along the traverses, but it is always advisable to keep the instrument switched on during a traverse in case there are any strong localized sources between stations. Instruments equipped with earphones or an audio signal are useful for this type of work as one can walk along without having to keep a constant watch on the meter. If there is any increase in the signal strength, a reading or readings can be recorded. In addition to carrying out surveys on foot, car-borne surveys can be conducted by mounting a detector on top of a car, provided that observations are not made over metalled roads. Ground radiometric surveys have been carried out in Scandinavia during the winter by mounting detectors on snow scooters.

Significant anomalies are considered as those which are at least two or three times greater than the background count rate which varies from area to area and has to be known before making any interpretations. Background is mainly due to cosmic rays, potassium-40 and minute amounts of thorium and uranium which occur in a wide range of rocks. Cosmic rays consist of atomic nuclei and protons from outer space with very high energies of the order of several thousand MeV. These particles collide with atoms in the atmosphere producing nuclear disintegrations which are accompanied by γ-radiation. Background can be determined by taking a number of readings just outside the area of interest or over ground in the survey area which is known not to contain any radioactive minerals or significant quantities of radioactive minerals. In the interpretation of radiometric surveys it must always be remembered that the magnitude of an anomaly is not necessarily a good guide to the amount of radioactive material present. For example, outcropping uranium mineralization can give rise to very strong anomalies of 0·5 mR/h or more, whereas the same mineralization buried under 1 m of overburden may only produce an anomaly of 0·05 mR/h or less.

Airborne surveys

Airborne radiometric surveys are superior to ground surveys for regional reconnaissance work because large areas can be covered quickly and relatively cheaply, but there is always the danger of missing important targets because of uneven coverage. There are numerous documented cases of good ground anomalies that were not picked up by an airborne survey owing to too wide a flight line spacing or rugged terrain which prevented the aircraft from maintaining a satisfactory ground clearance. In addition ground surveys have much greater resolution. For example, a uranium deposit consisting of a number of veins might produce a single large airborne anomaly, whereas a ground survey with a high density of readings will define the individual vein sources as a series of separate anomalies.

Nowadays airborne radiometric surveys are carried out with γ-ray spectrometers which monitor K, Th, U and total count radiation. The survey is flown at a height of 50–150 m, though it may be necessary to fly slightly higher than this in areas of rugged terrain with consequent loss of effectiveness. The altitude at which a survey can be flown depends upon the sensitivity of the instrument being used. Before the development of modern γ-ray spectrometers, it was necessary to fly at altitudes of 50–60 m to achieve meaningful results with the scintillometers and ratemeters available. The sensitivity of modern instruments is governed by the volume of the crystal or crystals used and with crystal volumes of 4500 cm³ and more, it is possible to fly at altitudes of 200 m. Since altitude is critical in assessing anomalies, the aircraft is fitted with a radar altimeter which 'contours' the ground to $\pm$ 10 m or better. Background determinations are made at high altitude or by flying over a lake and threshold values are usually taken at two to three times background. Slow aircraft are preferable as detection of the smallest radioactive zones depends upon the length of time the detector is exposed to a particular area. For total coverage the flight-line spacing should be $1\frac{1}{2}$ times the altitude, but this is rarely achieved in regional reconnaissance surveys owing to the cost, and line spacings of four times the altitude or greater are commonly used.

6.11   GEOTHERMAL METHODS

The average heat flow from the earth is approximately $1\cdot3 \times 10^{-6}$ cal/cm² s, but it can show wide differences owing to volcanic activity and the presence of fissures and cracks which permit the convective transfer of heat from depth by groundwaters. The average solar heat flow reaching the ground is approximately $1\cdot2 \times 10^{-2}$ cal/cm² s which is several orders of magnitude greater than the heat flow from the earth and it is clear that surface measurements of the earth's heat flow will be completely swamped by solar radiation. Owing to the extremely low thermal conductivity of soils and rocks, however, the effects of solar heat flow are barely noticeable at shallow depths below the surface. The solar heat flow can be considered to have diurnal and seasonal components. The diurnal effects can be detected to depths of about 1 m and, although the seasonal effects penetrate to depths of 20 m, they have a very long period with negligible day-to-day changes. Thus, temperature measurements at depths of 1–2 m below the surface can be used as a geophysical method for searching for anomalous heat flow patterns which may have some relation to subsurface structures.

The basic field procedure consists of drilling a series of short auger holes up to 2 m deep and placing temperature sensing devices at the bottom of the holes. The usual probe is made of plastic and has a temperature dependent platinum resistance sensor at the end which can measure temperature to an accuracy of 0·025°C. After inserting the probe, a time lapse of 1 h is normally allowed for conditions to stabilize. Measurements are preferred in dry ground, but, if conditions are wet, it is desirable that the water table is as high as possible. Measurements in the transition zone between saturated and dry ground tend to give unreliable readings. In addition, the presence of vegetation affects the temperature readings and corrections need to be applied in crossing from forest to open country. In the tropics readings in forest can be as much as 2–3°C higher than readings in open savannah. Readings can be compared from one season to another provided base readings are taken and a correction made for the seasonal variation.

The method has been used with some success in locating faults and salt domes (Poley and Steveninck, 1970). Thermal anomalies (both positive and negative) of the order of 1°C have been observed over some fault zones and salt domes may produce positive thermal anomalies of 1–2°C. Geothermal surveys have also been employed in prospecting for geothermal energy and groundwater. Since the oxidation of sulphides is an exothermic reaction, it is likely that thermal anomalies may occur over zones of mineralization and it is feasible that geothermal methods might have some application in mineral exploration. Very little work has been done in this field, however, and there is insufficient data available to assess the efficacy of the method.

6.12 WELL-LOGGING TECHNIQUES

Most of the geophysical techniques have been adapted for measurements in boreholes and there is a wide range of instruments available. The first borehole measurements were resistivity and SP surveys undertaken in the 1920's and known as electric logging. The petroleum industry made considerable use of such surveys and soon extended the methods to include seismic logging, neutron logging and γ-ray logging. In more recent years techniques which have become available include IP measurements, $Eh°$ and pH measurements, temperature measurements, borehole magnetometers and even borehole gravimeters.

Resistivity log
With this method a probe may be fitted with two current and two potential electrodes which press against the borehole sides as the probe is lowered down the hole, or one current electrode can be

placed on the surface at some distance from the borehole in which case one current electrode and two potential electrodes are used on the probe. The method is useful for correlating between holes as different stratigraphic units often show marked variations in resistivity.

SP log
SP logs are generally undertaken in conjunction with resistivity logs and are included under the term 'electric logging'. The presence of a self potential is usually indicative of permeable porous strata. In addition various economic minerals may give rise to self potentials. There is no way of predicting the type of SP profile that will be obtained, but it is a very simple technique and can sometimes give results which are useful in correlating from hole to hole.

Velocity log
Ordinary velocity logs are carried out by lowering a seismometer down a hole and firing a shot close to the surface. By measuring the travel times for various depths of the seismometer, the average velocities of material between different seismometer depths can be calculated.

Continuous velocity log (CVL) or sonic log
With this method seismic velocities are measured over very small intervals down the hole by using a special probe with a transmitter for producing an acoustic pulse and a receiver at a distance of 1–2 m from the transmitter. To eliminate the effects of travel time in the borehole fluid, two receivers are usually fitted to the probe. The conventional CVL measures travel times of compressional waves in microseconds/metre or metres/second and has become one of the routine tools of applied seismology. By integrating the logs with respect to depth, the total travel time between any two points in the hole can be obtained which allows reflection times to be converted to depths.

Neutron log
Hydrogen atoms are particularly efficient at capturing neutrons according to the reaction:

$$n + {}_{1}^{1}H \longrightarrow {}_{1}^{2}H + \gamma$$

Thus a neutron source can be used to measure the relative concentration of hydrogen atoms. In one method, known as the neutron-gamma method, a neutron source such as americium–beryllium is fitted to a probe in addition to a γ-ray detector some 40–50 cm away.

The γ-ray count is then a measure of the neutrons being captured, which in turn is related to the amount of hydrogen present in the rocks. This in turn closely reflects the amount of water or oil present, both of which have an abundance of hydrogen nuclei. Instead of measuring the γ-radiation to determine the amount of neutrons captured, the degree of scattering of neutrons by hydrogen nuclei can be measured by counting neutrons at the detector. This is known as the neutron–neutron method and is carried out by using a boron trifluoride counter as the detector. Neutron logs employed together with other logging techniques can be very useful in correlation studies.

γ-ray log

γ-ray logs, which measure the intensity of natural γ-radiation in the hole, are used principally in uranium exploration, but have wide application in ordinary correlation work. For example, argillaceous sediments, particularly older ones, tend to have high background γ-ray count rates, whereas coal, anhydrite and salt have low activities and dolomites, limestones and sandstones have intermediate ones. Of course, there is great variation depending on the local geology, and some sandstones, for instance, have very high background count rates.

Density log

In this method the attenuation of γ-radiation by the wall rocks is measured. A γ-ray source such as Co-60 or Cs-137 is mounted on a probe together with a detector 10–15 cm away. The count rate received by the detector is inversely proportional to the electron density (n) which is given by:

$$n = N\sigma Z/A$$

where N is Avogadro's number, σ is the density in g/cm^3, Z is the atomic number and A the atomic weight. Thus, the count rate is inversely proportional to the density in any system where Z and A are fixed. In practice the instrument is usually calibrated to give the density of a calcite–water system. A correction for Z/A is then required to give correct densities of other systems.

Caliper log

This is simply an instrument which measures variations in the borehole diameter. It is widely used in the petroleum industry where variations in the diameters of uncored holes can provide useful data in correlation studies. The softer and more readily eroded rock formations may result in a slightly larger diameter hole than the harder rock formations. Used in conjunction with other logging methods the caliper log is a useful tool in correlation work.

REFERENCES AND BIBLIOGRAPHY

Adams, J. A. S. and Gasparini, P. (1970). *Gamma-ray Spectrometry of Rocks*, Elsevier Publishing Co., Amsterdam, 295 pp.

Bergey, W. R., Clark, A. R., Frantz, J. C., Keevil, N. B. and Gordon Smith, F. (1957). Discovery of copper–nickel ore bodies at the Temagami Mine, Ontario. In *Methods and Case Histories in Mining Geophysics*, Mercury Press Co., New York, 168–175.

Bhattacharya, P. K. and Patra, H. P. (1968). *Direct Current Geoelectric Sounding*, Elsevier Publishing Co., Amsterdam, 135 pp.

Broughton-Edge, A. B. and Laby, T. H. (eds.) (1931). *The Principles and Practice of Geophysical Prospecting*, CUP, 372 pp.

Cagniard, L. (1953). Basic theory of the magnetotelluric method, *Geophysics*, **18**, 605–635.

Collet, L. S. (1965). The induced polarization and INPUT methods in geophysical exploration, *Geol. Surv. Canada*, Paper 65–6, 84–100.

La Compagnie Générale de Geophysique, (1955). Abaques de sondage electrique, *Geophysical Prospecting*, **3**, Supplement 3, 1–7, plus charts.

Cook, K. L. (1950). Quantitative interpretation of magnetic anomalies over veins, *Geophysics*, **15**, 667–686.

Crone, J. D. (1975). Pulse electromagnetic—PEM—ground method and equipment as applied in mineral exploration, paper presented at AIME meeting, New York, Feb. 16–20, 1975.

Darnley, A. G. (1970). Airborne gamma-ray spectrometry, *C.I.M. Bull.*, **63** (694), 145–154.

De Witte, L. (1948). A new method for interpretation of self-potential field data, *Geophysics*, **13**, 600–608.

Dobrin, M. B., (1960). *Introduction to Geophysical Prospecting*, McGraw-Hill Book Co., New York, 446 pp.

European Association of Exploration Geophysicists, (1958). *Surveys in Mining, Hydrological and Engineering Projects*, E. J. Brill, Leiden, 270 pp.

Faul, H. (ed.) (1954). *Nuclear Geology*, John Wiley & Sons, New York, 414 pp.

Geological Survey of Canada, (1970). Mining and Groundwater Geophysics/1967, *Economic Geology Report No. 26.*

Grasty, R. L. (1975). Uranium measurement by airborne gamma-ray spectrometry. *Geophysics*, **40**, 503–519.

Green, R. (1974). The seismic refraction method—a review, *Geoexploration*, **12**, 259–284.

Griffiths, D. H. and King, R. F. (1969). *Applied Geophysics for Engineers*, Pergamon Press, London, 223 pp.

Haalck, H. (ed.) (1953). *Lehrbuch des angewantes Geophysiks*, Teil 1, Berlin.

Hammer, S. (1939). Terrain corrections for gravimeter stations, *Geophysics*, **4**, 184–194.

Hammer, S. (1945). Estimating ore masses in gravity prospecting, *Geophysics*, **10**, 50–62.

Heiland, C. (1940). *Geophysical Exploration*, Prentice-Hall, New York, 1013 pp.

Heiskanen, W. A. and Vening Meinesz, F. A. (1958). *The Earth and its Gravity Field*, McGraw-Hill Book Co., New York, 470 pp.

Hinze, W. J. (1960). The gravity method in iron ore exploration, *Econ. Geol.*, **55**, 465–484.

Jacobs, J. A., Russell, R. D. and Wilson, J. Tuzo (1959). *Physics and Geology*, McGraw-Hill Book Co., New York, 424 pp.

Jacoby, W. (1971). Zur Berechnung der Schwerewirkung beleibig geformter dreidimensionaler Massen mit digitalen Rechenmachinen, *Zeit. f. Geophys.*, **33**, 163–167.

Jakovsky, J. J. (1957). *Exploration Geophysics*, Trija Publishing Co., Newport Beach, Calif., 1195 pp.

Jewell, T. O. and Ward, S. H. (1963). The influence of conductivity upon audio-frequency magnetic fields, *Geophysics*, **28**, 201–221.

Jung, K. (1937). Direkte Methoden zur Bestimmung von Störungsmassen aus Anomalien der Schwereintensität, *Zeit. f. Geophys.*, **13**, 45–67.

Kappelmeyer, O. and Haenel, R. (1974). Geothermics with special reference to application, *Geoexploration Monographs*, Series 1, No. 4, Gebruder Borntraeger, Berlin, 238 pp.

Kunetz, Geza (1966). *Principles of Direct Current Resistivity Prospecting*, Gebruder Borntraeger, Berlin, 103 pp.

Lasfarques, P. (1957). *Prospection Electrique*, Masson & Co., Paris, 290 pp.

Malmqvist, D. (1958). The geophysical case history of the Kankberg ore deposit in the Skellefte District, North Sweden. In *Surveys in Mining, Hydrological and Engineering Projects*, E. J. Brill, Leiden, 32–54.

McNeill, J. D. and Barringer, A. R. (1970). The airborne RADIOPHASE system—A review of experience: Presented at the 72nd annual meeting of CIMM, Toronto, April 22.

McNeill, J. D., Jagodits, F. L. and Middleton, R. S. (1973). Theory and application of the *E*-phase airborne resistivity method: *Proc. Sympos. Explor. Electromag. Meth.*, Univ. of Toronto, Toronto.

McQuillon, R. and Ardus, D. A. (1977). *Exploring the Geology of Shelf Seas*, Graham and Trotman Ltd., London, 234 pp.

Mooney, H. M. and Wetzel, W. W. (1956). *The potentials about a point electrode and apparent resistivity curves for a two-, three- and four-layer earth*, University of Minnesota Press, Minneapolis, 146 pp. and 243 sheets of reference curves.

Nagata, T. (1961). *Rock Magnetism*, Maruzen & Co., Tokyo, 350 pp.

Nagy, D. (1966). The gravitational attraction of a right rectangular prism, *Geophysics*, **25**, 203.

Nettleton, L. L. (1940). *Geophysical Prospecting for Oil*, McGraw-Hill Book Co., New York, 444 pp.

Nettleton, L. L. (1942). Gravity and magnetic calculations, *Geophysics*, **7**, 293–310.

Nininger, R. D. (1956). *Exploration for Nuclear Raw Materials*, Macmillan & Co., London, 293 pp.

Paal, G. (1965). Ore prospecting based on VLF-radio signals, *Geoexploration*, **3**, 139–147.

Palacky, G. J. and West, G. F. (1973). Quantitative interpretation of INPUT AEM measurements, *Geophysics*, **38**, 1145–1158.

Palacky, G. J. (1976). Use of decay patterns for the classification of anomalies in time-domain AEM measurements, *Geophysics*, **41**, 1031–1041.

Parasnis, D. S. (1962). *Principles of Applied Geophysics*, Methuen & Co., London, 176 pp.

Parasnis, D. S. (1966). *Mining Geophysics*, Elsevier Publishing Co., Amsterdam, 356 pp.

Pemberton, R. H. (1962). Airborne electromagnetics in review, *Geophysics*, **27**, 695–713.

Poley, J. Ph. and van Steveninck, J. (1970). Geothermal prospecting—delineation of shallow salt domes and surface faults by temperature measurements at a depth of approximately 2 m. *Geophysical Prospecting*, **18**, 666–700.

Rankama, K. (1954). *Isotope Geology*, Pergamon Press, London, 535 pp.

Richards, D. J. and Wabraven, F. (1975). Airborne geophysics and ERTS imagery, *Minerals Sci. Engng.*, **7**, 234–278.

Rocha Gomes, A. A. (1958). The discovery of a new ore body within the pyritic belt of Portugal by electromagnetic prospecting. In *Surveys in Mining, Hydrological and Engineering Projects*, E. J. Brill, Leiden, 97–109.

Roy, A. and Shikhar, C. Juin (1973). Comparative field performance of electrode arrays in time-domain induced polarization profiling, *Geophysical Prospecting*, **21**, 626–634.

Runcorn, S. U. (ed.) (1960). *Methods and Techniques in Geophysics*, vols. 1 and 2, Interscience Publishers Inc., New York.

Sato, M. and Mooney, H. M. (1960). The electrochemical mechanisms of sulphide self-potentials, *Geophysics*, **25**, 246–249.

Seigel, H. O. (1965). Three recent Irish discovery case histories using pulse-type induced polarization, *C.I.M. Bull.*, **58** (643), 1179–1184.

Seigel, H. O. (1974). The magnetic induced polarization (MIP) method, *Geophysics*, **39**, 321–339.

Seigel, H. O., Winkler, H. A. and Boniwell, J. B. (1957). Discovery of the Mobrun Copper Ltd. sulphide deposit, Noranda Mining District, Quebec. In *Methods and Case Histories in Mining Geophysics*, Mercury Press Co., Montreal, 237–245.

Seigel, H. O., Hill, H. L. and Baird, J. G. (1968). Discovery case history of the pyramid ore bodies, Pine Point, Northwest Territories, Canada, *Geophysics*, **33**, 645–656.

Sixth Commonwealth Mining and Metallurgical Congress, (1957). *Methods and Case Histories in Mining Geophysics*, Mercury Press Co., Montreal.

Society of Exploration Geophysicists, (1966). *Mining Geophysics*, vol. 1—case histories, Tulsa, Oklahoma, 492 pp.

Sumner, J. S. (1976). *Principles of Induced Polarization for Geophysical Exploration*, Elsevier Scientific Publishing Co., Amsterdam, 277 pp.

Tagg, G. F. (1934). Interpretation of resistivity measurements, *Trans. Am. Inst. Mining Met. Engrs.* **110**, Geophysical Prospecting, 183–200.

Tagg, G. F. (1964). *Earth Resistance*, George Neunes Ltd., London, 258 pp.

Telford, W. M., Geldart, L. P., Sheriff, R. E. and Keys, D. A. (1976). *Applied Geophysics*, Cambridge University Press, 860 pp.

Threadgold, P. (1969). Applications of well-logging techniques to mining exploration boreholes, Ninth Commonwealth Mining and Metallurgical Congress 1969, Paper 11, Instn. Min. Metall., London.

Van Nostrand, R. G. and Cook, K. L. (1966). Interpretation of resistivity data. *U.S.G.S. Prof. Paper* 499.

Waeselynck, M. (1974). Magnetotellurics: Principle and outline of the recording technique—a case history, *Geophysical Prospecting*, 22, 107–121.

Wait, J. R. (ed.) (1959). *Overvoltage Research and Geophysical Applications*, Pergamon Press, London, 158 pp.

Ward, S. H., O'Donnell, J., Rivera, R., Ware, G. H. and Fraser, D. C. (1966). AFMAG—applications and limitations, *Geophysics*, 31, 576–605.

Ward, S. H., O'Brien, D. P., Parry, J. R. and McKnight, B. K. (1968). AFMAG—interpretation, *Geophysics*, 33, 621–644.

Yüngül, S. (1956). Prospecting for chromite with gravimeter and magnetometer over rugged terrain in eastern Turkey, *Geophysics*, 21, 433–454.

Zonge, K. L., Sauck, W. A. and Sumner, J. S. (1972). Comparison of time, frequency and phase measurements in induced polarization, *Geophysical Prospecting*, 20, 626–648.

CHAPTER 7

Drilling Methods

In Chapter 5, a number of drilling techniques suitable for sampling unconsolidated overburden were described. Although it is not possible to categorize drilling methods into those suitable for rock and those suitable for unconsolidated material, since some drilling methods can be used in both environments, only those drilling techniques capable of penetrating solid rock will be described in this chapter.

There are four main methods used for drilling rock: percussion, churn, diamond and rotary. In percussion drilling compressed air is used to drive a hammer unit which imparts a rapid percussive action to a drilling bit at the end of a length of steel rods. In churn drilling, sometimes referred to as cable-tool drilling, a heavy chisel-like steel bit at the end of a wire cable is repeatedly jerked up and down. Since this is a percussive action, churn drilling is often referred to as percussion drilling. This may lead to much unnecessary confusion in describing two very different methods. In diamond drilling a cylindrical sample of rock or core is cut by a rotating annular drilling bit impregnated with diamonds. In rotary drilling a special non-coring bit is rotated at the end of a length of steel rods. There is much confusion in distinguishing rotary and diamond drilling since both involve a purely rotary action. Diamond coring bits are also used on rotary rigs from time to time and non-coring bits are often used in diamond drilling. The distinguishing feature is that the rotary motion in diamond drilling is imparted to the rods by a chuck, whereas in rotary drilling the rotary motion is applied by a device called the rotary table. Rotary rigs are generally very much bigger and more powerful than diamond-drilling machines and extremely large ones are used in oil-well drilling.

7.1 PERCUSSION DRILLING

Percussion drills consist basically of a hammer unit which is driven by compressed air. This hammer unit imparts a series of short, rapid blows to the drill steel or rods and at the same time slowly rotates them. The drills vary in size from small hand-held rock drills for drilling charge holes to large truck-mounted rigs capable of drilling large diameter holes. There are two main types of percussion drills: down-the-hole hammer and top hammer.

Down-the-hole hammer drills
As the name implies the hammer unit of this type of drill is lowered down the hole at the end of the rods. The cylindrical hammer unit is in effect an extension to the rod string although its diameter is usually somewhat greater than that of the rods. The bit on the end of the hammer unit contains a number of tungsten carbide inserts, either button-shaped (button bits) or chisel-ended (X-face bits). Only the percussive action is supplied to the bit by the hammer unit. Rotation of the rods is accomplished by a rotation unit on the rig itself. This rotation unit is track mounted and fed up or down by a chain feed. The rods generally vary in diameter from 85 to 115 mm and bits from 100 to 200 mm. The rigs vary in weight from 1·5 to 3·0 tonnes and entire truck mounted units complete with compressor and ancillary equipment may weigh 5 tonnes or more. Depth capabilities of up to 250 m are possible with some types, though most rigs are only capable of reaching depths in the range 100–125 m. Flushing of the drill cuttings from the hole is carried out using compressed air, though special foaming agents are available and are sometimes used to assist in flushing out holes in wet ground. Down-the-hole hammer drills are mainly used for shot hole and water well drilling and are not commonly used in mineral exploration. Figure 7.1 shows a typical down-the-hole hammer machine in action.

Top hammer drill
In this type of drill, both the percussive action and rod rotation are provided by a hammer unit which is track mounted on the rig and is moved up or down by a chain feed. The holes drilled by top hammer drills are smaller in diameter than those drilled by down-the-hole machines, with rods varying in diameter from 38 to 45 mm and bits from 64 mm to 102 mm. For use in mineral exploration the drills are usually mounted on trailers, trucks or large tractors. For use in quarries and open pits drills are sometimes mounted on track-laying vehicles and are then referred to as crawler drills. The latest machines are all hydraulically controlled and can easily be operated by one man and a helper.

Most percussion drills only use air for flushing the hole, but some machines use equipment designed for water circulation and flushing. For mineral exploration drills with water flushing facilities are far superior. If only air is used, sample return is virtually restricted to depths above the water table, for, although samples can often be blown clear for several metres below the water table, depending upon the depth and capacity of the compressor being used, eventually sample return ceases as it becomes impossible for the compressed air to blow the heavy, wet sludge up the hole. (This is not a problem for big down-the-hole hammer drills using high-capacity compressors.) Machines equipped for water circulation do not have this problem and sample return is unrestricted provided that no cavities are encountered. If the drill runs into cavernous ground, water circulation may be lost and no further samples can be collected. Casing can be run or drilled down to seal off cavities, but, if numerous cavities are present, percussion drilling may prove ineffective.

FIG. 7.1. A down-the-hole percussion drill in action.

When drilling with air, the samples may be collected by methods described for wagon drilling in Chapter 5. When drilling with water, samples may be readily and easily collected by attaching a T-piece to the top of the casing projecting above the hole collar. Containers placed beneath the side-projecting pipe will catch the sample as it flows from the hole. Since samples are bulky (a 102-mm hole will produce over 30 kg of sample over a 1·5-m run), it is necessary to split them. This is most conveniently done by having the sample run straight into a sample splitter as it comes from the hole. The simplest type of splitter consists of three Jones rifflers in series so that a one-eighth split is achieved.

The samples produced by percussion drills vary from fine dust to small chips depending upon the nature of rock being drilled. A coarse, friable grit, for example, may result in samples with a high proportion of coarse fragments, whereas samples from a massive limestone may be largely dust. Whatever rock is being drilled, however, it is usually possible to recognize rock types from the sample fragments as there are always a fair proportion ranging in size from 1·0 to 2·0 mm. To assist with logging and making correlations from hole to hole on a particular property, visual logs can be made up by glueing the sample powders on to stiff cards or boards. A convenient size for the cards or boards, which should be white, is 100 cm × 20 cm. They are then ruled into columns and rows 2·5 cm × 2·5 cm with each row representing a complete sample interval (e.g. 1·5 m). There should be a space at the top of the card for the property name and hole number. The sample depths can be written down the left hand column with the columns to the right reserved for differently treated sample powders. For instance, the first column might be for untreated sample material, the second for +16 mesh washed sample, the third for a heavy mineral separate, the fourth for a magnetic fraction, and so on. The powders are stuck to the card by applying a liberal amount of quick drying cellulose cement to the square and sprinkling the dry powder over the wet glue. Excess sample can be shaken off and the process repeated for other squares. To complete the cards it is also useful to include a graphical assay log down one of the columns.

When logging the holes, it is useful to be able to gauge the percentage of sample recovered. This can be done easily by comparing the actual volume of sample collected with the calculated volume of the hole. For example, let us suppose that the total volume of the dried one-eighth split of a sample collected from 21·0 to 22·5 m was 1300 cm^3. If the hole being drilled was 102 mm in diameter, it would have a volume of 12 260 cm^3 over a length of 1·5 m. Now the powdered material has roughly a 10% greater volume than the rock *in situ*

PERCUSSION DRILL LOG

PROPERTY WONDER ROCKS   HOLE No. WR33   TOTAL LENGTH 13.50m.   PAGE 1 OF 1

GRID REF. 4.5E 7.5N   BEARING —   INCLINATION -90°   LOGGED BY J.H.R.

BIT SIZE: casing 64mm, drill —   DEPTH: casing 6.0m, water table 12.9m   METRES: with air 13.50, with water —   DRILLED BY B.C.L.

Metres from	to	Description	Dry Vol.	Calc. Vol.	% Rec.	Sample No.	% Cu	% Zn	% Pb
0	1	red-brown soil	3·5l	3·5l	100	321	0·13	0·82	0·05
1	2	red-brown lateritic soil	3·5	3·5	100	322	0·18	0·93	0·07
2	3	red-brown laterite with pisoliths	3·5	3·5	100	323	0·19	2·31	0·05
3	4	as above with some needles of willemite	3·5	3·5	100	324	0·21	3·90	0·06
4	5	damp, dark-brown manganiferous laterite	0·7	3·5	20	325	0·22	2·36	0·09
5	6	as above to 5·7m then reddish lst. to 6·0m	1·4	3·5	40	326	0·28	4·28	0·02
6	7	white lst.	3·5	3·5	100	327	0·03	0·30	0·01
7	8	reddish lst.	3·5	3·5	100	328	0·03	0·30	0·01
8	9	as above	3·5	3·5	100	329	0·02	0·42	0·01
9	10	white lst. with minor willemite	3·5	3·5	100	330	0·05	1·32	0·02
10	11	as above	3·5	3·5	100	331	0·07	1·25	0·02
11	12	white lst.	3·4	3·5	100	332	0·02	0·25	0·01
12	13	reddish lst.	3·2	3·5	90	333	0·03	0·43	0·01
13	13·50	as above - no sample return below 13·30m.	0·4	3·5	20	334	0·01	0·72	0·01
		— END OF HOLE —							

FIG. 7.2. Example of a typical percussion drill-hole log.

so that the drill-hole should produce approximately $13\,500\ cm^3$ of sample for every $1\cdot5$-m advance. Therefore, overall recovery for this example is given by:

$$\% \text{ recovery} = \frac{8 \times 1300}{13\,500} \times 100\% = 77\%$$

Graphs can be plotted showing percent recovery against sample volume for different hole sizes so that actual recoveries can be read directly once the volume of the dried sample is determined. Figure 7.2 gives an example of a typical percussion drill-hole log.

For drilling overburden and cavernous formations, equipment has been developed which makes it possible to under-ream the hole and thus keep the casing close behind the advancing drill bit. This is achieved by using a special expanding bit which is lowered down the casing. Once through the casing, the bit expands and drills a hole slightly larger and just ahead of the casing which follows the bit down. The bit can be retracted and withdrawn up the casing at any time.

Percussion drills are excellent for testing targets to depths up to 100 m where core is not required. Since costs are one-third to one-quarter those of diamond drilling, they have proved especially useful for evaluating deposits which present more of a sampling problem than a geological one such as porphyry coppers. Drilling with percussion drills is very quick and penetration rates of up to 1 m/min in granite are possible with some machines. It is, therefore, quite feasible to complete two 60 m holes/day under good drilling conditions and still allow sufficient time for running casing, moving between sites and collecting samples satisfactorily. One disadvantage that is not often mentioned is that it can be rather unpleasant to work with percussion drills for any length of time as they are extremely noisy. To avoid impairment of hearing, people operating percussion drills on a regular day-to-day basis should wear ear plugs.

7.2   CHURN DRILLING

Churn or cable-tool drills were first used for drilling oil wells in the 1860's, but were later replaced by rotary drills which were much more efficient at drilling the deeper and deeper holes required as the petroleum exploration industry expanded. Today churn drills are mainly used for water well drilling, though in certain circumstances they may be useful in mineral exploration.

The basic lay-out of a churn drill is shown in the diagrammatic sketch in Fig. 7.3. A heavy drill stem, to which a sharpened steel bit is attached, is repeatedly raised and dropped by a cable which is paid

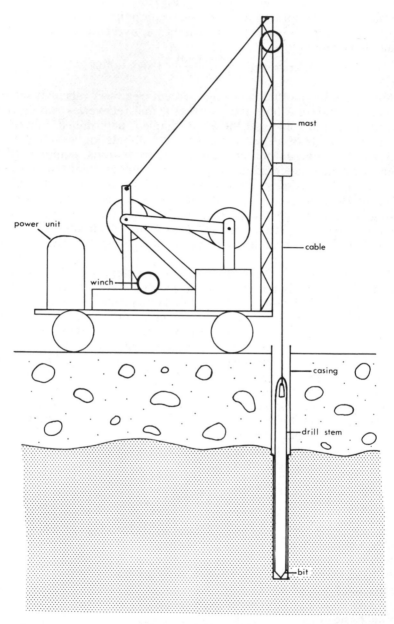

FIG. 7.3.   Sketch showing the main working parts of a churn drill.

out gradually by a winch as the hole deepens. The drilling action is supplied to the cable by a pulley on a beam which rocks up and down with a rectilinear motion at a rate of 30–60 strokes/min. The 'throw' of each stroke depends upon the size of the machine and the rock being drilled, but it usually varies between 40 and 100 cm. The striking edge of the bit is blunt, crushing rather than chipping the rock and, as the bit moves up and down, it twists round on the cable giving a 'churning' action. The correct tension on the cable is extremely important to maintain optimum drilling efficiency. The bit is allowed to fall and it should start its upward movement just before it hits the bottom of the hole. This results in the cable stretching slightly as the bit hits the bottom causing it to rebound quickly with a jerk or 'snap'. The speed of the drilling cycle and rate at which the cable is paid out is judged by the driller resting his hand against the cable to feel the correct 'snap'.

Cuttings and sludge are periodically removed by a bailer which is lowered to the bottom of the hole every 1–1·5 m of advance. A small amount of water is required both for drilling and bailing and, if the hole is dry, several bucketfuls of water are poured down the hole each time the bailer is used. Once the hole is below the water table it is usually unnecessary to add water.

Holes are generally 50–300 mm in diameter, but may be as small as 63 mm and very large machines can drill holes up to 1 m in diameter. The smaller machines have depth capacities of up to 200 m, while the larger machines can attain depths in excess of 500 m. In practice it is usual to start a hole at 250 or 300 mm and case off and progressively reduce the hole size as necessary. A number of different fishing tools are available to recover any equipment accidentally lost down the hole, which may happen if the cable breaks or the driller is careless. It is necessary to keep a good edge on the bit if efficient drilling progress is to be maintained. This is done by periodically dressing the bit by heating it with a blow torch or small forge, hammering an edge and retempering. Drilling rates per 12-h shift can be as much as 30 m in overburden or soft formations, but may be as little as 1 or 2 m in very hard rock.

Churn drills can be useful in exploration work for sampling soft formations up to depths of 100–150 m. Costs are comparable to and may be slightly less than percussion drilling. The main disadvantage is that churn drilling is very slow, but, if time is not particularly important and if only vertical holes are required, churn drilling is worth considering. Samples may be conveniently taken over lengths of 1–1·5 m, but it is important to ensure that the driller removes all the cuttings from the hole at the end of each sample interval. A common fault with churn drilling is that the bailer does not clean out all the cuttings and, if there are any high density minerals present, there will be a tendency for them to accumulate at the bottom of the hole. Since the minerals of interest in exploration generally have high densities, improper bailing may result in a false increase of values with depth. It is also important to case off any overburden to avoid contamination. If the overburden is thick, or if the formation being drilled is very friable, it is good practice to run the casing as the hole advances. By keeping the casing close behind the bit, contamination risks are reduced.

Plastic dustbins make convenient containers for receiving the samples as they are recovered by the bailer, but, since the samples are bulky (a 180-mm hole will produce over 100 kg of sample for every 1·5 m of advance), it is best to split them before bagging and removing from the site. A one-eighth split is convenient and may be carried out by passing the sample through a Jones riffler three times or, better still, by using a specially constructed splitter consisting of three Jones rifflers in series. Logging is generally not too difficult as a high proportion of the cuttings are usually coarse (3 mm).

7.3   DIAMOND DRILLING

Diamond drilling is by far the most important type of drilling undertaken in mineral exploration. The recovery of drill core enables details of the geology, ground conditions and mineralogy to be obtained that are not possible with any other method. The first diamond drill was built by a Swiss engineer in 1862 and by the early 1900's diamond drilling had become a well established technique. The earliest drills were powered by steam and, although steam power was still in use in the 1920's and 1930's, petrol and diesel engines had become the main source of power. Today the larger drills are usually powered by diesel engines, while petrol engines are most commonly used on the smaller machines. For underground drilling electric or air motors are used.

Machines and equipment
Figure 7.4 shows the basic layout of a diamond drill. The hollow drill rods pass through a tube in the swivelhead and are held firmly in place by a chuck. The swivelhead rotates the rods and has a feed mechanism for advancing the rods as drilling proceeds. The feed action may be achieved by a screw feed through a system of gears, it may be hydraulic (the most common system) or it may be manual through a leverage system. Manual feed is now obsolete, though it is still used on the small, lightweight, portable Winkie drill. With the

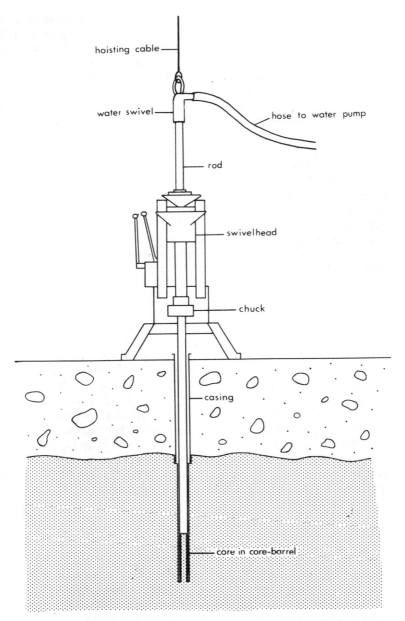

FIG. 7.4.   Sketch showing the main working parts of a diamond drill.

exception of the smallest machines, diamond drills have gearboxes so that drilling speed can be varied according to the rock being drilled. Drilling fluid (water, drilling mud, or air in special circumstances) is supplied to the rods by a water swivel which permits the rotating rods to be connected to a hose. The drilling fluid passes down the hollow rods, through the core-barrel to the bit where it cools and lubricates the cutting faces. The fluid then carries cuttings and sludge back up the hole outside the rods where it is discharged at the surface. To prevent the hole from collapsing in unconsolidated or soft, friable ground a pipe or casing of larger diameter than the rods is inserted to protect the hole. The rods and casing can be withdrawn from the hole by a winch on the machine operating through a sheave wheel on a shear-legs or tripod set above the rig. Some machines have a specially constructed mast for this, particularly truck-mounted machines, and on very large rigs a tall derrick is used. As drilling proceeds the drill core passes into the core-barrel where it is held firmly in place by the core spring or core lifter. The drill core is recovered by periodically withdrawing the rods and core-barrel from the hole.

Today there are two main types of diamond drilling: *standard* or *conventional* and *wireline*. In conventional drilling the rods have to be removed from the hole each time it is necessary to recover core from the core-barrel. In wireline drilling the core can be removed from the hole without withdrawing the rods. This is achieved by a special core tube which slots into the core-barrel and can be withdrawn by a device known as an *overshot* which is lowered down the inside of the rods on a cable and locks onto the top of the core tube. The action of the overshot locking onto the core tube causes latches which hold the core tube in the core-barrel to be withdrawn thus freeing the core tube so that it can be hoisted to the surface. After the core has been removed from the core tube, it is replaced in the core-barrel simply by dropping it down the rods. If the hole is very shallow or inclined upward, the core tube is pushed into position by pumping drilling fluid down the rods. Once the core tube is in the core-barrel, it automatically locks into place.

Hole sizes
Diamond drills have been built to a wide variety of sizes and recover core ranging from 19 mm in diameter to 300 mm and more. The most commonly used sizes in exploration drilling are shown in Table 7.1 which lists the standard North American and metric sizes.

Bits and reaming shells
Diamond drill bits, or crowns as they are sometimes known, come in a wide variety of shapes and sizes depending upon the drilling

TABLE 7.1
COMMONLY USED DIAMOND DRILL SIZES

North American Diamond Drill Sizes

	Standard			Wireline	
	Hole	Core		Hole	Core
XRT	30 mm	19 mm			
EX	38 mm	22 mm			
AX	48 mm	31 mm	AQ	48 mm	27 mm
BX	60 mm	42 mm	BQ	60 mm	36 mm
NX	76 mm	55 mm	NQ	76 mm	48 mm
HX	96 mm	74 mm	HQ	96 mm	63 mm
			PQ	123 mm	85 mm

Standard Metric Diamond Drill Sizes

Hole	Core		Approximate North American equivalent
	Thin bits	Thick bits	
36 mm	22 mm	—	EX
46 mm	32 mm	28 mm	AX
56 mm	42 mm	34 mm	
66 mm	52 mm	44 mm	BX
76 mm	62 mm	54 mm	NX
86 mm	72 mm	62 mm	
101 mm	—	75 mm	HX
116 mm	—	90 mm	
131 mm	—	105 mm	
146 mm	—	120 mm	

conditions for which they are required. A reaming shell is always fitted behind the bit to ensure that the hole size is maintained at the correct diameter. In most bits the drilling fluid simply passes down the inside of the bit, but for soft formations where it is necessary to protect the core in the bit from the eroding effect of the drilling fluid special bottom discharge bits with hollow walls and discharge ports in the cutting face have been designed. Bits are specified by their size, manufacturer's part, serial and type numbers and weight of diamonds in carats. In addition the number of diamonds/carat might also be given. In diamond bits large size diamonds (better for soft formations) range from 5 to 15 stones/carat and small size (better for hard formations) from 50 to 60 stones/carat. For example, two BX bits might both contain 12·0 carats, but one might have 60 stones and the other 300. In addition to normal diamond bits, impregnated bits consisting of diamond dust and very small stones within the metal matrix are made for special jobs; these are good for hard rock. Special non-coring diamond bits are sometimes used where core is not required, particularly when re-drilling a cemented section. For the drilling and coring of soft formations much cheaper tungsten carbide bits can be used in place of the more expensive diamond bits. Figure 7.5 shows some typical bits used in diamond drilling.

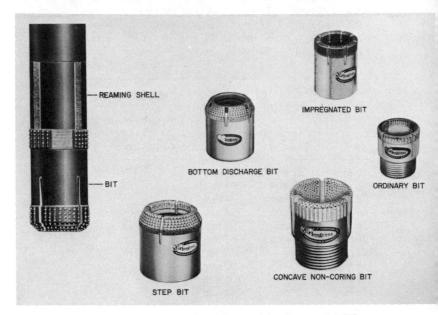

FIG. 7.5.   Some typical bits used in diamond drilling.

Core-barrels

The simplest type of core-barrel is a single tube 1·5–9 m long to which a reaming shell and bit are attached with a circular core spring or lifter fitted inside just above the bit to prevent core from falling out. Since the drill core is continually exposed to the eroding effect of the drilling fluid passing down to the bit, single-tube core-barrels are rarely used today. The double-tube core-barrel, which is the most widely used, contains an inner core tube for holding the core. This protects the core from the drilling fluid and improves core recovery in soft rock. The core tube is taken out of the core-barrel each time it is necessary to remove the core, which is then shaken out, pushed out with a thin metal rod or pumped out with water. For drilling very soft or friable formations, triple-tube core-barrels have been designed. These contain a further tube, split along its full length and nested inside the core tube. Core is protected in the split tube which is extracted from the core tube each time it is necessary to remove the core. Used in conjunction with bottom-discharge bits, these core-barrels give optimum protection to soft or friable core. The wireline core-barrel is essentially a double-tube core-barrel from which the core tube can be withdrawn without having to disassemble the core-barrel. The mechanism by which the core tube is removed has already been described. In addition wireline core-barrels contain a shut-off valve which is designed to minimize core blocking and grinding and hence improve core recovery. If the core starts blocking, the increase in water pressure causes the shut-off valve to close, thus cutting off the drilling fluid circulation and drawing the attention of the driller to the core blockage. The core tube can then be retrieved to remove the blocked core and allow drilling to commence until the core tube is full or a further blockage is observed.

Rods
Standard drill rods usually come in lengths varying from 1·5–6 m and can be joined together by threaded couplings. They are flush-jointed on the outside, but not on the inside where the couplings have narrower internal diameters than the rods themselves. Wireline rods have thinner walls and are flush-jointed internally as well as externally so that the core tube can be withdrawn without obstruction. Thread designs and specifications vary slightly according to the manufacturer.

Casing
Casing is seamless steel tubing which comes in a number of different lengths and can be screwed together with flush joints on the inside and outside. It has thinner walls than the rods and is designed to be slightly larger than the corresponding rod size. For example, AX rods will pass inside AX casing, AX casing will pass inside BX casing, BX rods will pass inside BX casing, BX casing will pass inside NX casing and so on. In addition, casing of a specific size will fit into a hole drilled by bits of the next size up. For example, NX casing shoes are the same size as HX coring bits, BX casing shoes are the same size as NX coring bits and so on. Casing can be driven down a pre-drilled hole by repeatedly raising and dropping a drive hammer on the top end of the casing or it can be drilled down if a special diamond bit known as a *casing shoe* is used. The running of casing in a hole can be a difficult operation and, if a lot of casing is required, it might be

necessary to make several size reductions. For example, a particular hole might have HX casing to 50 m, NX casing to 150 m and BX casing to 250 m.

After completing a hole, it is normal practice to recover the casing from the hole unless access to it is likely to be required again either for deepening or for geophysical logging. The casing often becomes tightly wedged in the hole and it is rarely possible to extract it with the hoist on the rig. For this reason large hydraulic jacks or a jarring hammer on the hoist have to be used; in many instances it is not possible to recover all of the casing.

Drilling fluids

Water is the most commonly used drilling fluid, but when drilling in overburden or soft, friable rocks, it is found to be too erosive. In such cases drilling mud, which consists of a bentonite–water mixture is used. This has a two-fold action. Firstly, it plasters the walls of the hole filling small cavities and cracks with clay. This reduces erosion and prevents caving into the hole. Secondly, the bentonite increases the density and viscosity of the drilling fluid which aids in the removal of cuttings and sludge. In certain instances the density of the drilling mud can be further increased by adding powdered barite. In addition special cutting oils may be used with the drilling water. For drilling very soft or friable rocks such as wet muds or coal seams, air is sometimes used as the drilling fluid. Bottom discharge bits have to be used and excellent core recovery is achieved. Air can only be used in very soft muds or rocks and, if any hard, abrasive material is present, the bit will be ruined very quickly.

Pumps

Except for use on the smallest machines, drilling pumps have to be capable of delivering water, mud or grouting cement. The size of the pump required will depend upon the size of the hole being drilled, but for most exploration drilling requirements pumps should be able to deliver 50–100 litres/min at pressures ranging from 30 to 50 kg/cm^2.

Fishing tools

If drill rods or tools are accidentally lost down a hole, the driller has a number of different fishing tools with which to retrieve the lost equipment. The most commonly used is the *recovering tap* which is tapered and threaded with 'V' threads for screwing into and gripping the top end of a lost rod or core-barrel. If recovering taps fail to work, a *bell tap*, which is designed to overlap and encase lost drill tools, can be used. To recover metal fragments magnetic extractors are sometimes used.

Cementing
If bad ground with cavities and caving material is encountered, it may be necessary to cement the hole to stabilize conditions so that drilling can continue. Neat cement and water are pumped down the hole until the bad section is completely filled. If quick drying cement has been used, it may be possible to re-drill the cemented section several hours after cementing, but, if ordinary cement is used, it usually requires 24–48 h before the cement sets hard. The drying time of cement can be speeded up by adding calcium chloride to the cement/water mixture. Core recovery of cement should be 100% if it has been carried out properly, though cemented sections are often re-drilled with non-coring bits.

Drilling rigs
The machines most commonly used in mineral exploration have depth capacities ranging from 200 to 1500 m depending upon the size of the machine and hole diameter being drilled. Much larger machines with depth capacities of 3000 m and more are also available, but holes of this depth are not common in mineral exploration. The controls on most machines include engine throttle, drill clutch, gear lever, hydraulic head feed control, swivel head up–down lever, hoist clutch and hoist brake. These are all centrally located so that the driller can maintain positive control of every operation. The chuck is usually manually operated with Allen keys, and rods and casing are joined and broken with pipe wrenches, which makes drilling a tough job. Automatic, hydraulically operated chucks are now available and even rod handling, breaking and joining is carried out with hydraulic controls on some of the latest machines, making the job of the driller much easier and greatly increasing productivity. A gauge to show the hydraulic feed pressure is usually part of the standard equipment on most machines. The productivity achieved depends on the size of the hole, the depth of the hole, the rock type being drilled and lastly, and probably most importantly, on the skill of the driller. For depths up to 300 m under good drilling conditions 20–30 m can be achieved with conventional equipment in a 12-h shift and 100–120 m with wireline equipment. These advances are exceptional and daily advance is usually very much less than this, but it should not fall below 5–6 m for conventional equipment and 20 m for wireline unless the holes are very deep or drilling conditions very bad. Drilling is a highly skilled business and a great deal of experience is necessary to know the best drilling speed and rate of penetration for different rock types so that optimum productivity and core recovery is achieved. Figure 7.6 shows a typical diamond drill in action.

FIG. 7.6. Exploratory diamond drill hole being drilled at −45° on a copper prospect in Zambia.

Portable diamond drills

Small machines that can be carried by one or two men have been built for drilling shallow targets. These machines include the Pak-sak drill and the 'X-ray' drill, which use XRT rods and bits, and the Winkie drill which uses EX size rods and bits.

The Pak-sak drill is roughly the size of a hand-held rock drill and is designed to be used in a similar manner. It is powered by a small petrol engine and has a separate water pump. Depths of up to 30 m are possible, but it is rarely practical to drill over 15 m. These drills are of little value except for collecting short core samples from rock outcrops.

The 'X-ray' and Winkie drills are somewhat larger and are really miniature versions of the larger machines. They can be carried by two men and mounted on heavy timbers for drilling. Depth capabilities of 50 m are possible, but this can be almost doubled if lightweight, magnesium–zirconium rods are used. The machines are very useful for testing shallow targets if drilling conditions are good. Owing to the small size of the XRT and EX bits, however, core recovery is very poor in bad ground and in areas with thick overburden or deep weathering they are almost completely ineffective.

Conventional versus wireline

The development of wireline methods in the 1960's was a major revolution in diamond-drilling techniques. Productivity and core recovery were dramatically increased and drilling costs reduced. Nevertheless, there are some disadvantages and conventional drilling may be preferable in certain cases. Wireline rods are thinner than conventional ones and their life is shorter and breakages more common. For a given hole size wireline core is smaller than conventional core with a wireline bit having to cut an annulus up to 40% larger than a conventional bit. This increases diamond wear and reduces penetration rates in hard rock. Hole deviation is also a problem. Owing to the thin rods and higher drilling pressures used in wireline, deviations are much greater than in conventional drilling and in many instances are completely unacceptable, though special minimum deviation reaming shell assemblies are now available to reduce this problem. Wireline has a distinct advantage in bad or caving ground, provided bit life is long. The core can be removed from the hole quickly each time there is a blockage and drilling can proceed without having to remove the rods and disturb a possible caving section in the hole. If bit life is short (less than five times a full core run), wireline becomes less attractive. In deep holes in excess of 500 m with hard rock formations and consequent short bit life, conventional drilling using long core-barrels of 6 or 9 m has advantages, particularly when using the latest all-hydraulic rigs.

Drilling overburden

When drilling through unconsolidated overburden, it is common practice to use a roller rock bit such as those used in rotary drilling until hard rock is reached. To reduce the chances of the open hole collapsing drilling mud is often used. The casing is then inserted in the open hole. If the overburden is very thick, it may be necessary to start the hole at a size well above HX so that several reductions in casing size can be made in order to case off the deep overburden section. Although casing shoes enable casing to be drilled down, the life of a casing shoe is often very short, particularly if the ground contains many quartz fragments and boulders. It is for this reason that open holes are often drilled with roller rock bits and the casing is only drilled down a few metres with a casing shoe if at all. If core recovery is required in overburden, it has been found that drilling

very slowly for short runs using tungsten carbide bits with little or no
water gives good core recovery of unconsolidated ground.

Reverse flow (dual tube) drilling
A new development in drilling, which in some ways is as rev-
olutionary as the introduction of wireline techniques, is that of
reverse flow or dual tube drilling. In this system special drilling rods
containing inner rods of smaller diameter are used. Drilling fluid is
passed down the outer tube and returns to the surface up the inner
tube, hence the term 'reverse flow'. A small proportion of the drilling
fluid returns up the hole annulus to provide adequate hole lubrication.
The system is particularly good for drilling and sampling areas with
thick overburden or deeply weathered rocks. The outer tube acts as a
casing and return of cuttings and sludge up the inner tube reduces
circulation loss due to cavities and fractures. In addition compressed
air can be injected in the drilling fluid to improve sample recovery in
bad lost circulation zones by reducing the hydrostatic head in the
inner tube. With roller rock bits penetration rates in overburden or
soft rock are very high and, since rock cuttings up to 2 cm across are
brought up, it is often unnecessary to core. When core is required, it
is a simple matter to change over to ordinary diamond drilling. The
Walker–Neer Company of the United States has introduced a system
which makes continuous coring possible with the reverse flow system.
With a special diamond bit in place of the roller rock bit core enters
the inner tube and is broken off in 12·5 cm lengths by a special core
breaker inside the tube and is pumped to the surface by the returning
drilling fluid. The diagram in Fig. 7.7 shows the basic principle behind
dual tube drilling.

Drilling costs
Diamond drilling is the most expensive type of drilling undertaken in
mineral exploration, generally being three times as costly as per
cussion or churn drilling and up to twice as expensive as rotary
drilling. The actual costs vary widely according to the part of the
world, size of the contract, depth of holes and diameter of core, but in
1977 normal drilling in different parts of the world varied anywhere
from US$20 to US$120/m for normal BX or NX drilling. In parti
cularly remote areas and/or in very difficult ground costs might be
much higher.

Time sheets and drillers' logs
It is common practice for diamond drilling contractors to use time
sheets to show the nature of work carried out during each shift. The
normal time sheet is divided into columns corresponding to half

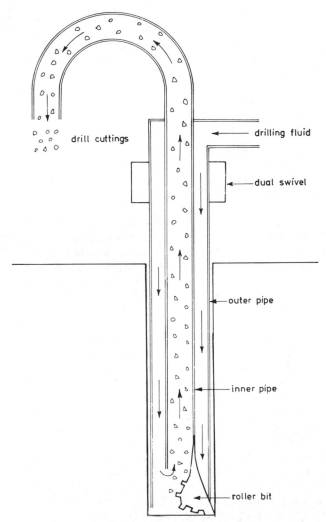

drill cuttings

drilling fluid

dual swivel

outer pipe

inner pipe

roller bit

FIG. 7.7.   Sketch to illustrate the principle of dual tube drilling.

hourly intervals of a shift and rows for the different jobs undertaken. These are: running rods, running casing, pulling rods, pulling casing, drilling mud, drilling rock, fishing, repairs, down time, surveying, rigging up, rigging down. The driller simply puts a tick against each half hour spent on a particular job. For example, a typical time sheet during a drilling shift might show 3 h running casing, 4 h drilling rock 1 h repairs, 1 h pulling rods, 1 h running rods. In addition the driller should enter the numbers of any new bits or casing shoes used. The time sheet should also show the amount and sizes of casing used and any items of equipment lost. The geologist in charge of the drilling should sign each time sheet he accepts as correct. The importance of the time sheet varies with the type of contract, but under some contracts it is extremely important that the time sheets are correct as it may affect considerably the drilling charges made by the contractor.

In addition to the time sheets the driller should also keep what is known as the driller's log. This should list such items as the drilling characteristics, circulation losses, cavities and cavings against the respective depths. This information is often very useful to the geologist. For instance, the absence of core over a section with a definite cavity will not be interpreted as core loss.

Sludge sampling

When drilling through unconsolidated overburden or in bad ground with very poor core recovery, it is good practice to collect sludge samples. This is quite easily done by digging a small trench near the hole collar with a gently sloping narrow channel way leading to the hole. A launder can be placed in the channel or it can be lined with cement or polythene sheeting. A small lip should project from the channel way out over the trench so that sludge will run into containers placed in the trench. The trench should have an outlet so that any drilling fluid passing into it can drain away. Samples may be collected in any suitable large container, or special sludge boxes can be used. Over each interval of sample collection the drilling fluid can be allowed to overflow from the collecting container while the sludge settles out. If it is considered necessary to collect as much of the fines as possible, several sludge boxes can be used in series to allow sufficient time for most of the fine material to settle out. When the sampling run has been completed, or when sufficient material has been collected, the container can be removed and replaced by an empty one to repeat the process. If the sludge samples are of a qualitative nature, such as the geochemical sampling of overburden, a small amount of sludge can be removed from the box and the rest thrown away. Samples required for quantitative purposes can be split on the site to facilitate handling and force-dried over a fire or gas burner if necessary. It should be remembered that it is not always possible to collect sludge samples because of loss of water return.

The drilling contract

As drilling is such an expensive undertaking, it is normal for the parties to enter into a formal contract. Essentially, there are two types of contract, *unit cost* and *cost-plus*, though the two may be combined. In a unit cost contract the contractor agrees to carry out the work at so much per metre and, regardless of the effort expended and materials used, the final cost is fixed by the total meterage drilled. In a cost-plus contract the contractor agrees to carry out the work at so much per metre plus additional costs of fuel and materials used, drillers' time, equipment rental, etc. The cost-plus contract should be avoided as it is always to the contractor's advantage and the final cost to the client is not dependent on results. For obvious reasons contractors are often unwilling to enter into a strictly unit cost contract and there is almost always a cost-plus element included. For example, if a contractor feels that drilling in a particular area may involve a lot of trouble with overburden, he may specify a rate of so much per metre plus an hourly charge together with additional costs of fuel and materials for drilling the overburden. The prospective client should always be extremely wary of any cost-plus items in a contract. For example, the final cost to a client from a contractor offering an attractive rate per metre in a contract together with a cost-plus clause may be much higher than the final cost from a contractor offering a less favourable rate per metre in a contract without a cost-plus clause.

Borehole surveying

Since a borehole rarely maintains its collar inclination and direction for any appreciable length and may show a marked deviation from the projected course, it is important to be able to survey the true path of the hole. The simplest surveying method is the acid etch from which the inclination can be measured. With this method a glass tube is half filled with hydrofluoric acid, stoppered and lowered down the hole in a special cylinder to the depth at which the measurement is required. It is left in the hole for 20–30 min to allow the acid to leave a clear etch mark on the glass. The bottle is then retrieved from the hole, the acid is washed out and the angle the etch mark makes with the tube measured to give the inclination of the hole. Owing to surface tension, however, the apparent angle of the liquid is always greater than the true dip of the hole and a meniscus correction has to be made for an accurate measurement. The amount of deviation from the true angle varies with the inclination of the hole, the size of the tube and the strength of the acid; correction graphs (Fig. 7.8) are used to determine the inclination.

As the *acid etch* test measures only inclination, a number of survey instruments, which measure both inclination and azimuth, are manufactured in different parts of the world. These include the Tro-Pari made in Canada, the Eastman borehole camera from West Germany and the Humphrey gyro survey instrument made in the

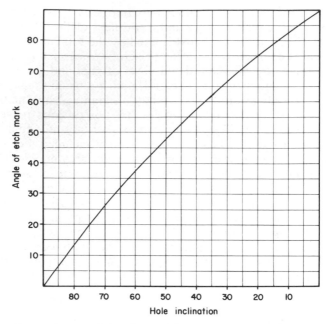

FIG. 7.8.   Correction graph for determining the true inclination from the etch mark in 25-mm tubes.

United States. The Tro-Pari instrument consists of a compass element and clinometer free to rotate until locked in place by a pre-set clockwork timer. The instrument is lowered down the hole in a special cylinder on the end of a brass or aluminium rod with the timer set to lock the instrument 5 or 10 min after reaching the required depth. A disadvantage of the Tro-Pari is that it is a single-shot instrument and only one measurement can be made each time it is lowered down the hole. This disadvantage is overcome in the rather more expensive multi-shot borehole camera such as the Eastman surveying instrument. This instrument takes a series of readings at fixed time intervals and records the azimuth and inclination on a miniature film. Both the Tro-Pari and Eastman instruments have the disadvantage that they are magnetic and cannot be used in the cased section of a hole or in strongly magnetic rocks. These disadvantages are overcome by the very much more expensive Humphrey gyro survey instrument which contains a miniature directional gyro for determining azimuth. This instrument can be lowered down the hole on a cable and a survey data are continuously displayed on a control box at the surface. A new borehole surveying instrument, produced by the

ABEM company in Sweden and known as the Reflex-Fotobor DD1, can also be used in cased holes and in magnetic formations. It consists of a 12-m rod or probe which contains three internal reflector rings set at distances of 3, 6 and 9 m respectively from a photographic recorder which uses 16-mm black-and-white film. The position of a vertical or bull's-eye bubble is also recorded. As the probe is lowered down the hole it flexes with the deviation of the hole and the degree of bending is measured by the offsets of the reflector rings and recorded on the film. The instrument continuously measures the deviation of the hole and is accurate to 10 cm in 100 m. The smallest holes that can be surveyed are 46 mm (AX) in diameter.

The importance attached to a borehole survey depends upon the nature of the job and depth of hole. Short holes of 100 m or less rarely show much deviation, but deeper holes may deviate considerably. There are known examples of 300-m wireline holes which show deviations of 60 m and more in plan. Surveying is not particularly important in cases where only one or two exploratory holes are drilled, but in the evaluation of a prospect the importance of surveying cannot be overemphasized. An acid etch test is very simple and takes very little time and for this reason it is a very useful method for checking vertical holes. If the acid etch indicates a bad deviation, a more time-consuming survey can be carried out with a Tro-Pari or similar instrument.

Plotting surveyed drill holes

The usual drill hole survey consists of a number of readings at different intervals down the hole and it is necessary to plot the actual continuous curve of the deviating drill hole from these few measurements. There is one simple method which gives a very good approximation for all but the most extreme cases. Figure 7.9(a) shows in section the hypothetical curve of a deviating drill hole which was vertical at the collar. The angles made by the tangents to the curve at points 1, 2, 3, 4 and 5 give the hole inclination at the respective points. These are tabulated below:

Survey point	Inclination	Depth down the hole
1	$-90°$	0
2	$-76\frac{1}{2}°$	100 m
3	$-66\frac{1}{2}°$	200 m
4	$-52°$	300 m
5	$-33\frac{1}{2}°$	350 m

The procedure is to take the mean value between successive inclinations which gives: $83\frac{1}{4}°$, $71\frac{1}{2}°$, $58\frac{1}{4}°$ and $42\frac{3}{4}°$. Then 0–100 m is

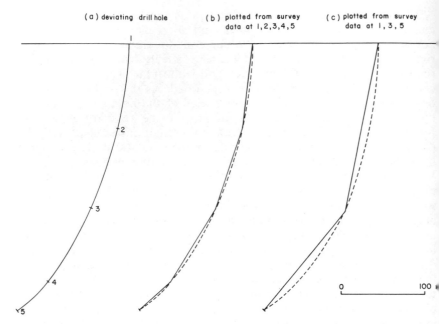

FIG. 7.9. Diagram to show how a deviating drill hole is plotted in section from survey data.

plotted at $83\frac{1}{4}°$, 100–200 m at $71\frac{1}{2}°$, 200–300 m at $58\frac{1}{4}°$ and 300–350 m at $42\frac{3}{4}°$ giving the result in Fig. 7.9(b). The dashed curve drawn through these plotted points gives almost a perfect fit with the original curve. Figure 7.10(a) shows in plan an extreme case of the hypothetical curve of a deviating drill hole. In this case the tangents to the curve at points 2, 3, 4 and 5 give the hole azimuth at the respective points. These are tabulated below:

Survey point	Azimuth	Horizontal distance
1	—*	0
2	10°	14 m
3	28°	33 m
4	$75\frac{1}{2}°$	54 m
5	$109\frac{1}{2}°$	37 m

*No azimuth as hole assumed to be collared vertically.

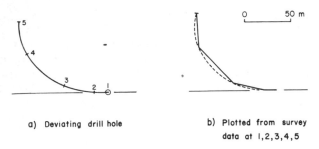

a) Deviating drill hole

b) Plotted from survey
data at 1,2,3,4,5

FIG. 7.10.   Diagram to show how a deviating drill hole is plotted in plan from survey data.

Then 14 m is plotted at 10° (there is no azimuth at point 1 so no mean value can be calculated between 1 and 2), 33 m at 19°, 54 m at $51\frac{3}{4}°$ and 37 m at $92\frac{1}{2}°$. The result in Fig. 7.10(b) is not such a good fit with the original curve as the example in Fig. 7.9, but this is an extreme case with a 90° change in azimuth. If there were more survey points, the plot would be much better. It is interesting to note that in the less extreme example shown in Fig. 7.9 a plot using only the data at points 1, 3 and 5 (Fig. 7.9(c)) still gives a fairly good approximation to the original curve.

In a deviating drill hole both the azimuth and inclination tend to change together (the hole often spirals) and the survey measurements for both inclination and azimuth have to be used in conjunction to plot the drill section and plan. Let us consider an actual example. A diamond drill hole X22 was inclined at −70° towards 259° magnetic and drilled to a final depth of 297 m. From the survey results given in Table 7.2 we are required to plot a plan view and a section through the collar position along 259° magnetic.

The required calculations are given in Table 7.3. To plot the borehole course in plan, the horizontal distance of 29·5 m is measured

TABLE 7.2
SURVEY RESULTS FROM BOREHOLE X22

Depth (m)	Inclination	Azimuth (magnetic)
collar	−70°	259°
102	−76°	312°
152	−78°	314°
251	−77°	305°
297	−75°	302°

along a bearing $285\frac{1}{2}°$. This is followed by 11·5 m on a bearing of 313°, 22·5 m on a bearing of $309\frac{1}{2}°$ and finally 12·5 m on a bearing of $303\frac{1}{2}°$ to give the result in Fig. 7.11. To plot the vertical section through the collar along 259°, the vertical distances in Table 7.3 are plotted down the hole and perpendiculars dropped to the line on the bearing 259° in the plan view give the respective distances of the hole from the vertical below the collar. The final result is shown in Fig. 7.11.

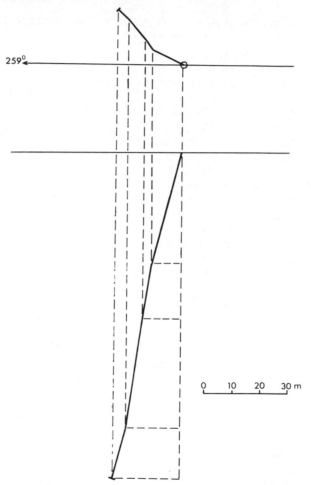

FIG. 7.11. Diagram to show how a deviating diamond drill hole is plotted in plan and in section from survey data.

TABLE 7.3
CALCULATIONS REQUIRED TO PLOT DRILL-HOLE X22 SHOWN IN Fig. 7.11

Depth	Interval	Inclination	Mean	Horizontal distance	Vertical distance	Azimuth	Mean
0	—	$-70°$	—	—	—	$259°$	—
	102		$73°$	$102 \cos 73° = 29{\cdot}5$	$102 \sin 73° = 97$		$285\frac{1}{2}°$
102		$-76°$				$312°$	
	50		$77°$	$50 \cos 77° = 11{\cdot}5$	$50 \sin 77° = 49$		$313°$
152		$-78°$				$314°$	
	99		$77\frac{1}{2}°$	$99 \cos 77\frac{1}{2}° = 22{\cdot}5$	$99 \sin 77\frac{1}{2}° = 96$		$309\frac{1}{2}°$
251		$-77°$				$305°$	
	46		$76°$	$46 \cos 76° = 12{\cdot}5$	$46 \sin 76° = 44$		$303\frac{1}{2}°$
297	—	$-75°$	—	—	—	$302°$	—

Deflections

Occasionally during the course of drilling a hole, it is necessary to change the direction of the drill hole. This procedure is known as a deflection and there are three main reasons why it may be necessary: (1) to correct the course of a deviating hole, (2) to obtain another intersection of a particular zone, and (3) to avoid an obstruction in the hole. Deflections are accomplished by setting a long tapering steel wedge in the hole just above the point at which the deflection is required. The wedge gradually forces the drill bit to one side until it is drilling a new hole slightly off the line of the old one. There are two types of wedges: non-directional and directional. Directional wedges have to be used in cases where the course of a deviating hole has to be corrected or where it is necessary to hit a restricted 'target'. Non-directional wedges can be used in cases where direction is not critical, such as avoiding an obstruction or obtaining an additional intersection of a particular zone. The typical wedge for a BX or NX hole is a solid steel rod 2–2·5 m long with a tapering facet 1·5–2 m long.

To set a non-directional wedge the hole is filled with cement or sand up to the point where the wedge is to be set, a block of wood is then placed at the bottom of the cemented hole and the wedge is pushed down on the wooden plug. Drilling off the wedge is carried out slowly and cautiously to make sure that the deflection is successful. Once the hole is on its new course drilling can be carried out as normal. The actual course taken by the deflected hole can be determined later by surveying.

The setting of a directional wedge (Fig. 7.12) is much more difficult and requires considerable skill on the part of the driller. The procedure is as follows. The hole is cemented and a wooden plug is set at the point where the deflection is required. A drive wedge, fastened by light copper rivets to a test shoe, is lowered down the hole on the end of the rods and driven down on the wooden plug. A spike at the bottom of the drive wedge locks it firmly into the wooden plug and the force of pushing it home shears the copper rivets. The rods and test shoe are withdrawn from the hole leaving the drive wedge in position after allowing sufficient time for an acid etch tube or survey instrument, contained in a compartment in the top of the test shoe, to measure the actual attitude of the drive wedge. The wedge itself is now joined to a wedge pilot shoe, which is rotated to the correct position, so that the wedge will be set at the correct angle for the required deflection when the wedge pilot shoe engages with the drive wedge in the hole. The wedge is now lowered down the hole until it slots into position on the drive wedge. If the whole operation has been carried out correctly, drilling off the wedge should result in the

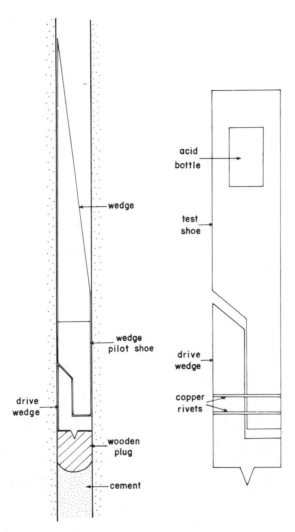

FIG. 7.12. The various parts of a directional wedge. Left: set in the hole.
Right: drive wedge attached to the test shoe before placing in the hole.

desired deflection. Correcting a deviating drill hole with directional
wedges can be a laborious and expensive process as it is generally
found that the hole has to be wedged a number of times (Fig. 7.13) to
keep it on course. The cost of wedging and possibilities of success
have to be carefully weighed against the cost of drilling a new hole.

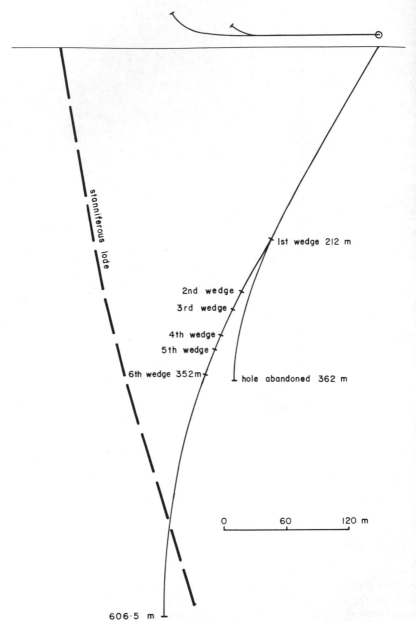

FIG. 7.13. Example of directional wedges used to correct a deviating diamond drill hole (after Walsham, 1967).

Orientation of drill core

It is possible to obtain a considerable amount of structural information from diamond drilling, but it may be ambiguous if the true orientation of the drill core is not known. In areas where there is good exposure and structures are relatively simple, the orientation of structures in the drill core can be deduced with little difficulty, particularly when several holes have been drilled. In areas with poor exposure and complex structures, however, it may prove impossible to ascribe true orientations to structures such as bedding, foliation, jointing, lineations, etc. intersected in drill holes.

A number of techniques for obtaining oriented core samples from drill holes have been developed in various parts of the world. One of the earliest methods was to obtain a wax impression of the end of a stub of core at the bottom of the hole by pushing a string of rods down the hole with a wax plug at the end. Knowing the orientation of the rod string, it was a simple matter to recover the core stub by drilling and then match it to the oriented wax impression. There are many difficulties with this method, one of the biggest being that of ensuring that the bottom of the hole is absolutely clean so that a satisfactory wax impression can be obtained.

A relatively simple method that is quite commonly employed today involves the use of an acid bottle. In this method an acid etch bottle containing hydrofluoric acid is placed in the upper part of a core tube which has a longitudinal line marked along its full length. The acid bottle also has a longitudinal mark along its side and it is placed in the core tube with its mark coinciding with the line on the core tube. The acid bottle is firmly fixed in place and a short run of core 20–30 cm long is drilled with the core barrel. The acid bottle is then left in the hole for about half an hour for a clear etch to form and is then withdrawn. Before the core is removed from the core tube a mark is placed on it coinciding with the line on the core tube. The core is then removed and a line drawn along its side with a marker pen. If the operation has been carried out correctly, this line on the core will correspond to the line on the core tube and hence the line on the acid bottle. The low point of the etch on the bottle is marked and its clockwise angular displacement from the longitudinal line looking down the hole is measured. This angular displacement is then transferred in the same sense to the core. This mark now lies in the vertical plane in which the hole is inclined. The angular displacement of the bottom of the points of structures such as bedding are then measured from the bottom of the etch mark (Fig. 7.14). The hole is then surveyed with a Tro-Pari or similar instrument which gives the inclination of the hole and true azimuth of the bottom point of the etch mark. This method assumes that the short length of core reco-

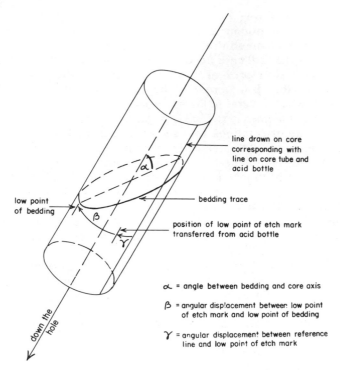

line drawn on core
corresponding with
line on core tube and
acid bottle

low point
of bedding

bedding trace

position of low point of etch mark
transferred from acid bottle

α = angle between bedding and core axis

β = angular displacement between low point
of etch mark and low point of bedding

γ = angular displacement between reference
line and low point of etch mark

down the
hole

FIG. 7.14.　Orientation of structures in drill core.

vered has not rotated in the core tube during drilling and for this reason swivel-type core barrels, in which the core tube is free and does not spin as the outer barrel is rotated, must be used.

Another instrument for obtaining oriented core samples is the Craelius core orienter made in Sweden (Roxstrom, 1961). This instrument operates on the same principle as the wax impression method, but achieves the result mechanically with six locator pins which protrude from the end of the instrument. When core presses against a central trigger piston extending beyond the locator pins, a spring pushes the locator pins against the top end of the core and the pins are pushed in by different amounts depending on the shape of the upper rough end of the core. A ball bearing indentation mark on an aluminium marker ring on the instrument indicates the lowest point in the plane of inclination of the hole (equivalent to the high point of an etch mark) and the recovered core can be oriented by matching its upper end against the locator pins.

A number of ingenious instruments have been designed and custom-built by different companies for their own specific needs. Although they differ somewhat in detail, they all work on the principle of placing an oriented mark or scratch on the core in the hole.

Once an oriented core sample has been recovered, the true orientation of structures measured in the core can be determined. One way of doing this is to use a two-circle goniometer for fixing the core in its original orientation and measuring the orientation of structures directly. It is not necessary to do this, however, and quick and accurate determinations can be made using stereographic projections. Let us consider an actual example of a drill hole inclined at $-60°$ to 50° from which an oriented core sample was obtained with bedding intersected at 42° to the core axis and the low point of the bedding 59° from the low point of the etch mark. The basic procedure (Fig. 7.15), which assumes that the hole is vertical and then rotates it to its correct inclination, is as follows:

1. Place a piece of tracing paper on a stereonet with a drawing pin through the centre so that the paper can be rotated around the centre point. Mark the north position (N). Measure off 50° for the hole azimuth and mark this position (H). Measure off another 59° from H and mark this position (A).

2. Rotate the tracing paper so that A lies on an E–W diameter. Measure 42° along the E–W diameter from the centre and trace the great circle through the poles.

3. Rotate the tracing paper so that H lies on the E–W diameter. Now move the hole 30° from the vertical position to its true inclination (H_1). The bedding intersection also has to be moved through 30° in the same sense. This is done by moving a number of points along the bedding plane great circle through 30° along small circles as shown.

4. Rotate the tracing paper until the plotted points all lie on the same great circle. Trace the great circle through the points. Mark the end of the E–W diameter on the opposite side of the centre to the trace of the great circle (B). Measure the number of degrees from the end of the E–W diameter to the trace of the great circle (39°). This is the true dip of the bedding.

5. Rotate the tracing paper so that N lies on north. The position of B now gives the true direction of dip of the bedding (140°). Thus, the beds dip at 39° to 140°.

A somewhat simpler procedure (Fig. 7.16) can be used by plotting the poles to the bedding planes. The procedure is as follows:

1. Proceed as in the first step for the example above.

2. Rotate A to the E–W diameter and plot A_1 42° from the equator.

3. Rotate H to the E–W diameter and plot H_1 30° from the centre at the correct position of the inclined hole. Also rotate A_1 30° in the same sense along a small circle to B_1.

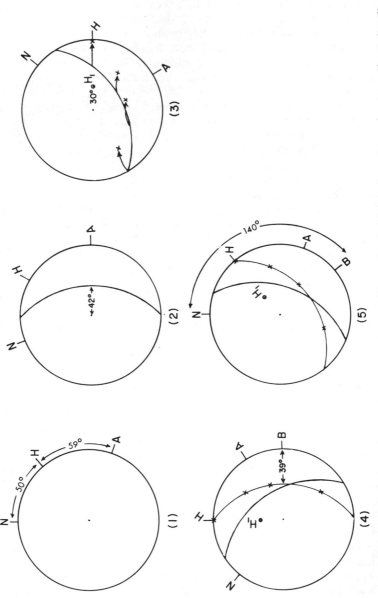

FIG. 7.15.   Diagram showing how an equal area stereographic projection is used to determine true dip and strike of bedding from an oriented core sample.

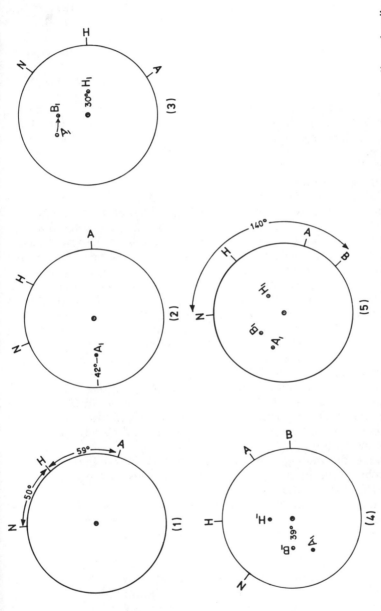

FIG. 7.16. Diagram showing how an equal area stereographic projection is used to determine true dip and strike of bedding of an oriented core sample by plotting poles to the bedding planes.

4. Rotate so that B_1 lies on the E–W diameter. Mark B on the equator on the same side of the centre as B_1. Measure the number of degrees from the centre to B_1 (39°). This is the true dip of the bedding.
5. Rotate N to north and read off the position of B on the equator (140°). This is the direction of dip.

To practise the method the reader may wish to solve the following problem.

An oriented core sample was retrieved from a borehole inclined at −45° to 225°. If the bedding angle is 35° and the low point of the bedding 70° from the low point of the etch mark, what is the true dip of the beds? (The answer is 53° to 330°.)

Determining structures without oriented core

In the absence of oriented core, it is still possible to determine structures if there are several drill holes with different orientations and if the structures are consistent over distances greater than the distances separating the boreholes. Let us consider an actual example. A diamond drill hole B_1 was drilled at an inclination of −45° due west and intersected bedding at 45° to the core axis. Two possible solutions that come to mind immediately are that either the bedding is horizontal or it is vertical. In fact, any dip from 0° to 90° is possible. A second drill hole B_2 was drilled at −60° to 195° and intersected the bedding at 25° to the core axis. Using the information from this second hole it is possible to reduce the number of solutions to two possibilities. The procedure shown in Fig. 7.17, using tracing paper over a stereonet is as follows:

1. Mark the north point N and plot B_1 at 45° from the centre on the west side of the E–W diameter. Mark a point (B_2) on the circumference at 195°. Rotate the tracing paper so that the azimuth of B_2 lies on the E–W diameter and plot B_2 30° toward the azimuth mark.
2. Rotate the tracing paper until B_1 and B_2 lie on the same great circle. Move both holes along small circles to the positions B_1' and B_2' on the principal diameter.
3. Rotate the tracing paper so that B_1' lies on the N–S diameter and trace small circles at 45° from each pole. This traces the locus of all possible positions of the poles to the bedding intersected in the hole.
4. Rotate the tracing paper until B_2' lies on the N–S diameter and trace small circles at 65° from each pole to plot the locus of all possible positions of the poles to the bedding intersected in the hole. The intersection points, P and Q, give the two possible solutions.

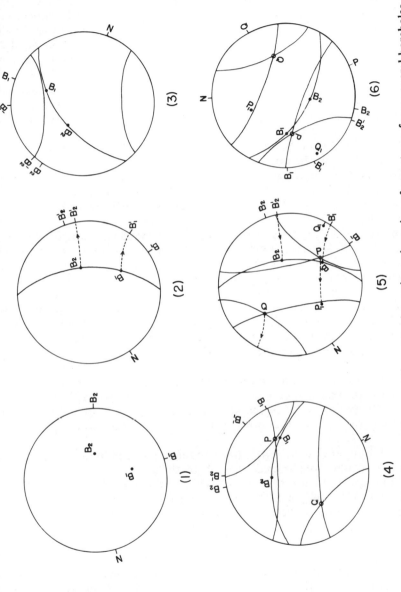

Fig. 7.17. Use of an equal area stereographic net to determine orientation of structures from several boreholes with different orientations.

5. The drill holes, B_1 and B_2, now have to be restored to their correct inclinations from the horizontal positions they were rotated into in step 2. Rotate the tracing paper until B_1 and B_2 lie on a great circle as in step 1 and move B_1' and B_2' along small circles back to their correct inclinations. Points P and Q also have to be rotated in the same sense and by the same amount along small circles to Q_1 and P_1.

6. Rotate P_1 and Q_1 to the E–W diameter in turn, mark the position at the end of the E–W diameter on the opposite side to the centre and measure the angular displacements from the centre. This gives the two possible dips, 38° and 84°.

Now, a third hole B_3 was drilled at −55° to 100° and intersected the bedding at 26° to the core axis.

7. This hole B_3 is paired with either of the first two and the procedure described above is followed. When this is done it is found that one of the points coincides with P giving the correct solution. Thus, the beds dip at 38° to 159°.

In areas with poor exposure it can be very useful to drill at least one hole with a different direction to the others to determine the true orientation of structures such as bedding, foliation, joints or veins. Sometimes two holes will give a unique solution, but it is generally necessary to have data from three holes. It is not necessary that the holes have different azimuths. In the example given, for instance, a third hole drilled due west at −70° would have had a bedding intersection at 50° to the core axis to give the correct solution.

To practise the method the reader may wish to solve the following problems.

1. A vertical drill hole intersected a number of quartz veins at 35° to the core axis. A second hole inclined at −70° to 120° also intersected the veins at 35° to the core axis. What is the dip and what are the two possible dip directions of the veins? (The answer is 55° to 22° or 55° to 216°.)

2. A hole angled at −45° south intersected bedding at 30° to the axis. A second hole angled at −70° to the northwest intersected the bedding at 31° to the core axis. What is the dip of the beds? (The answer is 50° to 84° or 28° to 250°.) A third hole is drilled southwest at −65° and intersects the bedding at 57° to the core axis. Which is the correct solution?

Core logging

Core logging forms an important aspect of an exploration geologist's job, as diamond drilling is usually the final and most important stage in the follow-up work to an exploration target. No special skills are required other than the ability to make competent mineral and rock identifications. The core should always be examined in good light, preferably natural light, though this may not always be possible. The core boxes should be laid out in correct order and, if a lot of core is being logged, it is best if the boxes are placed on trestles at a convenient height. Core that has been correctly placed in the boxes should run from left to right with the shallowest depth in the top left-hand corner and the lowest depth in the bottom right-hand corner.

Small wooden blocks with the depth clearly written on with a marker pen should be placed at the end of each drill run by the driller. For logging the geologist will require a notebook or logging sheets, a pen, a tape measure, a penknife, a clinometer or protractor and ruler for measuring angles, dilute HCl if there are any carbonate rocks, and possibly some test reagents (see following section). The core should be clean and it is the job of the driller to ensure that the core has been washed off before it is placed in the boxes. For easy identification of rock types it is always best to look at wet core so a bucket of water and paint brush or a watering can should be at hand during logging.

Besides noting lithologies and mineralization the geologist should also note core recovery for each run and it is important to ensure that the core has been placed in the boxes correctly and not 'stretched out' to give the appearance of better recovery than has been achieved. The percentage recovery can easily be determined by measuring each run with a tape measure. For example, if there is 1.52 m of core in the box within a run from 101.20 to 104.4 m, the recovery indicated is 54%. Any depths of interest should be interpolated if they fall within a run of less than 100% core recovery. For instance, in the example just quoted a lithological contact 56 cm from the 101.20 m marker would be calculated at a depth of 102.24 m and not 101.76 m.

As a core sample is a small cylinder of rock it is possible to misinterpret certain sedimentary or tectonic structures if the geologist does not try to visualize the rock as it might appear *in situ*. All structures such as bedding, foliation, joints, lineation, minor folds, etc. should be noted and dips measured with respect to the core wherever possible. There are two conventions for doing this: one measures the dip with respect to the core axis and the other with respect to the normal to the core. The core axis convention is the more common, but it is important to specify which convention is being used; if it is not specified, it is usually assumed to mean the core axis. For example, bedding in a particular hole might be measured as '30° to core axis' or '60° core normal'. Some workers prefer the core normal convention, which is sometimes referred to as the 'core dip', since it is equivalent to true dip in vertical holes.

In addition to going through the core centimetre by centimetre and box by box, it is extremely important to lay out in order all the boxes for a particular hole, so that the geologist can obtain an overall view. Very often major divisions that are not readily apparent if the core is simply examined box by box can be seen when the entire hole is laid out. In core logging it is important to break down the hole into a few major units. In areas with a clearly defined stratigraphic sequence this is generally straightforward, but even in new and unknown areas it is usually possible to define major rock units. This is easiest in a sedimentary succession, but major textural and compositional changes can also be noted in igneous rocks. A core log is easier to follow if it is broken up into a number of larger divisions. Nothing is worse than looking at page after page of meticulous logging without any breaks. The principle is illustrated below with a simple example from the hypothetical Bahati prospect, but it should not be considered as a good example of core logging as the descriptions are too brief and no structural information is given.

Core Log 1		Core Log 2	
0–5·3 m	argillaceous limestone	0–32·4 m	UPPER LIMESTONE
5·3–7·4 m	argillite		FORMATION
7·4–10·0 m	calcareous mudstone	0–5·3 m	argillaceous limestone
10·0–11·1 m	argillite	5·3–7·4 m	argillite
11·1–11·5 m	calcareous sandstone	7·4–10·0 m	calcareous mudstone
11·5–12·5 m	argillite	10·0–11·1 m	argillite
12·5–24·4 m	argillaceous limestone	11·1–11·5 m	calcareous sandstone
24·4–26·8 m	massive limestone	11·5–12·5 m	argillite
26·8–30·6 m	argillaceous limestone	12·5–24·4 m	argillaceous limestone
30·6–32·4 m	calcareous mudstone	24·4–26·8 m	massive limestone
32·4–50·1 m	fine-grained sandstone	26·8–30·6 m	argillaceous limestone
50·1–58·7 m	argillaceous limestone	30·6–32·4 m	calcareous mudstone
58·7–62·0 m	calcareous mudstone	32·4–50·1 m	BAHATI SANDSTONE
62·0–62·5 m	siltstone		FORMATION
62·5–64·9 m	argillaceous limestone		fine-grained
64·9–65·6 m	argillite		sandstone
65·6–66·0 m	sandy limestone	50·1–73·2 m	LOWER LIMESTONE
66·0–68·1 m	calcareous mudstone		FORMATION
68·1–73·2 m	argillaceous limestone	50·1–58·7 m	argillaceous limestone
		58·7–62·0 m	calcareous mudstone
		62·0–62·5 m	siltstone
		62·5–64·9 m	argillaceous limestone
		64·9–65·6 m	argillite
		65·6–66·0 m	sandy limestone
		66·0–68·1 m	calcareous mudstone
		68·1–73·2 m	argillaceous limestone

Various forms have been designed for diamond drill core logging and they may vary considerably depending on circumstances. For general purposes, it is useful to have a form that combines assay results with the core description. It is also preferable to have a separate column for mineralization as distinct from lithological description (Fig. 7.18). This does result in reduced space on the forms unless they are very wide, but it shows clearly the mineralized intervals which are of most interest. If mineralization is described within the main body of the litholigical description, as is done on some logging forms, it is not easy to abstract the mineralized intervals from the core log. Therefore, if a separate column is not used for describing mineralization, it is important to place the description of mineralization on separate lines below the lithological description, as shown below:

222·5–224·3 m   Light-grey, hard massive dolomite with abundant disseminated pyrite. Core rec. 100%.
224·3–227·4 m   White to brown, somewhat oxidized sandy dolomite. Bedding indistinct. Core rec. 100%.
Mineralization: weakly disseminated bornite to 226·0 m becoming much richer from 226·0 to 227·4 m.

In addition to ordinary logging forms special ones have been designed for computer plotting. This method is used more and more widely today, but it is of most value on mines where one is dealing with a narrow range of rock types and mineralization. For general exploration work computer plotting is of dubious value, though it should definitely be considered when an exploration target reaches the status of a prospect and is being drilled off.

Another aspect of core logging that should be considered is _geotechnical logging_ which is the noting of features relating to rock strength. This is not important in cases where a project is abandoned after drilling a few holes, but in cases where a project reaches the status of a prospect on which feasibility studies are undertaken, it is extremely important as the geotechnical properties have a direct bearing on the potential mining economics. There are cases of mines which proved to be uneconomic largely owing to poor rock strength which drastically affected the stability of stopes or pit walls; such expensive mistakes might have been avoided had more attention been paid to the geotechnical characteristics at the drilling stage. Geotechnical studies are part of the specialized field of engineering geology, but the features to note in core logging are fairly simple and include the following:

1. _Core loss_ is a good guide to rock strength. Strong, competent rocks tend to core well and soft, friable ones badly, though there are exceptions.

DIAMOND DRILL CORE LOG — SAMPLE RECORD

HOLE No MW3

PROPERTY MWAMBASHI

GRID REF 1100E 66+N

DRILLED BY Geomin.

STARTED 5-5-75 FINISHED 3-6-75

CORE SIZE 76mm to 162m, 66mm to 265m

LOGGED BY D.L. & J.H.R

BEARING 259° mag.

ELEVATION 1264+m

TOTAL LENGTH 265.0m

SURVEYS

collar	-70°	259°
	-68°	260°
160	-67°	263°
265		

| metres | | LITHOLOGICAL DESCRIPTION | MINERALIZATION | sample No | assays | | | from | to | rec. | rep. |
from	to				% Cu	% Pb	% Zn				
0	21.3	OVERBURDEN									
0	4.3	red-brown soil – no core									
4.3	21.3	Yellow-brown, pisoolitic laterite									
		core rec. 49%									

DIAMOND DRILL CORE LOG — SAMPLE RECORD

PROPERTY MWAMBASHI

HOLE No MW3   PAGE No 5

| metres | | LITHOLOGICAL DESCRIPTION | MINERALIZATION | sample No | assays | | | from | to | rec. | rep. |
from	to				% Cu	% Pb	% Zn				
		LOWER ROAN									
197.7	222.5	Black Shale									
		dark-grey, fissile, highly pyritic shale									
		somewhat leached and porous to 212m									
		Bedding varies from 68° to 75° to core axis									

FIG. 7.18. Example of part of a diamond drill log.

DIAMOND DRILL CORE LOG – SAMPLE RECORD

PROPERTY MWAMBASHI

HOLE No MW3   PAGE No 6

| metres | | LITHOLOGICAL DESCRIPTION | MINERALIZATION | sample No | assays | | | metres | | | |
from	to				% Cu	% Pb	% Zn	from	to	rec.	rep.
227.4	253.4	Footwall Formation									
227.4	228.6	Somewhat weathered, porous, poorly sorted	dissem. chalcopyrite and	516	2.67			231.0	232.0	1.0	1.0
		conglomerate with granite pebbles. core rec 100%.	chalcocite	517	2.23			232.0	233.0	1.0	1.0
				518	2.51			233.0	234.0	1.0	1.0
228.6	235.4	hard, massive, dolomitic quartzites with	dissem. chalcopyrite-weak	519	2.75			234.0	235.0	1.0	1.0
		occasional dark, biotite-rich granite cobbles.	to 230m. mod. good below	520	3.12			235.0	236.0	1.0	1.0
		core rec. 100%	230m.	521	3.91			236.0	237.0	1.0	1.0
235.4	253.4	conglomerate with abundant dark biot.-rich	dissem. cpy and rare	522	4.05			237.0	238.0	1.0	1.0
		granite pebbles – several calcareous	bornite-quite rich in places	523	4.44			238.0	239.5	1.5	1.5
		quartzite layers up to 50cm thick. Bedding	to 242m. virtually absent	524	3.00			239.5	240.5	1.0	1.0
		approx. 70° core axis. core rec. 100%	242-246, weak 246-252.	525	1.22			240.5	242.0	1.5	1.5
		– unconformity –		526	0.42			242.0	248.0	2.0	2.0
		BASEMENT		527	1.22			248.0	250.5	2.5	2.5
253.4	265.0	hard, fresh and poorly jointed leucocratic		528	0.67			250.5	252.5	2.0	2.0
		biotite granite									

FIG. 7.18.—contd.

2. *Drilling breaks per metre* should be noted and natural breaks (joints, bedding, faults) distinguished from drilling ones.
3. *Planes of weakness*, which include bedding planes, joints and schistosity, should be recorded, giving the core angles. Orientations are also important and, if true orientations are not known, as is usually the case, orientations relative to one another should be noted if possible.
4. *Rock strength* can be noted qualitatively using a scale that varies from the extremes 'impressions can be made with a fingernail' to 'broken with difficulty with a hammer'.
5. *Potential aquifers* should also be noted using porosity and fracture density as a rough guide.

Staining techniques

To assist with logging there are a number of staining techniques that can be applied to drill core for the identification and differentiation of various minerals. A common problem is to differentiate the various carbonate minerals, particularly calcite and dolomite. These can be roughly distinguished by the fact that calcite reacts readily with cold dilute HCl whereas dolomite reacts only very slightly and slowly, but there are two staining tests that are more diagnostic, one using alizarin-*S* and the other potassium ferricyanide.

Carbonate test no. 1. Etch the surface to be tested for a few minutes with dilute HCl. Then cover with cold reagent ($0 \cdot 1$ g alizarin-*S* in 100 ml $0 \cdot 2\%$ HCl) and leave to react for about 15 min. Any carbonates present will take on the following stains:

> calcite, aragonite, witherite—deep red
> ankerite, ferroandolomite, strontianite, cerrusite—purple
> dolomite, magnesite, siderite, smithsonite—unstained

Carbonate test no. 2. Combine equal parts of 2% HCl and 2% potassium ferricyanide to form the test reagent. Immerse sample to be tested in reagent for several minutes. Calcite and aragonite remain unchanged, ankerite and ferroandolomite turn dark blue with cold reagent and dolomite and siderite also turn dark blue, but only if the solution is hot and it is left to react for up to 5 min.

There are also some useful tests for sulphides. An old test (Gaudin, 1935) to differentiate pentlandite, pyrrhotite, chalcopyrite and pyrite is as follows:

Dissolve 20 g chromic oxide in 100 ml H_2O and then top up to 250 ml with concentrated HCl. This mixture is then diluted 1:2 with H_2O and left to age for half an hour before use. Immerse the sample in the test solution for 6 min at room temperature (20°C). Remove and wash off with water, then with ethyl alcohol and finally with diethyl ether. Allow to dry and the following stains should result:

> pentlandite—brilliant blue
> pyrrhotite—dark bronze
> chalcopyrite—brass yellow
> pyrite—unchanged

The temperature and time of contact are important for this test.

A useful stain for sulphides in general is known as *Mangula paint*. This is made up as follows:

Dissolve 40 g ammonium molybdate in 100 ml H_2O. Allow to stand for 2–3 h and add 20 g sodium pyrophosphate and dissolve. Any ammonium molybdate that remains undissolved will dissolve readily after addition of the phosphate. Pour into a polythene bottle and add 200 ml concentrated HCl followed by titanium dioxide which is used to give the white base. Up to 170 g titanium dioxide is recommended, but less than this may be sufficient to form a satisfactory white 'paint'. The indicator is now ready and needs to be used within a few hours of mixing. It is applied to the core or rock to be tested with a paint brush and will quickly turn blue if any sulphides are present. The test is useful for indicating the presence of fine-grained sulphides which are not readily visible, e.g. chalcocite in a dark argillite.

Some tests specific for certain metals may also be used. In the case of nickel the dimethyl-glyoxime test described in Chapter 2 is useful. Drops of *aqua regia* are first placed on the surface of the sample to be tested and then some drops of the DMG test solution are applied. A red colour indicates nickel.

A useful test for secondary zinc minerals is as follows:

Dissolve 30 g potassium ferricyanide in 1000 ml H_2O (solution 1). Dissolve 30 g oxalic acid in 995 ml 0·1 N HCl, add 5 ml *N,N*-diethylaniline and shake well (solution 2). For field use mix equal parts of solutions 1 and 2 to form test reagent. (The mixed solution deteriorates over a week or two so only enough for a few days use should be made up at a time.) When applied to the test sample any secondary zinc minerals present will turn orange.

Sampling core

Two types of samples are commonly taken from drill core: *chip samples* and *split core samples*. The more important are split core samples which are taken for accurate evaluation work of definite mineralized intervals. Before taking the samples, the geologist examines the core and marks off the intervals to be sampled by drawing a line along the core with a marker pen. When the intervals have been selected, the core is split in half by using a diamond saw or core splitter, which consists of a special press or hollow metal block for holding a short length of core. A long chisel blade is pressed against the core and struck with a hammer. If done correctly, the core should break longitudinally into two equal pieces, though the break is often uneven. If a diamond saw is available, it is preferable to a core splitter as it results in a perfect 'split', but, as most core splitting is done in the field, diamond saws are not commonly used. Once the core is split the geologist selects the individual sample lengths, taking care to note lithological boundaries and changes in grade and/or style of mineralization. The lengths chosen should normally vary from 1 to 3 m, unless sampling a definite narrow vein or mineralized bed with barren country rock on either side. Narrow sampling widths are sometimes taken for promotional reasons of rather dubious ethics to draw attention to a high grade. Assay tickets are made out and each sample placed in a numbered sample bag. Figure 7.19 shows a typical assay ticket which is usually placed in a book of 50 or 100 tickets and is divided into three portions by two rows of perforations. The end ticket is torn off and placed in the sample bag together with the sample, the second ticket is torn off and placed in the core tray or box with the split core that is retained for reference and the stub is left in the book. The bags of split core are then dispatched to a laboratory where they are crushed and a small split taken for analysis.

If there is a change in core size, it is important not to include both core sizes within a single sample, i.e. the depth where the core size changes should also mark the end of a sample. Assay results can be assumed to be proportional to length and weighted averages can be calculated for intervals spanning different core sizes, but samples for assay should not be made up from different core sizes. This principle can be illustrated by considering a simple example of changing from BX (44 mm diameter) to AX (28 mm diameter) core. If a 1-m sample immediately above the core size change assays 4·4% Zn and a 1-m sample immediately below the size change assays 9·6% Zn, the

Remains in book	Place in core tray	Place in sample bag
No. 0901	No. 0901	No. 0901
Property _____	Property _____	Assay for:
Hole No. _____	Hole No. _____	
From _____ to _____ metres	From _____ to _____ metres	
Assay for _____	Assay for _____	
Signed _____	Signed _____	

FIG. 7.19. Typical assay ticket used in diamond drill core sampling.

weighted average for the 2 m spanning the core size change is 7·0%. If, on the other hand, a single 2-m sample was taken over the same interval, the assay would produce a result of 5·9% Zn assuming the same s.g. Owing to the smaller size of the AX core, it has been under-represented in the total amount of material sent for assay. It is important to be aware of this problem when making up composite samples from existing sample material, e.g. using the coarse rejects of split core samples to make up a large sample for metallurgical testing. As an illustration, consider 25 m of AX (28 mm) core averaging 10·1% Zn, 35 m of BX (44 mm) core averaging 8·2% Zn and 15 m of NX (54 mm) core averaging 5·6% Zn, which are used to make up a composite sample for metallurgical tests. The weighted average, which represents the material *in situ*, gives 8·3% Zn, but when the samples are bulked a representative portion would assay only 7·3% Zn.

Chip samples are taken for semi-quantitative purposes, usually as a check against missing any mineralized intervals during logging or for rock geochemistry studies. The procedure consists in breaking off a small chip of core every 25–30 cm and making up composite samples corresponding to lengths of 2–3 m. Some companies take chip samples from all drill cores as a matter of routine whether any mineralization has been observed or not. This policy has much to commend it since the cost of the geochemical analysis is so small in comparison to the cost of drilling the hole, and it is a safeguard against missing any mineralization which may be difficult to identify, such as secondary zinc minerals in limestone or chalcocite in black argillite.

Drill sections

In any drilling job it is important to plot drill sections as the work progresses so that a full three dimensional picture can be obtained. Under the section on borehole surveying the method of plotting a drill hole in section from survey data was given. Since drilling is usually undertaken on an accurately surveyed grid, a series of sections is drawn along various grid lines across the strike. Figure 7.20 gives an example of a drill section across a stratiform copper deposit in Zambia.

To enable a deposit to be visualized in three dimensions it is useful to be able to show a number of sections together. One method of doing this is to plot the drill sections on transparent plastic or glass sheets which can be mounted in a wooden frame to produce a three dimensional model of the deposit being drilled. Instead of producing an actual three dimensional model, a three dimensional representation can be made on a two dimensional drawing. A common method of doing this is known as an *isometric section.* To plot an isometric

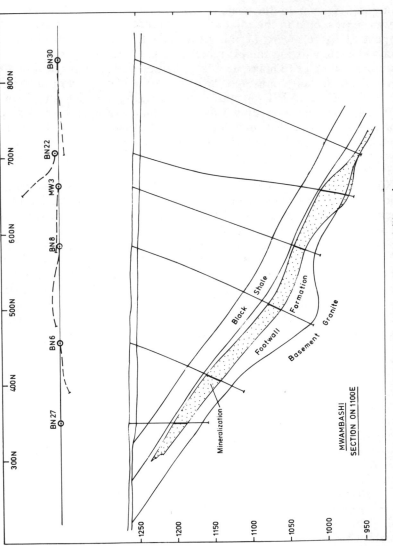

FIG. 7.20. Example of a drill section.

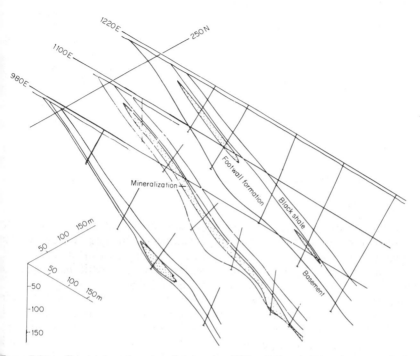

FIG. 7.21. Example of some isometric drill sections through part of the Mwambashi copper deposit on the Zambian Copperbelt.

section three rectilinear axes to represent two horizontal directions and a vertical direction are drawn on the paper, usually at 60° to each other and then all distances are plotted at true scale along the three axes. Figure 7.21 gives an example of some isometric sections through a deposit.

7.4   ROTARY DRILLING

Rotary drilling techniques are unequalled for drilling through overburden and soft rocks. The rigs vary in size from truck-mounted machines capable of a depth penetration of 600 m to extremely large machines used in drilling oil-wells to depths of 6000 m and more. The typical rotary bit used for most general work is shown in Fig. 7.22 and is known as a tricone or roller rock bit. Before 1956, it was not practical to drill hard rocks with such bits since the steel-toothed roller rock bits in general use were not capable of drilling igneous or

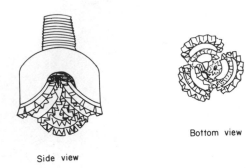

Bottom view

Side view

FIG. 7.22.   Sketch of roller rock bit used in rotary drilling.

hard metamorphic rocks and diamond coring or 'full-hole' bits had to
be used for penetrating hard rock formations. In more recent years
however, tricone bits with tungsten carbide button inserts have
become available and it is possible for a rotary drill to penetrate
almost all rocks without resorting to diamond bits.

Rotary drills are capable of an extremely high rate of penetration in
soft rocks and the larger machines used in oil-well drilling can drill
over 100 m/h, which is virtually as fast as the cuttings can be pumped
clear. In igneous rocks penetration with roller rock bits is very much
slower and may be as little as 1 m/h. For this reason rotary drills used
in mineral exploration or water-well drilling are combined with
down-the-hole percussion drills. Such combination rigs are available
mounted on large trucks complete with compressor, hoisting tower,
automatic rod handling facilities, mud pump and all ancillary equip
ment. These machines are very versatile and can drill holes quickly
through any type of formation by using roller rock bits for over
burden and soft rocks and the percussion drill for hard formations.
Some of these rotary rigs are equipped for dual tube drilling which
gives extremely good results in overburden or fractured and cavern
ous formations.

Roller rock bits are available in a wide range of sizes starting
from $2\frac{7}{8}$ in (73·9 mm), but most commonly used sizes in mineral
exploration are from $4\frac{1}{4}$ in (108 mm) to $6\frac{1}{4}$ in (158·8 mm). Large
sizes are used in oil-well drilling and may be up to 24 in
(609·6 mm), though somewhat smaller diameter holes are usually
drilled. The rods are normally 25 ft (7·62 m) in length, but in oil-well
drilling large derricks are able to handle rods four times this length.
Rotary drilling can be carried out at half the cost of diamond drilling
and in cases where core is not considered necessary, rotary drilling
has considerable merits in mineral exploration. It should also be noted
that diamond coring can be carried out with combination rotary rigs

but it is worthwhile only if short sections of the hole are to be cored. A major disadvantage of rotary drills is their large size which makes them totally unsuitable for drilling in remote locations. The much smaller diamond drill rigs can be manoeuvred into very awkward sites and can be easily slung beneath helicopters if necessary.

Sampling and logging

Rotary drill sampling is essentially a sludge sampling operation as described earlier in this chapter, the only real problem being the large volume of sample that is produced in a very short space of time so that facilities have to be available for handling large amounts of sludge. Since drilling muds are used, the samples collected will consist of a mixture of drill cuttings and mud, and for logging the sludge needs to be screened and washed to remove the drilling mud. After screening, the mud can be recirculated for further use in drilling. Roller rock bits drill by a crushing and breaking action so that cuttings tend to be fairly coarse, but in rock formations such as soft shales a considerable proportion of the sludge may be extremely fine and not separable from the drilling mud. This may result in sampling problems since contamination is unavoidable in such circumstances.

To assist with logging open holes a wide range of geophysical techniques is available (see Chapter 6) and can be of great assistance in identifying various rock formations and mineralized zones. In addition to these older established methods, new techniques have been developed which permit determination of metal content *in situ*. One such method, developed by Scintrex Ltd in Canada and known as Metalog®, makes use of the phenomenon of neutron activation (see Chapter 4). Most of the common base metals can be determined and the method has an advantage over a conventional assay in that the metal content of a much larger volume of rock is determined *in situ*. The equipment is very costly and is carried in the back of a specially equipped pick-up truck so that it is not a practical method for most exploration work. However, in the evaluation of a mineral deposit where a large number of drill holes are put down, the technique has considerable appeal. The cost saving in being able to drill open rotary holes instead of diamond drill holes may more than offset the cost of logging the holes by Metalog®.

REFERENCES AND BIBLIOGRAPHY

Campbell, R. C. (1963). Borehole surveying and directional drilling, *Quart. Col. School Mines*, **58**(4), 185–193.

Cumming, J. D. and Wicklund, A. P. (1976). *Diamond Drill Handbook*, J. K. Smit & Sons Ltd, Toronto, 541 pp.

Gaudin, G. (1935). Staining minerals for easier identification in quantitative mineragraphic problems, *Econ. Geol.*, **30**, 552–562.

Phillips, F. C. (1955). *The Use of Stereographic Projection in Structural Geology*, Edward Arnold Ltd, London, 86 pp.

Roxstrom, E. (1961). A new core orientation device, *Econ. Geol.*, **56**, 1310–1313.

Smail, E. L. J. (1966). Diamond drilling at Llanharry iron ore mine, South Wales, *Trans. Instn. Min. Metall.*, Lond., **75**, A175–181.

Walsham, B. T. (1967). Exploration by diamond drilling for tin in west Cornwall, *Trans. Instn. Min. Metall.*, Lond., **76**, A49–56.

Zimmer, P. W. (1963). Orientation of small diameter drill core. *Quart. Col. School Mines*, **58**(4), 67–82.

CHAPTER 8

Surveying

Surveying is the science of measuring man-made and natural features of an area of land and plotting these measurements to scale on a plan or section. *Geodetic surveying* is the science of determining the shape of the earth. It is the method used for surveying large areas such as entire countries and it is the main framework in which other surveys are located. *Topographic surveying* is the measurement and determination of physical features. *Cadastral surveying* is defined as measuring, defining and recording boundaries of properties. Very often topographic and cadastral surveys are carried out together. *Engineering surveying* is the accurate measurement of small areas for the purposes of constructing buildings, roads, railways and other engineering structures. Very often engineering work involves surveying in reverse, i.e. measurements have to be transferred from plans to the ground. With the exception of geodetic surveying, which involves spherical trigonometry, other types of surveying are based on simple trigonometrical relationships involving the measurement of angles and distances.

8.1 CHAINING

This is the simplest type of surveying and it is based purely on the measurement of distances. The land chain or surveying chain is made up of heavy gauge steel wire links each 20 cm long joined together in standard lengths of 20, 30 or 50 m. At either end of the chain are robust brass handles which enable the chain to be dragged along the ground and to be pulled taut between measurement points. Plastic tags are placed along the chain at intervals of 1 m with tags of a different colour every 5 m so that distances less than the full length of the chain can be measured with little difficulty. The length of a chain is always measured between the outside edges of the handles.

Field procedures

Chaining surveys are adequate for detailed surveys of small, relatively open areas with little topographic relief. The basic method of a chaining survey is to divide the area up into a series of triangles since the triangle is a geometric figure which can be determined uniquely simply from the measurement of its sides. This is done by selecting a number of points which will form the apices of an interlocking network of triangles within the area to be surveyed. These points should be selected so that the triangles formed are as close to equilateral triangles as is practically possible. Such triangles are termed *well conditioned triangles*. The various points or stations selected should be marked with *ranging rods* or pegs before the survey commences. Ranging rods are wooden or metal poles 2–3 m long, tipped with a steel point at one end and painted red and white in alternate bands 0·5 m long. In addition the approximate shape of the triangle network being used should be sketched in the field notebook and the stations correctly numbered or lettered. Ranging rods should be placed on the stations between which measurements are being made. This requires two people, a leading chainman and a follower. Each carries a ranging rod and in addition the leading chainman carries ten *surveying arrows*. These are straight lengths of heavy steel wire about 50 cm long, sharpened at one end with a ring for carrying at the other end. The leader pulls the chain out along the line of measurement while the follower holds the handle of his end of the chain against the survey station. When the chain is taut, the follower signals the leader exactly onto line by sighting along the ranging rods. The chain is shaken to make sure it is absolutely straight and the leader pushes an arrow into the ground against the outside edge of the chain handle. Then, while the chain is on the ground, offset measurements can be made to features of interest to either side of the line. When this is complete, another chain length is measured in the same manner as above, the leader places another arrow in the ground to mark the chain length and the follower removes the other arrow from the ground when the advance to the next chain length is made. The purpose of the arrows is not only to mark the chain lengths, but also to avoid any uncertainty in the number of chain lengths measured. On completion of the measurement between two stations the number of arrows carried by the follower will be equal to the number of full chain lengths measured. Offsets are made at right angles to the chain line and, if objects are fairly close (less than 6 or 7 m), the right angle can be estimated by eye. For larger offsets an *optical square* is very useful. This is a simple device which contains a pentagonal prism so that when it is held in the hand on a survey line a ranging rod to the side of the line can be lined up so that its image in the optical square coincides with that of a ranging rod sighted at the far end of the survey line. When this occurs, the ranging rod to the side of the line will be at right angles to the survey line at the point where the optical square is held. In addition to offsets, detail to the side of the chain line can be determined by measurements known as *ties*. The procedure is illustrated in Fig. 8.1 and, as this is a type of triangulation, points fixed by tie lines are very accurate. Field notes and measurements are written in special surveying notebooks which consist of good quality paper bound in hard, waterproof cloth covers. The pages are 20×10 cm and are bound along the short side in the same manner as a shorthand notebook. Two red lines about 15 mm apart to represent the chain line are ruled down the centre of the page. Bookings are always made from the bottom to the top. An example of a survey is given in Figs. 8.2 and 8.3 which illustrate the method of booking.

Although chaining surveys are usually carried out over areas where the slope of the ground does not need to be taken into account, it is sometimes necessary to chain up or down a fairly steep slope. When this happens, corrections have to be applied for the slope since the distance along the slope will be greater than the horizontal distance which is the measurement plotted on a plan. The simplest method to account for slope is known as 'stepping' the chain. This is done by measuring the slope in a series of steps with the up-slope end of the chain on the ground and the down-slope end above the ground so that the chain is approximately horizontal. The point of measurement on the ground on the down-slope side is then vertically below the end of the chain and is located with a plumb bob or drop arrow (a surveying arrow with a lead weight on the end). It will be appreciated that it is

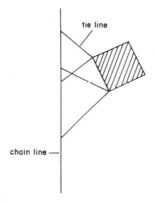

FIG. 8.1. The use of tie lines to locate detail to the side of a chain line.

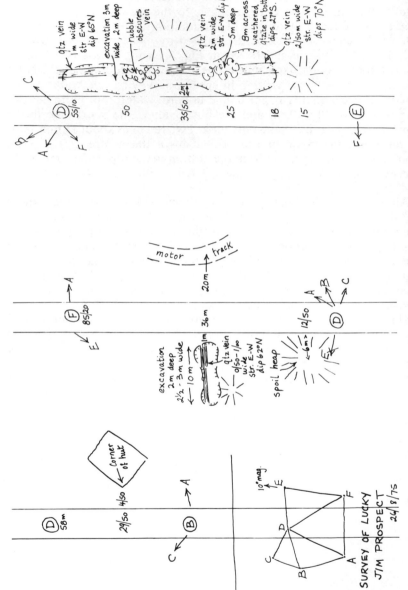

FIG. 8.2. Example of booking a chaining survey.

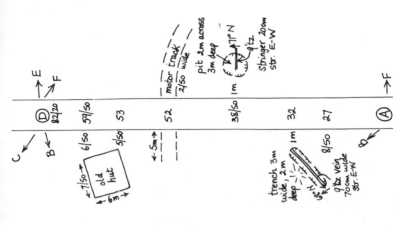

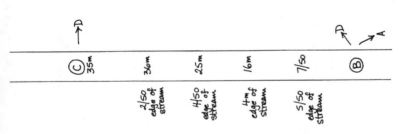

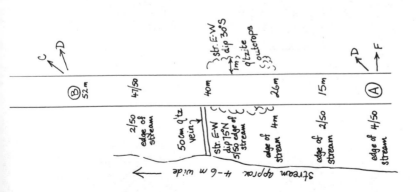

FIG. 8.2.—*contd.*

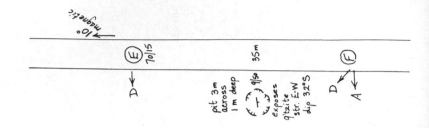

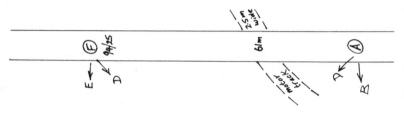

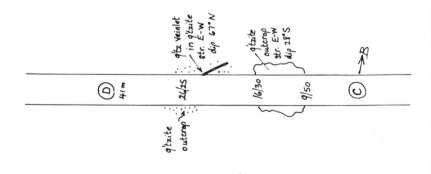

FIG. 8.2.—contd.

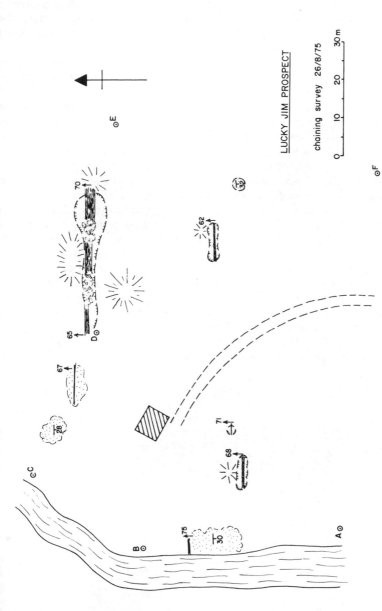

LUCKY JIM PROSPECT

chaining survey 26/8/75

Fig. 8.3.   Plot of the chaining survey booked in Fig. 8.2.

not possible to step a full chain length and it is usually necessary to step the chain in a series of lengths of 5–10 m. If too long a step is made, errors due to sag in the chain may well be greater than the error due to slope. It should also be appreciated that it is much easier to step a chain downhill than uphill.

Long even slopes can be corrected for by measuring the angle of slope with an instrument such as an Abney level. This simple instrument consists of a sighting tube to which a graduated arc is attached. A small bubble tube is mounted on and at right angles to an index arm pivoted at the centre of the graduated arc. An inclined mirror mounted in one half of the sighting tube enables the bubble to be seen at the same time as a target being viewed. To determine a vertical angle the target is sighted through the tube and the index arm turned left or right by a milled knob until the bubble appears in the centre of view and the knob is carefully adjusted until the target and bubble are exactly in line with a horizontal sighting wire in the tube. The angle of slope is then read from the graduated arc which is marked from 0 to +90° and −90° and a vernier enables angles to be determined to 10′. The principle of the Abney level is illustrated in Fig. 8.4 To make an accurate measurement of slope it is normal practice for the observer to use as a target a pole with a mark placed at the same height as the observer's eye.

A number of problems are often encountered during chaining surveys and solutions of two of the most common problems are given here. One problem (Fig. 8.5) is chaining between two stations A and B which are not intervisible owing to an intervening hill. A solution is

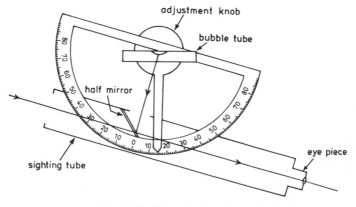

FIG. 8.4. Principle of the Abney level.

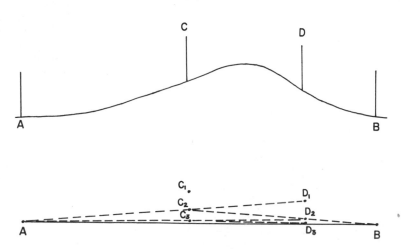

FIG. 8.5.   Method of ranging in a line between two points A and B which are
not intervisible.

for the leader to hold a ranging rod at point D_1 and the follower at C_1
such that C_1 and A can be sighted from D_1 and D_1 and B sighted from
C_1. Then the leader signals the follower from C_1 to C_2 where he is in
line with D_1 and A and the follower signals the leader from D_1 to D_2
where he is in line with C_1 and B. C_2 and D_2 are now closer to the
straight line between A and B than C_1 and D_1, but it is highly unlikely
that they will be on it exactly. To achieve this the process above is
repeated as many times as is necessary until both the leader and
follower find they no longer need to signal each other on to line.
When this happens, the two ranging rods will be on line between A
and B. Chaining of the line can now be carried out with the two
intervening ranging rods used as guides to keep on the true line.

Another problem is measuring across an obstruction such as a large
excavation or river or pond which is wider than a full chain length. A
procedure which can be used is illustrated in Fig. 8.6. The line being
chained is AB and the obstruction in this case is a river. Ranging rods
are placed at C and D so that they are both in line with A and B. Then
a right angle to AB is measured at C and a line chained to E. CE can
be any length but it is best to choose it roughly equal to the width of
the obstruction. Then a further distance EF is measured in line with
EC so that EF = EC. A right angle to FC is measured at F and a
ranging rod moved down the new line until a point G is reached which
is in line with E and D. Since triangle EFG is congruent with ECD,
the distance FG is equal to the unknown distance CD.

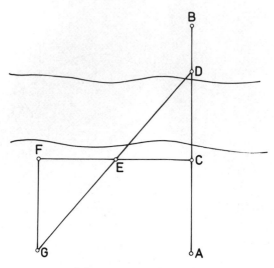

FIG. 8.6. Method of chaining across a river wider than a full chain length.

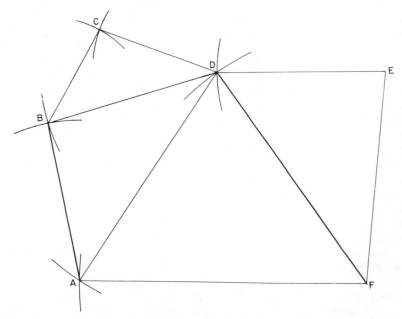

FIG. 8.7. Method of plotting the basic framework of a chaining survey.

Plotting the survey
Before plotting the survey, it is necessary to select a suitable scale. This will depend upon the amount of detail measured, but for most chaining surveys a scale between 1/200 and 1/2500 is usually used. Plotting should be done on good quality paper or permatrace to avoid shrinkage. Before starting the accurate plot, it is a good idea to sketch the framework of the survey to make sure it is orientated so that it will fit on the paper properly. 4H pencils can be used for lightly drawing in the chain lines and 2H pencils for plotting detail. The final plot should be drawn with India ink. Figure. 8.7 shows how the basic framework of chain lines is drawn using a compass to arc off the respective lengths of the different lines to locate the station positions at arc intersections. In the example given EF is scaled off accurately and all other station positions found at arc intersections. It is then a simple matter to plot detail from the field notes. As a check to plotting, it is always useful to measure a check line in the field across the survey area. In the example given (Fig. 8.3) a useful check line might have been AE or CF.

8.2   COMPASS AND TAPE SURVEYS

The compass and tape survey is probably the method of surveying most commonly used by exploration geologists for surveying small areas in detail. If carried out carefully, compass and tape surveys are suitable for quite accurate work. The only restriction is that they cannot be undertaken in areas with strongly magnetic rocks and consequent magnetic disturbance. In some parts of the world the remanent magnetization of rocks is so strong and variable that a magnetic compass is quite useless. These areas are exceptional, however, and compass and tape surveys can be carried out in most regions.

Many different types of compasses are available, but the most suitable for geological work are the standard military prismatic compass, the Brunton compass, the Swedish Silva compass and the Swiss Meridian compass. The Silva compass is not particularly suited to surveying, but it is excellent for geological mapping. It was designed for the Scandinavian sport of orienteering and the compass is mounted on a clear perspex plate in such a way that it can be used for measuring and plotting bearings on a map. It is oil damped and very quick to read, it can be adjusted for declination to give true bearings and can be supplied with a built-in clinometer which is essential for geological work. The Brunton compass is excellent for geological mapping as it incorporates an accurate clinometer and it is good for surveying. Its main disadvantage for survey work is that some people find it difficult to obtain steady readings when hand held. It can be mounted on a small tripod and, when so used, is probably the best compass for the most accurate work. The standard military prismatic compass is the best for ordinary survey work as it is very easy to sight and read accurately. Its main disadvantages are that it cannot be adjusted for declination and it does not have a clinometer. The Swiss Meridian compass, which is also a type of prismatic, is useful in this respect as it incorporates a clinometer, but it is not as robust as the ordinary military prismatic. Whichever compass is chosen, it is important to ensure that it is adjusted for the latitude range in which it is being used. The Brunton compass has a small copper weight on the needle which can be adjusted so that the needle remains horizontal whatever the latitude, north or south. The standard prismatic and Silva compasses, however, are sealed and cannot be adjusted. Thus, it is important to ensure that the region of the world in which they are to be used is stipulated to the makers when ordering a compass. For example, the compass card or needle of a compass set for a high latitude in the northern hemisphere will tilt off the horizontal when used in the southern hemisphere. In many instances this can be tolerated, but, if the tilt is too pronounced, it will prove impossible to obtain reliable bearings.

To undertake a survey a number of ranging rods are required in addition to a good compass and surveying tape. A second tape is also useful for tying in detail to either side to the line. The normal type of surveying is a traverse which should be closed on the starting point or an accurately fixed survey point such as a triangulation station. Compass readings should be taken with the observer standing over the survey station and a bearing taken by sighting on a ranging rod at the second station. As a check for magnetic deviations or errors in reading, back bearings should always be taken. The best bearing for the line can then be taken as the mean of the forward and back bearings. If slopes are more than a few degrees, an Abney level should be used to measure vertical angles so that slope corrections can be made. Detail close to either side of the line can be measured with offsets as in a chaining survey, but for more distant features a tie line with a compass bearing should be used. Figure 8.8 gives an example of how a survey should be booked. An ordinary notebook can be used, but a booking form such as that shown in the above example is a good idea.

Plotting the survey
The normal compass and tape survey can be plotted using a scale and protractor, since angles can be measured with a protractor as ac-

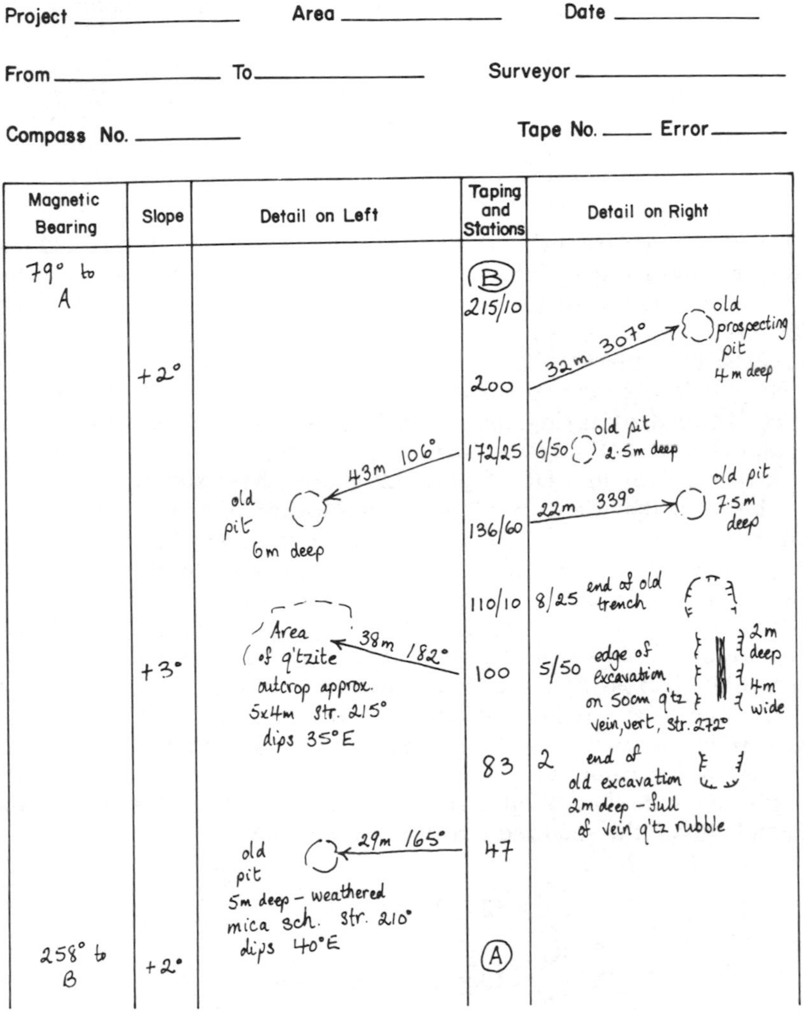

Project _____   Area _____   Date _____

From _____ To _____   Surveyor _____

Compass No. _____   Tape No. ____ Error_____

FIG. 8.8.   Method of booking a compass and tape survey.

curately as they can be read with a compass. If the survey is a closed
traverse, as it normally should be, it will be found that there is an
error of closure when the traverse is plotted and it is necessary to
correct for this. Let us consider the following example of a simple
compass traverse:

Line	Length (m)	Forward bearing	Back bearing
AB	82	56°	238°
BC	74	331°	151°
CD	91	282°	106°
DE	100	172°	356°
EA	45	130°	310°
	392		

It will be seen from this that there is a discrepancy in forward and back bearings for lines AB, CD, and DE. For line AB this discrepancy is 2° and we can take the mean bearing of 57° for plotting this line. The discrepancy for lines CD and DE is 4° in both cases and it suggests that there is probably a disturbance with the magnetic reading at D since the forward and back bearings of both BC and EA agree. Instead of taking the mean in this case we will discard the readings taken at D and only use the forward bearing for line CD and the back bearing for line DE. Figure 8.9 shows the resulting plot and it can be seen that there is an error of closure A'A of 13 m. This is rather an extreme error for such a short traverse, but it serves to illustrate a method of correction. For the traverse to close A' has to move 13 m to A along a bearing of 76°. It is assumed that the error is cumulative and therefore each station has to be corrected in turn. All the corrected points will be moved in the same direction (i.e. 76°) with a gradually increasing correction, smallest at B and greatest at A' (i.e. 13 m). The best method of apportioning the error of a compass traverse is known as *Bowditch's rule* which states that the amount of correction for each side is equal to the total correction times the length of that side divided by the total traverse length. In the example given the required corrections become:

$$\frac{82}{392} \times 13 = 2.72 \text{ for B}$$

$$\frac{74}{392} \times 13 = 2.45 \quad 5.17 \text{ for C}$$

$$\frac{91}{392} \times 13 = 3.02 \quad 8.19 \text{ for D}$$

$$\frac{100}{392} \times 13 = 3.32 \quad 11.51 \text{ for E}$$

$$\frac{45}{392} \times 13 = 1.49 \quad 13.00 \text{ for A'}$$

The corrected traverse is shown in Fig. 8.9.

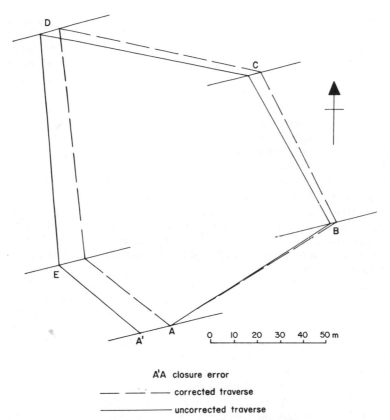

FIG. 8.9. Plot of the compass traverse given in the text.

8.3 DETERMINATION OF ELEVATION

The elevation of a point is defined as its vertical distance above or below a datum surface. For international comparison of elevations the datum surface adopted is mean sea level. For small restricted areas an arbitrary datum plane may be used.

There are three main methods for determining heights or elevations: (1) levelling, (2) measurement of vertical angles and (3) barometric heighting.

Levelling

This method is based on determining a horizontal reference line or series of lines and comparing vertical departures of points of unknown elevation from them. The instrument for doing this is known as a *level* and consists essentially of a sensitive bubble tube attached to a telescope. The level is mounted on a tripod and has three levelling screws for adjusting the telescope to a precise horizontal position. Once adjusted, the telescope can be rotated on a horizontal circle so that readings can be taken in any azimuth. In addition the more accurate levels have a fine setting screw near the eyepiece for tilting the telescope. With this arrangement the bubble can be centred precisely for each reading. Instruments for very precise work contain optical micrometers which enable readings to be estimated to hundredths of a millimetre. For less precise work self-levelling or *automatic levels* are available. These usually have a bull's-eye level for quick and approximate levelling on the tripod head and a pendulum compensator in the instrument automatically ensures that the line of sight is horizontal without further adjustment. These instruments are particularly suited to engineering surveys and generally contain stadia hairs for tacheometric measurement of distance in addition to a horizontal circle which can be read to a tenth of a degree.

To carry out a levelling survey the level is set up near a point of known elevation, for example an Ordnance Survey bench mark, a *levelling staff* is held absolutely vertical by an assistant on the known point and the observer reads the point on the staff intersected by the horizontal cross hair in the telescope. Levelling staffs are usually made of mahogany or aluminium alloy and are telescopic to facilitate transport. With the telescopic extensions fully extended they are 4–5 m long. Figures and 10 mm graduations are clearly marked in red and black for alternate metres on a white background. Some staffs have bulls-eye levels mounted on them to ensure that they are held vertically for accurate readings. In addition staffs for very precise work have figures and 5 mm graduations marked on an invar strip. Once the first reading, known as the *backsight*, has been taken and booked, the telescope can be swung around on its horizontal circle and readings taken to points of unknown elevations (*intermediate sights*). It will be appreciated that these intermediate sights are restricted to an upper elevation range equal to the height of the instrument above the ground and a lower elevation range equal to the total staff length less the instrument height above the ground. To accommodate other points outside this range the instrument must be moved to a more advantageous site. This is done by taking a last reading known as a *foresight* near the extreme of the elevation or sighting range and moving the instrument to a new site where the foresight point is re-read as a new backsight and the whole process is repeated. The procedure is illustrated in Fig. 8.10. If there is a big rise or fall in elevation between points being levelled, it may be necessary

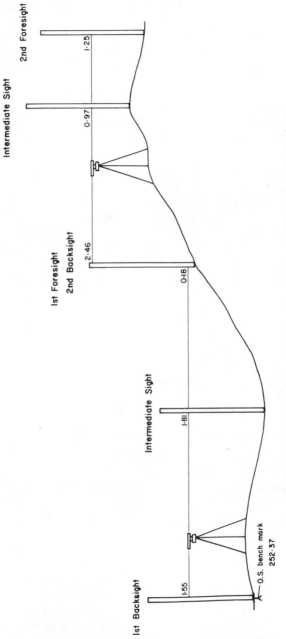

Fig. 8.10. Method of using a level to determine elevations (see Table 8.1).

to make a number of backsight and foresight observations before the required points can be levelled.

There are two methods of booking a levelling survey, the *rise and fall* method and the *height of collimation* method. These are shown in Table 8.1 for the example in Fig. 8.10. In the height of collimation method the first backsight reading is added to the reduced level to give the height of collimation or line of sight. All reduced levels for intermediate sights are obtained by subtracting the staff readings from the height of collimation. This is also done for the foresight. When the new backsight is taken, a new height of collimation is obtained and the

TABLE 8.1

TWO METHODS OF BOOKING A LEVELLING SURVEY. THE EXAMPLE IS ILLUSTRATED IN FIG. 8.10

	Constant Height of Collimation				
Station	Height of collimation	Backsight	Intermediate sight	Foresight	Reduced level
OS bench mark	253·92	1.55			252·37
No.1			1·81		252·11
No.2				0.18	253·74
	256·20	2·46			
No.3			0·97		255·23
				1·25	254·95

		Rise and Fall				
Station	Backsight	Intermediate sight	Foresight	Rise	Fall	Reduced level
OS bench mark	1·55					252·37
No.1		1·81			0·26	252·11
No.2			0·18	1·37		253·74
	2·46					
No.3		0·97		1·49		255.23
			1·25	1·21		254·95

procedure is repeated. In the rise and fall method differences between the backsight and intermediate and foresight readings are determined. If the backsight reading is greater, the difference is booked as a rise and, if the backsight reading is smaller, the difference is booked as a fall. The reduced levels of the intermediate sights and foresight are then obtained by adding a rise and subtracting a fall from the reduced level of the backsight. In both methods booking errors can be checked since the difference between the sum of the backsights and sum of the foresights should be equal to the difference between the reduced levels of the first backsight and last foresight.

For accurate work a levelling traverse should be closed, either to the starting point or to another precisely levelled point such as an Ordnance Survey bench mark. There are a number of sources of error in levelling and these include: inaccurate staff readings, instrument out of adjustment, instrument not precisely level and staff not vertical. Most errors can be minimized by ensuring a high standard of field work. Errors in reading the staff can be eliminated by practice and by keeping sight lengths short. The clarity with which the staff can be seen depends upon the telescope, but with most instruments readings can be estimated reliably to 2 mm at a distance of 60 m. The accuracy with which the instrument is levelled depends upon the instrument being correctly aligned and on the sensitivity of the bubble and accuracy with which it can be centred. This is largely governed by the quality of the instrument, but it should be possible to keep the error within ±5 mm/km for most instruments.

Other errors in levelling are caused by the curvature of the earth and by refraction of light rays in the atmosphere. Consider two points A and B in Fig. 8.11 at the same elevation on the earth. A level with

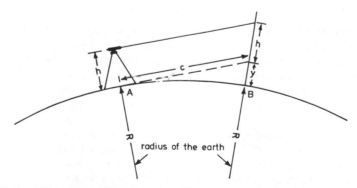

Fig. 8.11. Diagram to show how levelling errors are caused by the curvature of the earth.

instrument height h at A should give a staff reading of h at B, but owing to the curvature of the earth, the line of sight, which is tangential, will give a reading of $h + y$ (an error) at B. Thus the reduced level of B will be calculated lower than its true elevation. If R is the radius of the earth and the distance AB is c (the arc distance on the ground can be taken as equal to the sighting distance), then

$$c^2 + R^2 = (R + y)^2$$
$$2Ry = c^2 + y^2$$

Since y is very small, y^2 will be even smaller and compared to c^2 can be considered negligible. Therefore, the error due to the curvature of the earth is given by:

$$y = \frac{c^2}{2R} \tag{8.1}$$

Owing to refraction in the atmosphere light rays tend to curve downwards towards the earth and this has the effect of reducing the curvature error. The refraction error is given by:

$$\frac{kc^2}{R} \tag{8.2}$$

where k is a coefficient of refraction and varies from 0·055 to 0·081. A value of 0·068 can be adopted for general use, but values of k can be computed for any area by taking reciprocal readings from known stations. The combined curvature and refraction error is:

$$y = \frac{c^2}{R} \frac{(1 - 2k)}{2} \tag{8.3}$$

For most general purposes this may be expressed as:

$$y = 0·068d^2 \text{ metres} \tag{8.5}$$

where d is the line of sight in kilometres. From this it can be seen that a line of sight of 120 m results in an error of 1 mm. These formulae can be used to make corrections for curvature of the earth and refraction, but in levelling it is normal practice to keep lines of sight shorter than 100 m to avoid any errors. In any case precise staff readings can not be made at sight lengths over 100 m.

Recently *electronic levels* have been developed to determine automatically accurate elevations over a small area. The instrumentation works on the principle of measuring pressure differences in a long, narrow, flexible, liquid-filled plastic tube. Electronic signals from sensitive transducers at either end of the tube indicate the degree of pressure difference which is related to the height difference between the ends of the tube. A sensor head is attached to one end of the tube and the meter with the measuring circuitry and digital display is attached to the other end. To operate the equipment the sensor and instrument are placed at a point of known elevation and the reading is set to the correct value. With the sensor left at the known point the instrument can be carried about automatically displaying the elevations of unknown points on which it is placed. The density of the liquid and hence pressure differences vary with temperature, but the instrument is designed to correct automatically for this variation. The instrumentation can detect elevation differences as small as 3 mm, but its overall accuracy is somewhat less than this. The equipment is unlikely to replace conventional levels for really precise work and its principal use at present is probably in detailed gravity surveys where there is a distinct advantage in being able to determine elevations accurately at the same time as the gravity readings are being taken.

Measurement of vertical angles

If the horizontal distance between two points is known, the elevation of one of them relative to the other can be determined by measuring the vertical angle between the horizontal and a line of sight between the two points. The instrument required for this is known as a *theodolite* (see Section 8.6). Figure 8.12 shows a theodolite with instrument height h set up at a point A to measure a vertical angle to a flag on a hill at point C. If AB is the horizontal distance between A and C, α the vertical angle, h' the height of the flag and H the elevation of A, then the elevation of C is given by:

$$H + AB \tan \alpha + h - h' \tag{8.6}$$

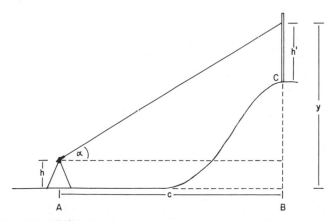

FIG. 8.12.   Theodolite set up at A to measure the elevation of C knowing the horizontal distance AB between A and C.

If the theodolite is on the hill at C with elevation H and the vertical angle negative, the elevation of A is given by:

$$H - AB \tan \alpha - h + h' \qquad (8.7)$$

Since determination of elevation by measuring vertical angles generally involves much greater distances than normal level sights, corrections for refraction and curvature of the earth (formula 8.5) should be made for accurate work. Let us consider an example with the following data:

$$\alpha = + 10° 35' 20''$$
$$AB = 10\,359 \text{ m}$$
$$h = 1·54 \text{ m}$$
$$h' = 6·20 \text{ m}$$
$$H = 152·31 \text{ m}$$

then,

$$152·31 + 10\,359 \times \tan(10° 35' 20'') + 1·54 - 6·20 = 2084·2 \text{ m}$$

Using formula 8.5 the correction for refraction and curvature of the earth becomes:

$$y = 0·068(10·359)^2 = 7·30 \text{ m}$$

Thus, the corrected elevation of C is 2091·5 m.

In the examples given above the horizontal distance is known as would be the case between points fixed by triangulation. Often, as in theodolite traverse work, it is the slope distance that is measured. In this case the elevation of a higher point (Fig. 8.13) is given by:

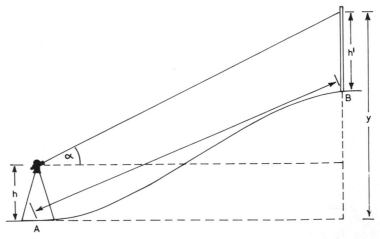

FIG. 8.13.   Theodolite set up at A to measure the elevation of B knowing the slope distance AB.

$$H + AB \sin \alpha + h - h' \qquad (8.8)$$

and the elevation of a lower point is given by:

$$H - AB \sin \alpha - h + h' \qquad (8.9)$$

Note that in the formulae given the slope distance is assumed to be the same as the sighting distance. This is only true if the instrument and target heights are the same. If they are not equal, a small correction should be applied. However, this is so small in all but extreme cases that it can be ignored.

Barometric heighting

Elevations can be measured with a barometer or altimeter by virtue of the fact that atmospheric pressure decreases with altitude. Assuming an isothermal atmosphere it can be shown from Boyle's and Charles's laws and a consideration of the equilibrium of a column of air under the action of gravity that:

$$h_2 - h_1 = K \log_e \frac{p_1}{p_2}$$

where p_2 is the atmospheric pressure at elevation h_2, p_1 the atmospheric pressure at elevation h_1 and K a constant depending on the Gas Constant, the absolute temperature and the value of the acceleration due to gravity. At 0°C (273 K) the following general formula can be used:

$$h_2 - h_1 = 42\,221 \log_e \frac{p_1}{p_2}$$

or

$$h_2 - h_1 = 18\,336 \log_{10} \frac{p_1}{p_2} \qquad (8.10)$$

As the constant in formula 8.10 is dependent on the value of the acceleration due to gravity, it will vary with latitude and altitude. At sea level and 0°C it varies from 18 385 at the equator to 18 292 at the poles. A correction for temperature needs to be applied and, if θ_m is the mean absolute temperature between stations at h_1 and h_2, the temperature correction is given by multiplying formula 8.10 by $\theta_m/273$ or:

$$\left(1 + \frac{t_1 + t_2}{546}\right) \qquad (8.11)$$

where t_1 and t_2 are the temperatures at h_1 and h_2 in degrees Centigrade.

The above formulae are based on the assumption of an isothermal atmosphere which is not strictly correct since temperature decreases with altitude in the troposphere (the first 10 km of the atmosphere) until it stabilizes in the stratosphere (above 10 km) at about $-40°C$. This fall in temperature, known as the lapse-rate, is about $6·5°C/1000$ m. An analysis of this has been made by various workers and a series of formulae derived which relate to a 'standard atmosphere'. Figure 8.14 shows the relationship between pressure and temperature as applied to a 'standard atmosphere' at a latitude of 45° and with a temperature of 15°C and a pressure of 760 mm of mercury at sea level. The pressure values vary slightly with the value of the acceleration due to gravity and hence latitude. Atmospheric pressure is measured in Newtons per square metre (N/m^2) or millibars (mb) or millimetres of mercury (mm Hg).

$$1·333 \text{ mb} = 1 \text{ mm Hg}$$
$$100 \text{ N/m}^2 = 1 \text{ mb}$$

The classical mercury barometer is not used in survey work as it is too awkward and delicate to transport in the field. Rather aneroid barometers are used which are calibrated directly in feet or metres derived from a formula similar to 8.10, so that direct pressure measurements are not made. The better instruments can be read to 1 m or slightly less. When taking readings humidity corrections are necessary in addition to temperature corrections, as the theoretical

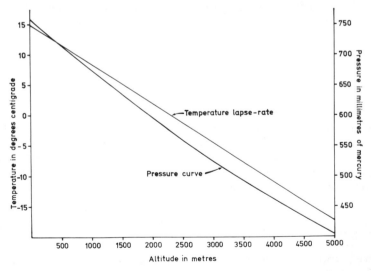

FIG. 8.14. Relationship between pressure and temperature of a 'standard atmosphere' at a latitude of 45° with a temperature of 15°C and a pressure of 760 mm Hg at sea level.

considerations are based on dry air. Correction tables or charts for both temperature and humidity are usually supplied with the various available instruments by the makers. Although the altimeters used for survey work give direct readings of elevation, it is not feasible to take isolated spot readings as this does not take into account pressure variations due to weather changes in the troposphere (*tropos* is Greek for 'to turn'). Although weather changes may be highly irregular, they often follow a diurnal pattern with a steady decrease in atmospheric pressure during the morning followed by an increase towards nightfall. The mornings are generally the best time for surveys. The errors caused by weather changes may put readings out by 30 m or more! There are three basic methods used to overcome this: (i) one-base method, (ii) two-base method and (iii) 'leap-frogging'.

One-base method
With this method two instruments are required, a base instrument and a field instrument. A base station is selected in the survey area at a point of known elevation roughly in the middle of the elevation range being surveyed. Both instruments are read at the base at the start of the survey and the readings compared. It is generally found that the readings are slightly different. This difference is noted as a plus or minus amount necessary to make the field instrument's reading the same as the base instrument's. This is known as *indexing*. Then an observer at the base reads his instrument every 15 min or so, noting the time and making the necessary temperature and humidity corrections (makers generally supply a hygrometer for this). The second observer moves around the area taking readings, noting the time of each reading and making temperature and humidity corrections. At the completion of the circuit, the field observer returns to the base and readings are compared again. Since weather changes can be very localized, reading the field instrument outside a 5–10 km radius of the base is not recommended for accurate work. The variations in the base instrument readings with time are then used to make corrections to the various field readings. An example of some typical survey readings is given below assuming corrections have been made for temperature and humidity.

Indexing at base with elevation 270 m:

Base instrument: 273 m
Field instrument: 275 m

The base instrument correction is $270 - 273 = -3$ m
The field instrument correction is $270 - 275 = -5$ m

	Base Instrument			Field Instrument	
Time	Reading	Time variation	Time	Reading	Station
7:45	273	—	8:00	257	1
8:00	274	+ 1	8:20	263	2
8:15	276	+ 3	8:35	242	3
8:30	275	+ 2	8:50	286	4
8:45	273	—	9:10	292	5
9:00	272	− 1			
9:15	271	− 2			

Then the correct elevation at each station is given by:

$$\text{field reading} + \frac{\text{field instrument}}{\text{index correction}} + \frac{\text{correction for time}}{\text{variation at base}}$$

Station 1: $257 - 5 - 1 = 251$ m
Station 2: $263 - 5 - 3 = 255$ m
Station 3: $242 - 5 - 2 = 235$ m
Station 4: $286 - 5 + 0 = 281$ m
Station 5: $292 - 5 + 2 = 289$ m

Two-base method
This method was designed to obviate the need to make humidity and temperature corrections. While this is not strictly valid, the errors are very small provided that the elevation range is not more than 500 or 600 m. With this method two bases are established, one near the top end of the elevation range being surveyed and the other at the bottom end. Before commencing the survey all three instruments are indexed at one of the bases and any discrepancies noted. Readings at both bases are then plotted against time and the times of the various field readings noted as in the case of the one-base method. An example of some typical survey readings is given below:

High base elevation: 563 m
Low base elevation: 186 m

Indexing at low base:

Low base instrument: 185 m
High base instrument: 185 m
Field instrument: 184 m

Thus, low base correction: + 1
high base correction: + 1
field correction: + 2

Low Base Readings		High Base Readings		Field Readings		
7:30	185	7:30	564	7:45	231	No. 1
7:45	186	7:45	564	8:05	342	No. 2
8:00	186	8:00	565	8:35	407	No. 3
8:15	187	8:15	565	8:55	482	No. 4
8:30	188	8:30	566	9:10	501	No. 5
8:45	189	8:45	566	9:31	396	No.6
9:00	189	9:00	566			
9:15	188	9:15	567			
9:30	188	9:30	567			

Then the correct elevation at each station is given by:

$$h = h_L + \Delta H \frac{r_F - r_L}{r_H - r_L} \tag{8.12}$$

where h_L is the true elevation of the lower base, ΔH the height difference between the low and high bases, r_F the field reading, r_L the low base reading and r_H the high base reading, all made at, or adjusted to, the same time.

$$\text{Station 1: } 186 + 377 \times \frac{233 - 187}{565 - 187} = 231 \text{ m}$$

$$\text{Station 2: } 186 + 377 \times \frac{344 - 187}{566 - 187} = 342 \text{ m}$$

$$\text{Station 3: } 186 + 377 \times \frac{409 - 189}{567 - 189} = 405 \text{ m}$$

$$\text{Station 4: } 186 + 377 \times \frac{484 - 190}{567 - 190} = 480 \text{ m}$$

$$\text{Station 5: } 186 + 377 \times \frac{503 - 189}{568 - 189} = 498 \text{ m}$$

$$\text{Station 6: } 186 + 377 \times \frac{398 - 189}{568 - 189} = 394 \text{ m}$$

'Leap-frogging' method
Since both the one- and two-base methods require the use of more than one instrument and observer (unless self-recording base instruments are used), this method can be used for less precise work if only one instrument is available. A reading is taken at a point of known elevation (station 1) and then the observer moves quickly to station 2, takes a reading and returns quickly to station 1 to note if there is any change in reading. If so, a correction is applied to the reading at station 2. Then the observer moves to station 3, takes a reading and quickly returns to

station 2 to note if there has been any change. The observer then moves to station 4 and the operation is repeated by always checking against the previously occupied station. At the end of the traverse it is good practice to tie in to another point of known elevation. An example of some readings corrected for humidity and temperature is given below:

Station 1 (known elevation): 278 m

Readings	Stations
277	1
256	2
277	1
311	3
257	2
228	4
313	3
269	5
227	4

Thus, the corrected readings are:

Station 1	278 m
Station 2	257 m
Station 3	311 m
Station 4	227 m
Station 5	271 m

8.4   PLANE TABLE SURVEYING

Plane tabling is probably the method of surveying most widely used by geologists. It is unique among surveying techniques in that a map is prepared directly in the field as the survey progresses without measuring angles and without calculations (unless determining elevations and making tacheometric measurements of distance). It can be used for both large and small scale surveys, though its importance has greatly diminished with the general availability of aerial photographs. It is still a useful technique, however, for detailed surveys of small areas. The basic equipment is very simple and consists of a drawing board which can be fitted to a robust tripod, a spirit level for levelling the board, a plumbing fork, which permits a plumb bob to hang below the table directly under any point on the drawing, a compass for orientating the table magnetically if so desired and a sight rule known as an *alidade*.

To carry out a survey a sheet of high quality drawing or cartridge paper is placed on the board and fastened around the edges by special clips or drawing pins to the under surface of the board. The paper should be stretched smoothly and tightly over the board without any wrinkles. Then, equipped with a number of well sharpened 4H and 2H pencils and the necessary ancillary equipment mentioned earlier, one is ready to undertake a survey. The plane table can be used for making a complete survey with control points and detail or it can be used from existing control points to fill in detail. The technique for plotting detail will be described first.

For the purposes of describing the technique let us suppose that we are required to survey features of detail to one side of a line between two pegged stations, A and B. First point A is plotted at a convenient point on the board with a suitable scale selected so that the distance AB can be scaled off on the drawing. Then the plane table is set up over peg A, levelled and with the alidade rule touching point A, the peg B is sighted through the alidade vanes. A line is ruled along the alidade edge and the distance AB scaled off and point B plotted. Then the alidade is rotated still with the edge of the rule on point A and a series of sights is taken to points of interest and the rays to each lightly ruled on the paper. It is a good idea to write some identifying remark very lightly alongside each ray for easy identification. Then the plane table is moved to peg B and set up over this point in such a manner that the ruled line BA is lined up exactly with point A by sighting on peg A through the alidade with the edge of the alidade rule alongside the line BA. A series of rays is then drawn to all the points of interest by taking sights with the alidade to each point in turn while the alidade rule touches point B. The positions of the various points of detail are then located at the intersections of respective rays. The procedure is illustrated in Fig. 8.15. For accurate work it is necessary that the points A and B on the drawing are located vertically above the respective points on the ground. This is achieved by using the plumbing fork.

In addition to surveying detail, the plane table can be used to locate control points, either by a *plane table traverse* or by *plane table triangulation*. In a plane table traverse the positions of the various stations are plotted in turn by occupying each station, aligning the plotted ray on the board with the previously occupied station and drawing a new ray to the next station. It will be appreciated that the distance between stations has to be measured on the ground to plot them. The procedure illustrated in Fig. 8.16 is as follows: (1) With the distance between stations A and B scaled off on the board, station B is occupied and the ray BA aligned with station A. (2) Station C is sighted through the alidade, a ray is drawn on the board, the distance BC scaled off and point C plotted. (3) Station C is now occupied and

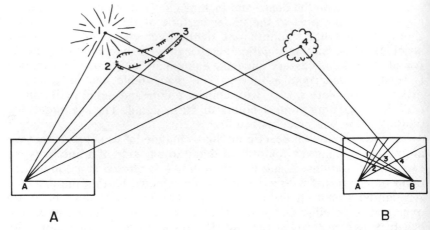

A   B

FIG. 8.15.   Method of locating points of detail with the plane table.

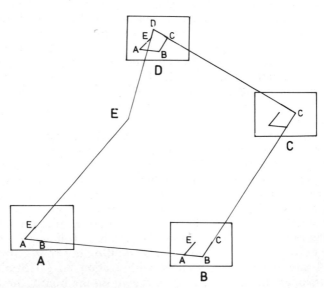

FIG. 8.16.   Method of carrying out a plane table traverse.

the process above continued to locate station D. This procedure is continued right round the traverse until it eventually closes on station A.

To avoid having to measure distance between stations on the ground, a plane table triangulation can be carried out. With this method a series of control points is build up from two initial stations whose distance apart is known by successively occupying other stations located by intersecting rays in the manner described for locating points of detail.

In addition to the ordinary survey methods described above, elevations can be measured by using a *telescopic alidade*. This is simply an alidade rule with a telescope mounted on its centre. The telescope has a vertical circle with a micrometer for accurate measurement of vertical angles and it can be levelled independently of the table. The telescope also contains stadia hairs which enable distance and levels to be measured tacheometrically if so desired (see Section 8.7) thus obviating the need to locate points with intersecting rays as described for the ordinary alidade.

In addition to the techniques described above, a point can be fixed by resection if a minimum of three other points have already been located and plotted accurately. The board is set up at the unknown point X and roughly oriented with respect to the three known points. Then with the edge of the alidade rule touching each known point in turn a series of rays is drawn backwards while sighting on each point. If the board has been correctly oriented, the ray will intersect in a point which is the correct position of the point X. Usually this does not happen and the three rays form a small triangle known as the *triangle of error* (Fig. 8.17). There are a number of methods which can be used to solve this triangle of error or *three point problem* as it is often known. The simplest method is by trial and error and involves turning the board slightly and repeating the resection until the rays meet in a point or the triangle of error becomes very small. A resection cannot be carried out if the unknown point and three known points lie on a circle or if two of the known points are diametrically opposed.

Plane tabling is a very good method for geologically mapping small areas in detail (scales 1/500 to 1/2500). This is most efficiently carried out by having a field assistant to plot the rays on the board. This allows the geologist to move around and select the most useful points of detail to be fixed. The geologist carries a surveying rod and holds it at each point selected while the field assistant sights on them in turn and plots the respective rays. Each point is numbered consecutively and the geologist can make detailed notes on the geological features at the same time that the rays are plotted. It is often useful to use

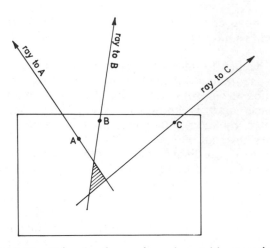

FIG. 8.17. Triangle of error in a plane table resection.

pegs, paint or flagging ribbon to identify each point on the ground for accurate relocation when the second set of rays is plotted from the second control point. A survey can be carried out by the geologist alone, but is is very much slower, particularly if there is a lot of detail to be surveyed accurately.

In summary, the main advantages of plane tabling are that it is quick, the technique is easy to acquire and the map is produced directly in the field with the rapid sketching in of as much detail as required. The main disadvantages are that the scale of the map must be known before the survey commences and field work is not possible in wet or windy weather.

8.5 SURVEYING CALCULATIONS

Most surveying calculations are based on elementary co-ordinate geometry and trigonometry, though the actual computations are tedious and laborious. To ensure a sufficient degree of accuracy log tables to at least seven places used to be used. Since the advent of the small scientific electronic calculator, however, calculations using log tables are now obsolete. For large numbers of surveying observations, which used to take days and even weeks of tedious computation, small progammable electronic calculators are available to do the work in a matter of minutes. Before considering some actual examples, a review of some elementary relationships and formulae in co-ordinate geometry and trigonometry is given below.

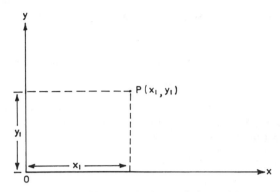

FIG. 8.18. Rectilinear grid for defining grid co-ordinates.

(1) *Grid co-ordinates.* In a reference grid defined by two rectilinear axes x and y (Fig. 8.18), a point P is fixed by two co-ordinates x_1 and y_1. In conventional co-ordinate geometry the point P can be located anywhere with respect to the origin O, but in surveying it is normal practice to select the origin outside the survey area so that the co-ordinates of any point will be positive, i.e. the survey grid is restricted to the first quadrant. In surveying the co-ordinates of a point are known as the *eastings* and *northings*, an easting corresponding to a positive x co-ordinate and a northing to a positive y co-ordinate. Thus, a point A on a survey grid may be defined by its easting, E_A, and northing, N_A, i.e. its co-ordinates are (E_A, N_A).

(2) *Length of a line.* The length of a line between two points A and B can be determined from the co-ordinates by the following formula:

$$AB = \sqrt{(N_B - N_A)^2 + (E_B - E_A)^2} \qquad (8.13)$$

(3) *The bearing of a line.* The grid azimuth of a line between two points A and B can be calculated from the co-ordinates by the following formula:

$$\theta = \arctan \frac{(N_B - N_A)}{(E_B - E_A)} \qquad (8.14)$$

where θ equals the acute angle AB makes with an E–W line. Then

If $N_B - N_A$ and $E_B - E_A$ are both positive,

$$\overrightarrow{AB} = 90° - \theta$$

If $N_B - N_A$ is negative and $E_B - E_A$ positive,

$$\overrightarrow{AB} = 90° + \theta$$

If $N_B - N_A$ and $E_B - E_A$ are both negative,

$$\overrightarrow{AB} = 270° - \theta$$

If $N_B - N_A$ is positive and $E_B - E_A$ negative,

$$\overrightarrow{AB} = 270° + \theta$$

(4) *Determination of co-ordinates from a length and bearing.* Given the co-ordinates of a point A, the bearing to a point B and the length AB, the co-ordinates of B can be calculated.

$$E_B = E_A + AB \sin \overrightarrow{AB} \qquad (8.15)$$

$$N_B = N_A + AB \cos \overrightarrow{AB} \qquad (8.16)$$

(5) *Computation of a triangle.* In surveying we are generally concerned with the computation of a triangle knowing one of its sides and the three angles. This computation is based on the well known sine rule.

$$\frac{AB}{\sin C} = \frac{BC}{\sin A} = \frac{CA}{\sin B} \qquad (8.17)$$

(6) *Intersected point.* A point C can be located from two known points A and B by measuring the angles to C at A and B (Fig. 8.19). Such a point C is known as an intersected point. This is really a form of triangulation, except for a true triangulation fix, the angle at C should also be measured. Let us consider an actual example. We are given the following data and are required to calculate the co-ordinates of C:

	eastings	northings
A	114 126·32	79 361·18
B	110 735·40	68 864·62

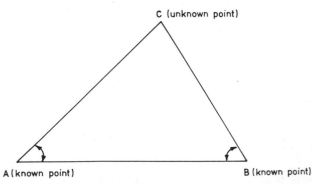

FIG. 8.19.   Point C fixed by intersection from A and B knowing the distance AB.

$$C\hat{A}B = 46° 37' 20''$$

$$C\hat{B}A = 62° 08' 40''$$

Then,

$$AB = \sqrt{(N_B - N_A)^2 + (E_B - E_A)^2} = 11\,030{\cdot}69$$

$$\overrightarrow{AB} = 270° - \arctan \frac{N_B - N_A}{E_B - E_A} = 197° 54' 11''$$

$$A\hat{C}B = 180° - 46° 37' 20'' - 62° 08' 40'' = 71° 14' 00''$$

$$AC = \frac{AB \sin B}{\sin C} = 10\,300{\cdot}14$$

$$\overrightarrow{AC} = \overrightarrow{AB} - C\hat{A}B = 151° 16' 51''$$

$$E_C = E_A + AC \sin (151° 16' 51'') = 119\,075{\cdot}71$$

$$N_C = N_A + AC \cos (151° 16' 51'') = 70\,328{\cdot}11$$

(7) *Resected point.* A point P can be located by measuring the angles to three known points, A, B and C (Fig. 8.20) Such a point P is known as a resected point. For a good resection none of the rays should

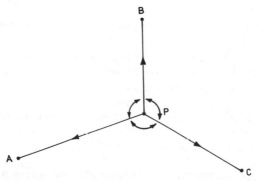

Fig. 8.20. Fixing unknown point P by resection to known points A, B and C.

intersect at less than 30° at P. The computation of the co-ordinates at P is quite involved and one of the methods, known as a *Collins resection*, is given below. There are two possible cases: (i) point P inside the triangle formed by A, B and C, and (ii) point P outside the triangle formed by A, B and C.

(i) *P inside triangle ABC (Fig. 8.21 (a)).* Join the points A, B and C to form the triangle ABC. Plot rays from P to A, B and C. Draw a ray from C making an angle 180° − α with CB. Draw a ray from B making

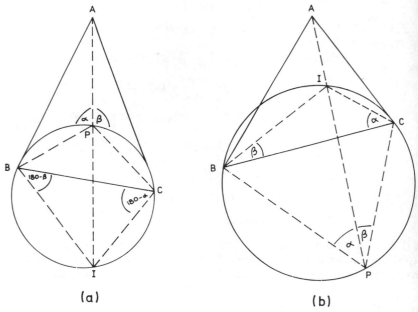

(a)   (b)

FIG. 8.21.   Construction for calculating a Collins resection.

an angle $180° - \beta$ with BC. These rays meet in a point I. Draw the ray from A through P which will be found to intersect BI and CI at I. The points I, B, P and C now lie on a circle.

(ii) *P outside triangle ABC (Fig. 8.21 (b))*. Proceed with construction as before, but draw a ray from C making an angle α with CB and a ray from B making an angle β with BC. These rays intersect at I which is on the ray drawn from A to P. Once again the points P, B, I and C lie on a circle.

The geometric constructions above make the solution of the point P possible and the procedure is illustrated by an example below:

$$\alpha = 119° \ 15' \ 00'' \qquad 180 - \alpha = 60° \ 45' \ 00''$$

$$\beta = 137° \ 30' \ 00'' \qquad 180 - \beta = 42° \ 30' \ 00''$$

	eastings	northings
A	11 920·20	12 370·30
B	10 730·10	11 051·00
C	12 069·90	10 434·80

(i) Compute length and bearing of BC:

$$BC = \sqrt{(E_C - E_B)^2 + (N_C - N_B)^2} = 1474·71$$

$$\overrightarrow{BC} = 90° + \arctan \frac{N_C - N_B}{E_C - E_B} = 114° \ 41' \ 55''$$

(ii) Solve triangle BCI:

$$B\hat{I}C = 180 - (180 - \alpha + 180 - \beta) = 76° \ 45' \ 00''$$

$$CI = \frac{BC \sin (180 - \beta)}{\sin B\hat{I}C} = 1023 \cdot 55$$

$$\overrightarrow{CI} = \overrightarrow{CB} - (180 - \alpha) = 294° \ 41' \ 55'' - 60° \ 45' \ 00'' = 233° \ 56' \ 55''$$

$$E_I = CI \sin \overrightarrow{CI} + E_C = 11 \ 242 \cdot 07$$

$$N_I = CI \cos \overrightarrow{CI} + N_C = 9832 \cdot 43$$

(iii) Compute bearings IA (IP) and IB:

$$\overrightarrow{IP} = \overrightarrow{IA} = 90° - \arctan \frac{N_A - N_I}{E_A - E_I} = 14° \ 57' \ 37''$$

$$\overrightarrow{IB} = 270° + \arctan \frac{N_B - N_I}{E_B - E_I} = 337° \ 12' \ 39''$$

(iv) Solve triangle BIP:

$$B\hat{I}P = \overrightarrow{IP} - \overrightarrow{IB} = 37° \ 44' \ 58''$$

$$BI = (E_I - E_B)^2 + (N_I - N_B)^2 = 1321 \cdot 75$$

$$P\hat{B}I = 180 - (180 - \alpha) - B\hat{I}P = 81° \ 30' \ 02''$$

$$IP = \frac{BI \sin P\hat{B}I}{\sin (180 - \alpha)} = 1498 \cdot 27$$

$$E_P = IP \sin \overrightarrow{IP} + E_I = 11 \ 628 \cdot 85$$

$$N_P = IP \cos \overrightarrow{IP} + N_I = 11 \ 279 \cdot 92$$

8.6 MEASUREMENTS WITH A THEODOLITE

A theodolite is an instrument designed for the precise measurement of vertical and horizontal angles. It consists of a telescope which can be rotated both on vertical and horizontal axes. Measurements of the degree of rotation are made on two graduated circles, one horizontal and the other vertical. Fractions of a degree may be measured with a vernier on the less precise instruments, but an optical micrometer is generally used. Theodolites vary from small instruments capable of measuring angles to 5' to extreme precision instruments capable of a 0·1" accuracy. The less precise instruments are often referred to as transits. Some instruments incorporate a compass and can be used for observing magnetic bearings in addition to being used as a normal theodolite. A good theodolite with an accuracy of 10″–20″ is adequate for most jobs in exploration, and the more precise ones with a 1″ or better accuracy are only necessary for primary triangulation surveys and extremely precise astronomical observations.

Figure 8.22 shows the main parts of an ordinary theodolite. Both the vertical and horizontal circles have clamping screws which can be released to permit full rotation of the instrument about its two axes. When the clamping screws are locked, fine adjustment tangent screws can be used for slow and precise movement of either circle. The theodolite can be moved around two horizontal plates. With the lower one clamped, the upper one can be rotated with respect to the graduated horizontal circle on the lower plate giving the degree of rotation. When the upper plate is clamped, the theodolite can be

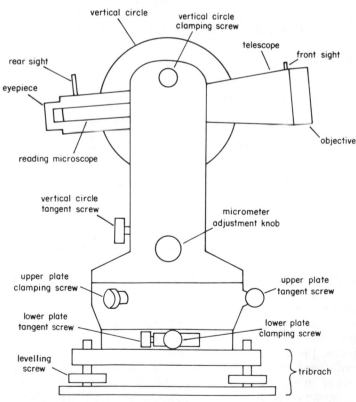

FIG. 8.22. Diagram showing the basic parts of a theodolite.

rotated around its horizontal axis by means of the lower plate so that the reading of the graduated horizontal circle remains fixed. To take a reading the theodolite is set up firmly on its tripod and levelled precisely with its levelling screws. With the lower plate clamped and upper plate and vertical circles unclamped the telescope is swung round and aimed at the target using the sights on top of the telescope. The horizontal and vertical circles are clamped and accurate sighting on the target is accomplished with the fine adjustment screws so that the cross hairs in the telescope are centred precisely on the target. If the instrument has a separate vertical bubble this should be centred exactly before the final lining up of the cross hairs on the target is carried out.

The readings of both vertical and horizontal circles are made by looking down the reading microscope which is usually placed close to the telescope eyepiece. Illumination of the graduated circles is made by a small hinged mirror which should be adjusted to allow as much light into the instrument as possible. Various systems of reading are used. Some of the larger theodolites have separate viewing for the horizontal and vertical circles, but most theodolites display the two circles together with the micrometer scale within the same viewing window. Such an arrangement is shown in Fig. 8.23 which shows the

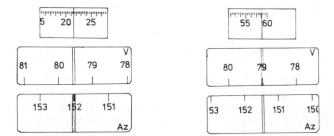

FIG. 8.23. Reading system used in the Wild TIA theodolite. The left-hand figure shows the micrometer adjusted to read the horizontal circle (152° 21′ 30″) and the right-hand figure shows the micrometer adjusted to read the vertical circle (79° 59′ 00″).

system used on the Wild TIA theodolite. The vertical circle is graduated from 0 to 360° with 90° being horizontal with the telescope in its normal position and 270° being horizontal with the telescope turned through 180° (transitted as it is known). Positive elevations are less than 90° or 270° and negative angles are greater than 90° or 270°. In the figure the vertical angle shown is + 10° 01′ 00″ (90° − 79° 59′). The horizontal angle shown in the example in Fig. 8.23 is 152° 21′ 30″. In both cases the correct reading is obtained in turn by turning the micrometer adjustment knob until the lower figure is centred (152° and 79° in the example)*.

If it is required to line up a target with a particular horizontal reading, the lower plate is clamped and the upper plate turned until the required angle is seen in the reading microscope. Then the upper plate is clamped and the final centring to the required angle made with the fine adjustment screw. Now with the upper plate left firmly clamped, the lower plate can be unclamped and the theodolite turned and sighted on the target. When roughly lined up on the target, the lower plate can be clamped and the telescope brought precisely in line with the target by using the fine tangent screw on the lower plate. Different sightings made from this station by rotating the upper plate while the lower one remains firmly clamped will all be made with reference to the reading set on the first target. For exact centring over a station the theodolite has an attachment for fastening a plumb bob underneath.

Various accessories are available for most theodolites. These include small battery-powered plug-in lamps to replace the illuminating mirror for night work, eyepiece prisms for sights up to $+65°$, diagonal eyepieces for sights up to $+90°$, solar prisms for fitting over the telescope objective so that sun observations can be made, and laser eyepieces which project a laser beam out of the telescope coinciding exactly with the line of sight for use in alignment work.

Theodolite traversing
Extremely accurate traversing surveys can be carried out by measuring the angles with a theodolite. Figure 8.24 shows a closed traverse with six stations, A to F. To carry out a traverse in a clockwise direction as shown in the figure, the theodolite is set up at each station in turn and readings taken to forward stations followed by the back stations. This results in measurement of the internal angles as shown in the diagram. At each station the theodolite is read both with the telescope in its normal position and transitted (turned through 180° on its vertical circle). This is known as 'face left' and ' face right'. For sighting, a surveying staff can be held at each forward and back station, but for really precise work special tripod-mounted targets should be used. These targets consist of black metal plates on which three or four white or yellow triangles are painted with their apices pointing towards the centre point of the target. The target plates are mounted on special carriers which can be locked into the

---

*It should be noted that some theodolites are graduated in grads and not degrees ($400^g \equiv 360°$).

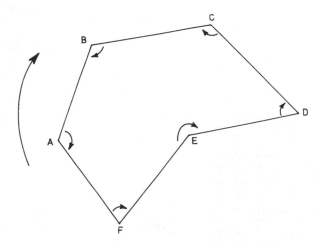

FIG. 8.24. Example of a closed theodolite traverse.

tribrach secured to the tripod head. (The theodolite base with its three levelling screws is known as a tribrach (Fig. 8.22) and the theodolite can be detached from its tribrach by loosening a securing screw.) This enables very precise work to be undertaken by allowing targets and theodolite to be interchanged between stations by detaching from the tribrachs which are left undisturbed and precisely centred and levelled on the tripods.

Figure 8.25 shows the booking of a closed theodolite traverse. The column for staff reading is for the determination of elevation using the vertical angles and, if targets were being used, the target height would be entered instead. To plot a theodolite traverse, co-ordinates should be calculated instead of using a protractor since there is no point in measuring angles in the field at far greater precision than they can be plotted. The following steps are undertaken in computing a theodolite traverse for final plotting:

1. Correct lengths for elevation.
2. Abstract internal (or external) angles, correct and compute bearings.
3. Compute co-ordinates.
4. Correct co-ordinates for closure error.

The lengths are corrected by using the vertical angles or height differences if there is more than a very small difference between instrument and target heights. Using the vertical angles in the example given in Fig. 8.25 the corrected lengths are as follows:

AB $268 \cdot 45 \cos(2° 13' 30'') = 268 \cdot 25$ m
BC $320 \cdot 52 \cos(7° 21' 10'') = 317 \cdot 88$ m
CD $329 \cdot 91 \cos(0° 00' 45'') = 329 \cdot 91$ m
DE $299 \cdot 23 \cos(9° 35' 15'') = 295 \cdot 05$ m
EF $295 \cdot 89 \cos(3° 18' 05'') = 295 \cdot 40$ m
FA $267 \cdot 04 \cos(4° 32' 00'') = 266 \cdot 20$ m

The internal angles are abstracted as follows using the mean horizontal angles in Fig. 8.25.

	160° 15′ 05″	
A	−35° 14′ 55″	
	125° 00′ 10″	125° 00′ 10″
	239° 56′ 15″	
B	−121° 11′ 10″	
	118° 45′ 05″	118° 45′ 05″
	166° 58′ 25″	
C	−42° 13′ 20″	
	124° 45′ 05″	124° 45′ 05″
	131° 58′ 50″	
D	−75° 29′ 15″	
	56° 29′ 35″	56° 29′ 35″
	22° 36′ 40″	
E	−162° 37′ 15″	
	219° 59′ 25″	219° 59′ 25″
	90° 11′ 30″	
F	−15° 11′ 25″	
	75° 00′ 05″	75° 00′ 05″
		−719° 59′ 25″
		+720° 00′ 00″ (six-sided figure)
		+35″

Corrections: $\dfrac{35}{6} = +6$

Thus, the corrected internal angles are:

A	125° 00′ 16″
B	118° 45′ 11″
C	124° 45′ 11″
D	56° 29′ 41″
E	219° 59′ 31″
F	75° 00′ 11″
	720° 00′ 01″

The corrections above are derived from the rule that in any closed figure of n sides the sum of internal angles equals $n \times 180° - 360°$, i.e. $6 \times 180 - 360 = 720$ for the above example.

The co-ordinates are calculated from formulae 8.15 and 8.16, knowing the co-ordinates of the starting point and the bearing of A to B. If the bearing and co-ordinates are not known, arbitrary co-ordinates and a bearing can be assigned for the purposes of the calculation. In the example being considered the bearing of A to B is given as 20° 30′ 00″. Then the bearings of the other lines are calculated from the internal angles as follows:

A to B	20° 30′ 00″	
	+180° 00′ 00″	
	200° 30′ 00″	
	−118° 45′ 11″	(B)
B to C	81° 44′ 49″	
	+180° 00′ 00″	
	261° 44′ 49″	
	−124° 45′ 11″	(C)
C to D	136° 59′ 38″	
	+180° 00′ 00″	
	316° 59′ 38″	
	−56° 29′ 41″	(D)
D to E	260° 29′ 57″	
	+180° 00′ 00″	
	440° 29′ 57″	
	−219° 59′ 31″	(E)
E to F	220° 30′ 26″	
	+180° 00′ 00″	
	400° 30′ 26″	
	−75° 00′ 10″	(F)
F to A	325° 30′ 16″	
	+180° 00′ 00″	
	505° 30′ 16″	
	−125° 00′ 16″	(A)
A to B	380° 30′ 00″ = 20° 30′ 00″	

If the co-ordinates of A are given as 1202·35 E and 8572·10 N, the co-ordinates of other stations are as follows:

Eastings

B	$1202·35 + 268·25 \sin 20° 30′ = 1296·29$
C	$1296·29 + 317·88 \sin 81° 44′ 49″ = 1610·88$
D	$1610·88 + 329·91 \sin 136° 59′ 38″ = 1835·90$
E	$1835·90 + 295·05 \sin 260° 29′ 57″ = 1544·90$
F	$1544·90 + 295·40 \sin 220° 30′ 26″ = 1353·02$
A	$1353·02 + 266·20 \sin 325° 30′ 16″ = 1202·26$

Stations	Horizontal Angles			Vertical Angles			Staff Reading	Instrument Height	Length	Remarks
	F.L.	F.R.	Mean	F.L.	F.R.	Mean				
A to B	35°15'00"	215°14'50"	35°14'55"	+2°13'40"	+2°13'40"	+2°13'40"	1·40	1·37	268·45	Station A is trig. pt. "The Rocks"
A to F	160°15'10"	340°15'00"	160°15'05"	-4°32'10"	-4°32'10"	-4°32'10"	1·45		267·08	
B to C	121°11'10"	301°11'10"	121°11'10"	+7°21'30"	+7°21'30"	+7°21'30"	1·35	1·45	320·50	
B to A	239°56'10"	59°56'20"	239°56'15"	-2°13'10"	-2°13'30"	-2°13'20"	1·40		268·45	
C to D	42°13'20"	222°13'20"	42°13'20"	+0°01'00"	+0°00'40"	+0°00'50"	1·40	1·34	329·88	
C to B	166°58'30"	346°58'20"	166°58'25"	-7°20'40"	-7°21'00"	-7°20'50"	1·38		320·54	
D to E	75°29'30"	255°29'00"	75°29'15"	-9°35'10"	-9°35'50"	-9°35'30"	1·41	1·39	299·20	
D to C	131°58'40"	311°59'00"	131°58'50"	-0°00'40"	-0°00'40"	-0°00'40"	1·50		329·94	
E to F	162°37'20"	342°37'10"	162°37'15"	-3°17'50"	-3°18'30"	-3°18'10"	1·39	1·48	295·88	
E to D	22°36'30"	202°36'50"	22°36'40"	+9°34'40"	+9°35'20"	+9°35'00"	1·41		299·26	
F to A	15°11'30"	195°11'20"	15°11'25"	+4°31'40"	+4°31'50"	+4°31'50"	1·38	1·43	267·00	
F to E	90°11'30"	270°11'30"	90°11'30"	+3°17'40"	+3°18'20"	+3°18'00"	1·40		295·90	

FIG. 8.25.   Typical booking of a theodolite traverse.

Northings

B	$8572 \cdot 10 + 268 \cdot 25 \cos 20° 30' 00" = 8823 \cdot 36$
C	$8823 \cdot 36 + 317 \cdot 88 \cos 81° 44' 49" = 8868 \cdot 99$
D	$8868 \cdot 99 + 329 \cdot 91 \cos 136° 59' 38" = 8627 \cdot 73$
E	$8627 \cdot 73 + 295 \cdot 05 \cos 260° 29' 57" = 8579 \cdot 03$
F	$8579 \cdot 03 + 295 \cdot 40 \cos 220° 30' 26" = 8354 \cdot 43$
A	$8354 \cdot 43 + 266 \cdot 20 \cos 325° 30' 16" = 8573 \cdot 82$

Thus, the difference in the A co-ordinates from the starting and calculated values is:

$$E = 1202 \cdot 35 - 1202 \cdot 26 = +0 \cdot 09 \text{ m}$$

$$N = 8572 \cdot 10 - 8573 \cdot 82 = -1 \cdot 72 \text{ m}$$

The overall closure error is given by:

$$1 \bigg/ \frac{L}{\sqrt{(E^2 + N^2)}}$$

where L is the total length of the traverse. In the example given the closure error is 1/1029, which is poor for a theodolite traverse and indicates that there are probably errors or inaccuracies in the length measurements. There are a number of ways of adjusting the closure error, but the simplest method is probably Bowditch's rule, which is adequate for most work. Using the rule, corrections are made separately for the eastings and northings.

Triangulation surveys

Triangulation is the basis of all major surveying work and is the method used for surveying entire countries. It is based on the fact that a triangle of given dimensions has a unique shape and any triangle can be determined by knowing the lengths of its three sides or by knowing two angles and the length of one of the sides. In surveying a large area a network of triangles is built up from a single precisely measured base line by taking angular observations with a theodolite. Figure 8.26 shows how this is done. XX' is an accurately measured base line and from it points A and B are accurately fixed. Then the other triangles are built by measuring the angles at A, B and C and calculating the distance AB. Triangle ACD is fixed by measuring the three angles and calculating the distance AC from the earlier determined triangle. This process can be continued as far as desired. Primary triangles have sides of approximately 50 km, secondary triangles have sides of 15–25 km and tertiary triangles have sides less than 10 km. Ideally, the triangles should be selected so that they are as near to equilateral triangles as possible. This is often hard to achieve, but no angle should be less than 30° for best results.

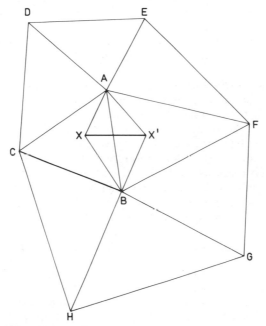

FIG. 8.26. Triangulation network built up from the base line XX'.

Readings are taken from all three apices of the triangle and sights are made to survey targets or specially constructed beacons at the other stations being observed. Weather conditions have to be ideal for long shots and surveyors often have to wait many days or even weeks to obtain satisfactory readings. Very often readings are made at night. In addition to horizontal angles, vertical angles are also observed, but, as they are of lesser importance, they are often made at a different time from horizontal angles if conditions are not right. To be reliable vertical angles should only be observed when refraction is at a minimum and varying least, which is generally from late morning to early afternoon. The normal procedure of observing angles is to select one beacon as the RO (referring object) and then move around in a clockwise direction (known as face left) to all the other stations in turn, eventually closing on the RO. Then the telescope is transitted and starting from the RO all stations are observed again going around in the opposite sense for the face right readings. In approaching the targets it is important to do so from the same side and bring the cross hairs into line very slowly with the tangent screw. If the target is overshot, it is normal procedure to repeat all readings again and not accept a reading for which the telescope had to be brought back on target. Having completed a series of face left and face right observations, the upper and lower plates are moved with respect to each other to set new readings on the horizontal circle and the entire process is repeated several times. The angles are corrected for any differences, which should be very slight, and then the included angles are determined.

In calculating a triangulation survey the three angles determined for each triangle are added and corrections made so that they total exactly 180°. If the triangles being observed are large, it will be found that the sum of the three angles is greater than 180°. This is because the angles have been measured in three different planes owing to the curvature of the earth. This difference is known as the spherical excess and corrections have to be made for it in primary triangulation. The spherical excess varies with the size of the triangle and slightly with latitude since the earth is not a perfect sphere. Some average values for the spherical excess are given below:

Area of triangle (km^2)	Spherical excess
94	0·5″
188	1·0″
937	5·0″
1877	10·0″

8.7 MEASUREMENT OF DISTANCE

Tapes

The earliest measurements of distance were made with bars or rods of known length and even in early 1800's the base lines for triangulation surveys were still being measured with special steel or bimetallic bars. Very accurate measurements are possible with bars, but, as it is a cumbersome and time consuming method, metal tapes replaced bars and have been used widely since the late 1800's. For the most accurate work invar tapes are hung in catenary between special straining trestles. Wires attached to rings at either end of the tape pass over pulleys on the straining trestles and standard weights attached to the wires keep the tape in constant tension. Readings are obtained from special index heads mounted on tripods just below the stretched tape. Points are transferred to the ground by an optical plummet, which may also be mounted on the tripods, and alignment between several measuring points is achieved by using a theodolite. Corrections need to be applied if conditions differ from those at which the tapes were standardized. These include a tension correction and temperature correction. A correction also needs to be applied for sag if the tension is different from that at which the tape was standardized or if the tape was standardized on the flat and not in catenary. Corrections may also have to be applied for slope and lack of precise alignment. In addition, for triangulation surveys all measurements are reduced to mean sea level, so the following altitude correction has to be applied:

$$\frac{- L \times H}{6\,367\,000} \text{ metres,}$$

where L is the measured length in metres and H the elevation in metres.

For less precise work, such as theodolite traverses or small local triangulation surveys, measurements can be made with metal tapes, or surveying bands as they are often known, laid flat on the ground and pulled tight between measuring marks. As in the case of supported tapes, alignment between measuring points is maintained with a theodolite. Corrections for slope should also be applied if the ground slope is more than a few degrees. An approximate slope correction, which is sufficiently accurate for all but the steepest slopes, is $-h^2/2L$, where h is the height difference between the measured points and L the taped distance. When measuring distance with tapes on the ground, it is good practice to measure the distance between stations in two directions. For example, in measuring the distance between two stations, A and B, one would start taping at A and finish at B and then repeat the process at B and finish at A. This is a convenient check in theodolite traverse work as the two distances should agree within a few centimetres over distances of several hundreds of metres. The mean of the two distances obtained is adopted as the correct measurement. Metal tapes should be treated carefully. Good ones are expensive and they are not designed for rough work like the much more robust surveying chains. After use in wet or damp conditions, steel tapes should be dried and lightly oiled to prevent rusting.

The accuracies that can be obtained with the various linear measuring methods are compared below:

Invar tapes in catenary	1/500 000–1/1 000 000
Steel tapes on ground	1/10 000–1/20 000
Chaining	1/500–1/1000

Tellurometer

This is an instrument developed in South Africa which uses radio waves to measure distance. Since long radio waves cannot be directed, high frequency waves of about 3-cm wavelength, which can be concentrated in a narrow beam, are used. The equipment consists of a master instrument and slave instrument, both being equipped with parabolic reflector aerials. A signal is transmitted from the master to the slave instrument which receives it and then re-transmits it back to the master instrument. The time delay between reception and re-transmission is so small that it can be ignored. Instead of measuring the time delay between sending and receiving a signal to determine the distance, as is done in radar measurements, the instrument measures the phase difference in nanoseconds between sending and receiving. Since the carrier wave has too short a wavelength to make effective use of the phase shift for measurement of distance, it is pattern modulated at 10 MHz. This gives the same effect as using a carrier wave of 30-m wavelength. By varying the pattern modulation and observing the different phase shifts, it is possible to calculate the distance between master and slave. The instrument has a range of about 200 m–80 km, but its optimum range is 25–50 km. Temperature, pressure and humidity corrections need to be applied and spurious readings may be caused by reflection of the signals from the ground; this is particularly bad over water. The instrument has an accuracy of 1/500 000 over the longer range, but at distances of less than 2 km its accuracy is only of the order of 1/10 000.

Geodimeter

This is an instrument developed in Sweden which operates on the same basic principle as the tellurometer, but uses visible light instead of radio waves. The light carrier wave is pattern modulated by varying the intensity electronically at a frequency of 10 MHz giving an effective wavelength of 30 m. As in the case of the tellurometer different pattern modulation frequencies are used to observe the different phase shifts which enable the distance between the instrument and its reflector to be calculated. Observations need to be taken at night or twilight and the stations have to be intervisible which is not necessary in the case of the tellurometer. Distances up to 30 km can be measured with an accuracy of 1/200 000–1/300 000.

Tacheometry
Any theodolite fitted with stadia hairs, which are two short horizontal lines equidistant above and below the main cross hairs, can be used for tacheometric measurement of distance. The principle is illustrated in Fig. 8.27 which shows a theodolite set up at point A to read a staff

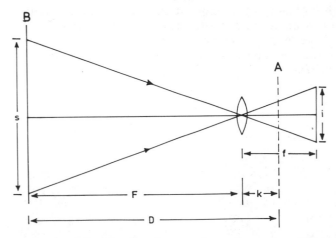

FIG. 8.27. Principle of tacheometric measurement of distance.

at point B. The difference between the upper and lower stadia readings is s and i is the distance between stadia hairs. Now, from similar triangles the following relationship holds:

$$f/i = F/s$$

so that,

$$F = (f/i)s$$

The distance to be measured, D, is the distance between the staff and the vertical axis of the theodolite and it is given by

$$D = F + k$$

so that,

$$D = (f/i)s + k$$

The term f/i is a constant for a particular instrument designated by c so that the final formula is:

$$D = cs + k$$

The theodolite is usually designed so that the instrument or multiplying constant c is equal to 100. In addition many tacheometers are fitted with an anallatic lens which eliminates the other instrument or additive constant k, so that the distance D is simply given by

$$D = 100(u - l)$$

where u is the upper stadia reading and l the lower stadia reading.

Figure 8.27 shows the simplest case where the theodolite and staff are on the same level. The situation is complicated when they are not on the same level as shown in Fig. 8.28. If the staff were inclined towards the theodolite at an angle α, the horizontal and vertical distances H and V would be given by:

$$H = 100s' \cos \alpha$$

$$V = 100s' \sin \alpha$$

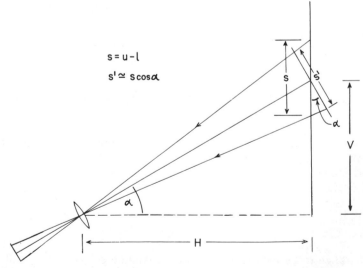

$$s = u - l$$
$$s' \simeq s \cos \alpha$$

FIG. 8.28.   Tacheometric determination of distance for inclined shots.

The normal surveying practice is to keep the staff vertical as this is easily done on a routine basis and it would not be possible for the staff man to judge the correct angle of tilt. Thus, the corrected formulae for the horizontal and vertical distances H and V become

$$H = 100(u - l) \cos^2 \alpha$$

$$V = 100(u - l) \sin \alpha \cos \alpha$$

Ordinary tacheometric measurements involve many calculations, but tables such as Redmond (1961) are published and enable the user to read the horizontal and vertical distances directly from them. The tables are published for every 20' of vertical angle, so the normal practice is to line up on the staff and adjust the vertical circle fine adjustment to read at one of the exact intervals and then read the staff rather than the other way round which would probably result in an odd angle reading not given in the tables. Figure 8.29 shows an

Station	Instrument Height	Horizontal Angle	Vertical Angle	Mid-point	Bottom Reading	Top Reading	Diff.	H	V	Remarks
A to ①	4·63	35°25′10″	+5°20′	10·40	12·80	8·00	480	475·9	+44·42	
②		67°37′20″	+6°40′	9·05	11·92	6·20	572	564·3	+65·96	
③		143°21′30″	−2°00′	8·22	11·52	4·75	677	676·2	−23·61	
④		232°18′00″	+9°40′	6·48	6·85	6·10	75	72·9	+12·41	
B		301°04′50″	+1°20′	8·13	10·77	5·50	527	526·7	+12·26	O.S. trig. pt.
⑤	4·54	82°53′30″	−3°40′	7·56	10·51	4·90	561	557·7	−35·74	
⑥		164°12′10″	+1°40′	5·95	6·76	5·10	166	165·9	+4·83	
⑦		171°01′40″	+4°20′	5·20	6·44	3·95	249	247·6	+18·76	

FIG. 8.29. Example of booking a tacheometric survey (units in feet).

example of the booking of a tacheometric survey with values for H and V taken from Redmond's tables (an instrument constant of 100 is assumed).

Some tacheometers, known as self-reducing tacheometers, have curved stadia lines which approach or recede from the zero curves as the line of sight is inclined, giving continuous solutions to the standard tacheometric formulae so that horizontal and vertical distances can be read directly without tables or calculations. Another type of tacheometer, known as a reduction tacheometer, achieves the same result as a self-reducing tacheometer by means of prismatic wedges in front of the objective lens which adjust the line of sight as the telescope is inclined so that the horizontal and vertical distances can be read directly.

To obviate the need for tedious calculations some instruments are fitted with special graduated scales on the vertical circle which allow quick and simple calculation of the horizontal and vertical distances to be made. This does not give a direct reading as in the case of self-reducing or reduction tacheometers, but the instruments are considerably cheaper. The most widely used of these systems is known as *Beaman's arc* which is generally fitted to most telescopic alidades. There are two graduated scales, one giving a value for V and the other for H. As an example of the Beaman arc let us suppose that with the telescope sighted on the staff at a station the arc readings are $V = 22$, $H = 5$ with a mid-staff reading of 2·45 and top and bottom stadia difference of 1·35, then

$$\text{vertical} = sV - h = 1·35\,(22) - 2·45 = 27·25$$
$$\text{horizontal} = 100s - sH = 135 - 1·35\,(5) = 128·25$$

In addition to the tacheometers described, there is also a class of tacheometers known as movable-hair or tangential tacheometers, in which the separation of the stadia hairs can be adjusted to give a constant staff difference reading. The distance is then determined by the amount of separation of the stadia lines. These tacheometers are rarely used now, however, and will not be described.

Tacheometry is extremely useful for surveying small areas where great detail and height control are required as many measurements can be taken from one instrument position. The accuracy is only of the order of 1/500 so it is not a method to use where great precision is required, but this is not a problem for most exploration work or evaluation studies in geology. It is also a useful method for traverse work in areas where terrain might make taping very difficult. Its usefulness, however, is limited by sight lengths which are generally restricted to 300 m and less with most instruments.

Infra-red tacheometers (distomats)

A number of manufacturers in different parts of the world now produce automatic instruments using an invisible beam of infra-red light to measure distance on the same principle as the geodimeter. They are either built as integral units incorporating a theodolite or are available as accessories to convert ordinary theodolites to distomats. A GaAs diode is used to provide the carrier wave and a special prismatic target is used to reflect the beam back to the instrument. Distances up to 1000 m can be measured under normal circumstances, but the range can be extended by using a larger number of prism reflectors. Operation is extremely simple. The target is sighted in the telescope and a start switch triggers the computer controlled measuring cycle. After a time lapse of about 10 s the slope distance is displayed to the millimetre with an accuracy of ± 5 mm. By entering the vertical angle into the built-in calculator, the vertical and horizontal distances are automatically displayed. In addition station coordinates can also be computed and displayed by entering the horizontal angle into the computer. For setting out work the instruments can also be used in an automatic measuring mode which displays the slope distance every few seconds for tracking a moving prismatic target.

Distomats have revolutionized engineering site surveying and have made the older methods virtually obsolete. The instruments are expensive, however, and for undertaking small infrequent surveys, such as in mineral exploration work, conventional tacheometers are still important survey instruments.

8.8 ASTRONOMICAL SURVEYING

The use of astronomy in surveying forms the basis of classical navigation and enables the observer to fix his position accurately anywhere on the earth. A full understanding of astronomical surveying requires a thorough grounding in basic astronomy and a knowledge of spherical trigonometry, but it is easy to learn the main principles so as to be able to carry out some of the simpler determinations which are described below.

Figure 8.30 shows the earth with centre O at the centre of a very much larger imaginary sphere known as the *celestial sphere*, which is a sphere on which all the heavenly bodies can be considered to lie. N and S are the celestial north and south poles respectively and they are located directly above the earth's true north and south poles. EE' is the celestial equator which may be considered as an extension of the earth's equator. If an observer is on the earth's surface at o, point Z on the celestial sphere vertically above him is known as the observer's zenith, hh' is his horizon and ϕ his latitude. Since the celestial sphere can be considered to have an infinite radius (the nearest star is some 58×10^8 earth radii from the earth's surface), the observer's horizon hh' can be considered to pass through the centre of the celestial sphere along HH'.

Keeping the concept of the celestial sphere in mind some further definitions are necessary:

Observer's meridian. This is a great circle on the celestial sphere which passes through the observer's zenith and celestial pole.

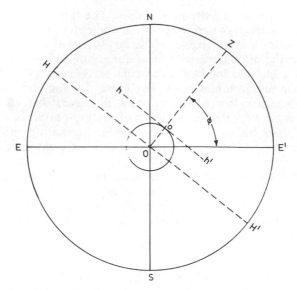

FIG. 8.30.   Earth at the centre of the celestial sphere.

Hour circle. A great circle through the celestial poles.

The hour angle of a heavenly body is the angular distance from the observer's meridian westwards to the hour circle through the heavenly body.

Ecliptic. This is the path of the sun described as a great circle on the celestial sphere. It is inclined at an angle of 23° 27′ to the celestial equator.

Altitude of a star (α). This is the vertical angle from the observer's horizon to the star along the arc of a great circle on the celestial sphere which passes through the observer's zenith and position of the star.

Declination of a star (δ). This is the vertical angle from the celestial equator to the star along the arc of an hour circle passing through the position of the star

Transit of a star. Owing to the rotation of the earth, the stars appear to move along arcs on the celestial sphere. If a star is observed for a period of several hours, its altitude will increase until it reaches a maximum on the observer's meridian. Then it will decrease until it reaches a minimum once again on the observer's meridian. These two points where it crosses the meridian are known as the upper and lower transits. It should be noted that the lower transit of many stars cannot be observed as it takes place below the observer's horizon. Stars whose upper and lower transits can be observed are known as *circumpolar stars*.

Elongation of a circumpolar star is when it is at the extreme east (east elongation) or west (west elongation) of the pole.

Obliquity. This is the inclination of the earth's axis (23°27') to the normal to the plane of the earth's path around the sun.

First point of Aries. This is the point on the celestial equator which is crossed by the ecliptic at the *vernal equinox* when the sun passes from south to north of the celestial equator. The point of Aries moves approximately 50" of arc a year along the celestial equator. This is known as the 'precession of the equinoxes'.

The right ascension (RA) of a heavenly body is the angular distance eastwards along the equator from the point of Aries to the hour circle through the heavenly body.

To make use of the methods of field astronomy surveyors have to refer to the *Star Almanac* which is published annually. This lists all the pertinent data about the sun and the stars which are commonly observed so that the various calculations can be made.

Time and longitude

Owing to the rotation of the earth the sun appears to revolve around it making $365\frac{1}{4}$ such 'revolutions' *per annum*. The stars on the other hand appear to make $366\frac{1}{4}$ 'revolutions' *per annum*. Thus, there is a difference between a solar day and a sidereal day with a sidereal day being about 3 min 55 s shorter. This difference (R) between solar and sidereal time is listed in the Star Almanac and the two times are the same once a year on about the 22nd September. Longitude can be determined if the time is known accurately. The zero meridian of longitude passes through Greenwich, England, and normal time measurement, which is based on a mean solar day of 24 h, is made with reference to Greenwich Mean Time (GMT) or Universal Time (UT) as it is often known. To determine longitude at any locality the precise moment in terms of GMT when the sun reaches its highest point (upper transit) is measured giving mid-day in terms of Local Apparent Time (LAT). This can be converted to Greenwich Apparent Time (GAT) by the following relationship:

$$GAT = GMT - E$$

where E is known as the equation of time and can be found in the Star Almanac. Longitude may be defined by:

$$Longitude = LAT - GAT$$

As an example, let us suppose that the upper transit of the sun was observed on 15th April 1977 to take place at 15 h 15 min 20 s GMT then,

$$
\begin{array}{ll}
15 \text{ h } 15 \text{ min } 20\!\cdot\!0 \text{ s} & \text{(GMT)} \\
-12 \text{ h } 00 \text{ min } 00\!\cdot\!2 \text{ s} & (E \text{ from Star Almanac}) \\
\hline
3 \text{ h } 15 \text{ min } 19\!\cdot\!8 \text{ s} & \text{(GAT)}
\end{array}
$$

$$
\begin{array}{ll}
12 \text{ h } 00 \text{ min } 00\!\cdot\!0 \text{ s} & \text{(LAT)} \\
-3 \text{ h } 15 \text{ min } 19\!\cdot\!8 \text{ s} & \text{(GAT)} \\
\hline
8 \text{ h } 44 \text{ min } 40\!\cdot\!2 \text{ s} & \text{(longitude)}
\end{array}
$$

Now, in terms of time 1 h is equivalent to 15°, 1 min to 15′ and 1 s to 15″ so that the above longitude converts to 131°10′3″ west. (If the longitude is less than 12 h, it is west and, if it is more than 12 h, it must be subtracted from 24 h to give longitude east.) The accurate determination of GMT is quite easy since a number of special radio stations in various parts of the world broadcast time signals for use by land surveyors.

It is often difficult to determine the precise moment of the upper transit of the sun and it is easier to sight the theodolite on the sun just before upper transit and note the time when the upper or lower limb crosses the horizontal cross hair on its ascent and when the same limb touches the cross hair on its descent with the vertical angle of the theodolite firmly fixed. The mean time will be the moment of upper transit. In fact, a correction needs to be applied, but this can be ignored if the sun is observed very close to its upper transit and the time interval is small.

Determination of azimuth and latitude
The simplest method of determining azimuth would be to observe a star at its upper transit when the vertical circle of the theodolite will lie in a true north–south direction. However it is not possible to obtain an accurate horizontal fix on the upper transit since the vertical component of a star's motion becomes very slow near its transit and the star appears to be moving horizontally. To avoid this problem the star has to be observed on the ascent side of its transit when it can be accurately fixed by the cross hairs. Then it is observed on its descent with the vertical angle set at the same reading and followed by the horizontal circle until it is bisected by the vertical cross hair as it crosses the horizontal cross hair. The mean of the two horizontal angles obtained on either side of the transit will give the correct horizontal reading of the true north–south line. If the latitude is known, the azimuth can be determined by observing a circumpolar star at eastern or western elongation. This is ideal for very accurate work since there is so little change in azimuth near elongation that the star will appear to ascend or descend the vertical cross hair depending

on which elongation is being observed. This gives adequate time for precise face left and face right readings. The azimuth of the reference line is then given by

$$\sin^{-1} \frac{\cos \delta}{\cos \phi} + \theta$$

where δ is the declination of the star, ϕ is the observer's latitude and θ is the horizontal angle between the reference line and the point of elongation.

Latitude can be determined by measuring the altitude of a star at its upper transit. Then, latitude is given by

$$90° - \text{altitude } (\alpha) - \text{declination } (\delta)$$

When observing the altitude of a star it is necessary to apply a correction for refraction in the atmosphere. The correction varies with altitude, but for altitudes greater than 20°, an approximate value for the correction is $-58'' \cot \alpha$. For really precise work adjustments have to be made for air temperature and pressure and reference needs to be made to refraction tables in the Star Almanac.

As an example, the upper transit of γ Ursae Majoris (each star is designated by a Greek letter and the name of the constellation on the star charts) was observed at an altitude of 20° 49' 50'' (average of face left and face right) on the 25th April 1977 in Lusaka, Zambia. Correcting for refraction the altitude becomes:

$$20° \, 49' \, 50'' - 58'' \cot(20° \, 49' \, 50'') = 20° \, 47' \, 18''$$

In the Star Almanac the declination for γ Ursae Majoris is 53° 49' 19'' on the 25th April 1977. Thus, the latitude is:

$$90° \, 00' \, 00'' - 53° \, 49' \, 19'' - 20° \, 47' \, 18'' = 15° \, 23' \, 23'' \text{ South}$$

Latitude can also be determined by observing the sun, but corrections need to be applied for parallax since the sun is much closer to the earth than any of the stars and a measurable angle is subtended at the sun by the earth's radius, so that the observer's horizon cannot be assumed to pass through the centre of the celestial sphere as in the case of the stars. This correction is $+8.80'' \cos \alpha$ where α is the altitude corrected for refraction. In addition it is not easy to observe the sun's centre. Unless a special solar prism is fitted to the telescope objective, which permits direct aiming at the sun's centre, it is normal practice to sight on the edge of the sun with the cross hairs. In this case a correction for the semi-diameter needs to be applied. This is approximately 16', but it varies slightly and the exact correction can be determined from the Star Almanac. To illustrate the procedure suppose that the altitude of the sun's lower limb was observed as

42° 26′ 55″ at upper transit on the 22nd June 1977 at 6 h 30 min GMT and it is required to find the latitude. Firstly, the altitude is corrected for refraction which results in a corrected angle of 42° 25′ 52″. Secondly, the correction for parallax is applied which gives an angle of 42° 25′ 58·5″. Thirdly, the correction for the semi-diameter is looked up in the Star Almanac for June 1977 and added to the altitude giving a final corrected altitude of 42° 41′ 46·5″. Fourthly, the declination is looked up in the Star Almanac for 6 h 30 min GMT on the 22nd June 1977. This gives a value of 23° 26′ 18″ so that the latitude becomes:

$$90° - 23° 26′ 18″ - 42° 41′ 46·5″ = 23° 51′ 55·5″$$

For most astronomical observations it will be necessary to use eyepiece prisms or, better still, diagonal eyepieces. In addition, for work at night, illumination will be necessary both for reading angles and for seeing the cross hairs. For solar observations it is absolutely necessary to use a sun filter over the telescope objective.

REFERENCES AND BIBLIOGRAPHY

Bannister, A. and Raymond, S. (1973). *Surveying*, Pitman Publishing, London, 548 pp.
Bouchard, H. and Moffit, F. H. (1965). *Surveying*, International Textbook Co., Scranton, Pennsylvania, USA, 754 pp.
Clendinning, J. and Olliver, J. G. (1966). *The Principles of Surveying*, Blackie and Son Ltd., Glasgow, 463 pp.
Dugdale, R. H. (1960). *Surveying*, MacDonald and Evans Ltd., London, 120 pp.
Her Majesty's Nautical Almanac Office, (1976). *The Star Almanac for Land Surveyors for the Year 1977*, Her Majesty's Stationery Office, London, 80 pp.
Ministry of Defence (1965). *Textbook of Topographical Surveying*, Her Majesty's Stationery Office, London, 388 pp.
Redmond, F. A. (1961). *Tacheometric Tables*, The Technical Press, London, 256 pp.

CHAPTER 9

Ore Reserve Calculations

9.1 ORE AND ORE RESERVES

Ore is defined as a mineral or mineral aggregate more or less mixed with a gangue, from which a metal or metals can be won at a profit or in the hopes of a profit.* From this definition it follows that local production and transportation costs are a prime factor and what may constitute an ore body in one region may not do so in a more remote part of the world. The term 'ore', therefore, should not be applied indiscriminately to any mineral deposit and the term 'ore zone' should be avoided when referring to mineralized intersections in boreholes and trenches unless they are clearly part of a deposit which has distinct possibilities of being economic. On the other hand, one should not be too pedantic about the term 'ore', otherwise mining engineers would forever be adding to and subtracting from ore reserves on the basis of metal price fluctuations. In extreme cases, of course, this does happen when sharp price rises may bring material that was considered as waste into the ore reserves and, vice versa, when rising costs and depressed prices may result in ore being reclassified as waste. As regards deposits not undergoing exploitation, it is acceptable to refer to the mineralization as 'ore' when a reasonable tonnage has been shown to exist and when there is a distinct hope that it might be worked at a profit.

Before it can be decided whether or not a mineral deposit might be economically viable, it is necessary to determine the tonnage available together with the overall grade. It is obvious that the degree of confidence with which this can be done depends upon the amount of data available. There are three standard categories for quoting ore reserves:

*Strictly ore refers only to metallic minerals, but non-metallic minerals such as fluorite are often included.

1. Proven or measured
2. Indicated or probable
3. Inferred or possible

The proven or measured category is only applied when there is a high degree of certainty, usually only in mines where the ore is blocked out and thoroughly sampled. This is often referred to as *ore-in-sight*. This category should not normally be used in exploration work where the sample information is all from drill-holes. An exception to this might be material which is very uniform and predictable such as a bed of limestone or phosphate rock.

A fairly high degree of confidence is implied by the indicated or probable category and it should only be used for ore that has been clearly outlined in three dimensions by drilling, pitting, trenching, etc. This category is often referred to as 'drill indicated ore'.

The inferred or possible category implies a degree of uncertainty and is used in cases where there is good geological evidence for continuity, but only a limited amount of sample data such as a few widely spaced boreholes.

In quoting ore reserves it is generally accepted that the first two categories may be added together, but many people object to the inclusion of possible ore in the total reserves. Although there are no hard and fast rules for classifying ore reserves and the division between the three classes is arbitrary, it is extremely important not to be too liberal in applying the term 'ore reserves'. For instance, the geologist who finds a mineralized outcrop and then quotes reserve figures by extrapolating along strike and down dip is not only foolhardy but dishonest. In cases where mineralized rock has only been defined in two dimensions or to shallow depths and there are good indications that it extends to depths below that seen, it is common practice to quote probable or possible reserves in the form: 'there are x tonnes for every n metres of depth'. For mineralization in outcrop the depth given might be 3 or 5 m. For mineralization located in pits or shallow drill-holes, the depth given might be 30 m or more.

In ore reserve calculations we are concerned with calculating a volume and an average grade. The tonnage is derived from the volume by multiplying by the tonnage factor. If the volume is determined in cubic metres, the *tonnage factor* is simply the specific gravity of the ore. The volume is commonly determined by calculating an area in two of the dimensions and then multiplying by the third dimension to determine the final volume. To determine total area it is usually possible to divide the area under consideration into a number of regular geometric figures such as squares, rectangles, triangles, trapezia, etc. (Fig 9.1).

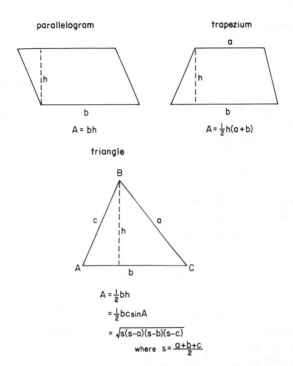

FIG. 9.1. Formulae for the areas of some simple figures.

The average grade is determined by making use of weighted averages. Let us consider an example in two dimensions. Figure 9.2 shows a vertical section through two drill-holes and we are required to determine the area of copper ore between the drill-holes and assign an average grade to it. In this case, the area is readily determined

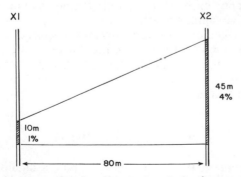

FIG. 9.2. Section through two adjacent boreholes intersecting copper ore.

from the trapezium formula:

$$(10 + 45)80/2 = 2200 \text{ m}^2$$

The grade could be determined by taking the mean of the two intersections which gives a value of 2·5%. However, this does not take into account the much longer and richer intersection in X2 which would contribute more to the grade of the block than the shorter intersection. One way of weighting the grades is known as the metre-% method where the average grade $\bar{g}$ is given by:

$$\bar{g} = \frac{\Sigma PT}{\Sigma T}$$

where P is the grade of an intersection and T its true thickness. For the example above the mean grade is 3·45%. Another method is known as the percentage method and the mean grade is determined by the formula:

$$\bar{g} = \left(\frac{\Sigma PT}{\Sigma T} + \Sigma P\right)\Big/3$$

which gives 2·82% for the example above.

In the example in Fig. 9.2 there is a positive correlation between grade and thickness, i.e. the longer intersection has a higher grade. If we swap the grades between X1 and X2, we will have an example of a negative correlation, i.e. the longer intersection has a lower grade. The straight mean value is 2·5%, the same as in the case for a positive correlation, but the metre-% method gives 1·55% and the percentage method 2·18%. The ordinary mean should never be used because it gives the same answer whether there is positive or negative correlation. Although the metre-% method is widely used, it can be shown that it over-evaluates for positive correlation and under-evaluates for negative correlation. For this reason the percentage method is preferred. In cases where there is poor correlation, i.e. grade and thickness are variable, both methods give similar results.

In exploration geology we are normally required to calculate ore reserves from a number of drill-hole intersections. For tabular bodies with a low dip (up to 20°) a number of different methods can be used. These can be divided into two types: (i) plan methods and (ii) cross-sectional methods.

9.2 PLAN METHODS

Some of the plan methods are based upon assigning an 'area of influence' around each drill-hole. Three of these are shown in Fig. 9.3.

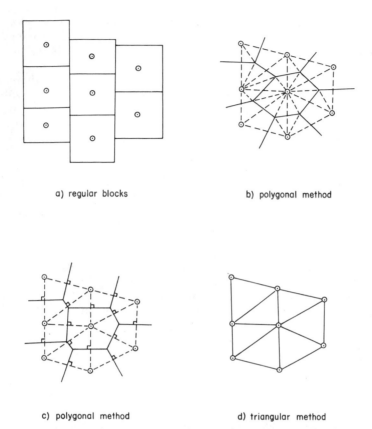

a) regular blocks

b) polygonal method

c) polygonal method

d) triangular method

FIG. 9.3. Some different methods for calculating ore reserves of low dipping ore bodies in plan.

The first method (a) is based on dividing the ore body into rectangular blocks. If the holes are regularly spaced, the lines separating the blocks are drawn halfway between the holes. If the holes are irregularly spaced, the size of the blocks will be arbitrary. The other two methods (b) and (c) are two variations of the polygonal method. In the first (b) the sides of the polygons around each hole are located by joining the points at the intersections of the bisectors of the angles between the lines joining the holes. In the second (c) the sides of the polygons are the perpendicular bisectors of the lines between holes. In the 'area of influence' methods each block or polygon is assigned the grade and thickness of the hole at its centre. The area of each block or polygon is determined and then multiplied by its thickness to determine the volume. The sum of the individual volumes gives the total volume of the ore body. The average grade is determined by summing the products of each block volume and its grade and dividing this sum by the total volume.

These 'area of influence' methods over-evaluate when there is a positive correlation between thickness and grade and under-evaluate when there is a negative correlation. This problem is overcome by the triangular method (Fig. 9.3 (d)). With this method, the area is divided into triangles by drawing lines between the holes. The thickness and grade for each triangle is determined as a weighted average of the values in the holes at the corners of the triangle. The grade can be determined by either the metre-% or percentage method. In the case of the percentage method the formula for the three dimensional case becomes:

$$\bar{g} = \left(\frac{\Sigma PT}{\Sigma T} + \Sigma P\right)\Big/ 4$$

The average thickness can be determined as a simple mean of the three thicknesses at each corner. Let us consider the example of the triangular method for the three drill-hole intersections given in Fig. 9.4. By the metre-% method the grade assigned to the block is:

$$\frac{8 \times 2 \cdot 9 + 25 \times 5 \cdot 2 + 15 \times 3 \cdot 8}{(8 + 25 + 15)} = 4 \cdot 38\%$$

and the thickness is:

$$\frac{8 + 25 + 15}{3} = 16 \text{ m}$$

By the percentage method the grade is:

$$\left(\frac{8 \times 2 \cdot 9 + 15 \times 3 \cdot 8 + 25 \times 5 \cdot 2}{8 + 25 + 15} + 2 \cdot 9 + 5 \cdot 2 + 3 \cdot 8\right)\Big/ 4 = 4 \cdot 07\%$$

and the thickness is 16 m as before.

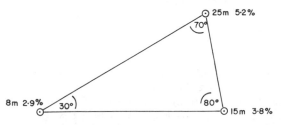

FIG. 9.4. Plan view of three adjacent boreholes intersecting copper ore joined for ore reserve calculation by the triangular method.

We can refine this further by weighting the values at each corner according to the size of the included angle. The weighting factor is determined by dividing the angle in degrees at each corner by 60. If we do this the thickness becomes:

$$\left(8 \times \frac{30}{60} + 15 \times \frac{80}{60} + 25 \times \frac{70}{60}\right)\Big/ 3 = 17 \cdot 72 \text{ m}$$

and the grade becomes 4·50% by the metre-% method and 4·27% by the percentage method.

The triangular method is better than the 'area of influence' methods and, if there is either positive or negative correlation between grade and thickness, it is better to use the percentage method in calculating the mean grade. If there is no correlation, the metre-% method is adequate and less laborious to calculate. Whether the metre-% or percentage method is used it is best to weight the values according to the size of the included angles as shown above.

Another plan method, known as the contour method (Gilmour, 1964), is very simple to use and gives good results. In this method two contour maps of the ore body are drawn, one showing variation in thickness (isopachs) and the other metre-% values. To calculate ore reserves the areas between contours on the isopachyte plan are determined with a planimeter and each area obtained is multiplied by the respective mean value of the two bounding contours. The sum of all the values so obtained gives the total volume of the ore body and this multiplied by the correct tonnage factor gives the total ore reserves. To determine the average grade the metre-% map is treated in the same manner as the isopachyte map to give a total cubic metre-%. This value divided by the total volume determined from the isopachyte map gives the average grade. If one has the use of a planimeter, this is probably the best of the plan methods to calculate ore reserves as it avoids the problems of over- and under-evaluation that sometimes occur with some of the other plan methods.

In any of the plan methods it is extremely important to use the true positions of the ore intersections and not the collar positions of the drill-holes as these will only be the same for perfectly vertical holes.

9.3 CROSS-SECTIONAL METHODS

These are based on drawing sections across the ore body, determining the areas of ore in each section and computing the final volume by using the distance between sections. In addition to being an alternative method to plan methods, cross-sectional methods are useful for determining reserves of irregular bodies or bodies of variable dip

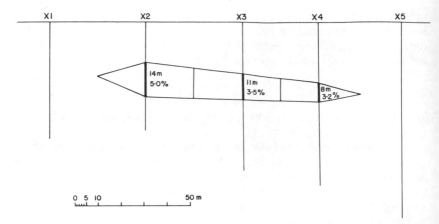

FIG. 9.5. Example of a drill section through a copper ore body to illustrate ore reserve calculations by the cross-sectional method.

which cannot be calculated by ordinary plan methods. Figure 9.5 shows an example of a drill section and we are required to determine its area and average grade. In this example the calculations are quite easy as the section through the ore body has been drawn as a regular figure. First the ore body section is divided into compartments at the mid-points between holes and the grade in each hole is assigned to its respective compartment. Next the area of each compartment is calculated and multiplied by its grade. These are summed and divided by the total area to give the average grade for the section. In the example given the average grade becomes:

$$\frac{425 \times 5 + 418 \times 3 \cdot 5 + 216 \times 3 \cdot 2}{425 + 418 + 216} = 4 \cdot 04\%$$

over a total area of 1059 m².

This area was readily determined by dividing the figure into triangles and trapezia, but it is not always possible to draw the sections as regular figures and we need to consider methods of determining areas of irregular figures. One method is to measure it directly with a *planimeter*. This is a mechanical integrating instrument which consists of a pole block, a pole arm, a tracing arm and a measuring unit. To use the instrument the pole block is placed on the plan outside the area being measured and the tracing point on the end of the tracing arm is moved round the outline of the figure being measured. The area is determined by reading the measuring unit before tracing the outline of the figure and after returning to the commencing point. The difference between the two readings multiplied by a scale factor gives the area. The measuring unit consists of a graduated drum with a vernier scale and another indicator to show the number of complete revolutions of the drum. Planimeters with an adjustable pole arm can be adjusted for the scale of plan being used, whereas fixed pole arm planimeters give the area in square millimetres. Areas can be determined by resting the pole block inside the figure, but this is more complicated and requires a correction.

Another method is known as 'give and take' and consists of fitting regular figures (triangles, rectangles, trapezia, etc.) into the figure making sure that all small areas of the irregular figure left out along boundary lines ('give') are made up by approximately equal areas outside the figure ('take'). An example of this is shown in Fig. 9.6(a) which gives a total area of $1708 \cdot 5 \text{ m}^2$.

A third method, known as 'counting squares', consists of placing a squared overlay on the figure and counting whole and part squares contained within the boundaries of the irregular figure (Fig. 9.6(b)).

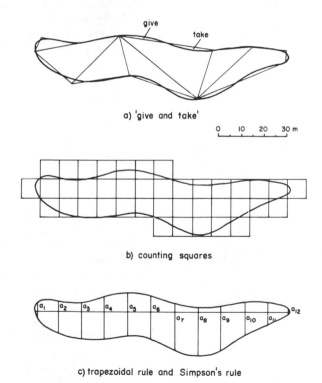

a) 'give and take'

b) counting squares

c) trapezoidal rule and Simpson's rule

FIG. 9.6. Different methods of determining areas of irregular figures.

The total area is then obtained by summing the fractions of the squares and whole squares.

Another method is known as the trapezoidal rule. This is done by drawing a longitudinal line within the figure and dividing it into a number of equal compartments with cross lines drawn at right angles to the longitudinal line (Fig. 9.6(c)). The total area is given by:

$$A = d\left(\frac{a_1 + a_n}{2} + a_2 + a_3 + \cdots + a_{n-1}\right)$$

where d is the width of the compartments and a_1, a_2, ... a_n the lengths of the cross lines. In the example shown in Fig. 9.6(c) the area becomes:

$$A = 10\left(\frac{5 + 2\cdot5}{2} + 13\cdot5 + 16 + 18 + 18\cdot5 + 20 + 21 + 19 + 14 + 10\right)$$
$$= 1722\cdot5 \text{ m}^2$$

Another method which gives greater accuracy than the other methods is known as Simpson's rule. The figure is divided up in the same manner as for the trapezoidal rule, but it is essential that there is an *uneven* number of equal strips. According to the rule, the area is given by:

$$A = \frac{d}{3}[a_1 + a_n + 2(a_3 + a_5 + \cdots + a_{n-1}) + 4(a_2 + a_4 + \cdots + a_{n-2})]$$

In the example above the area becomes:

$$A = \frac{10}{3}[5 + 2\cdot5 + 2(16 + 18\cdot5 + 20 + 19 + 10)$$
$$+ 4(13\cdot5 + 18 + 18\cdot5 + 21 + 14)]$$
$$= 1713.3 \text{ m}^2$$

Once the areas of the individual cross-sections have been calculated, there are three ways of determining the total volume. One method is to multiply the area of each cross-section by the distances midway between the adjacent cross-sections. An arbitrary distance can be extended beyond the end cross-sections. This is usually a quarter to a half of the distance to the adjacent barren or 'non-ore' section. The sum of the individual volumes so obtained gives the total volume. A second method is known as the trapezium method for volumes. If the sections are equidistant, the formula becomes:

$$V = D\left(\frac{A_1 + A_n}{2} + A_2 + A_3 + \cdots + A_{n-1}\right)$$

where A_1, A_2, ... A_n are the corresponding areas and D the distance between them. If A_1 and A_n are the end 'ore' cross-sections, an arbitrary volume of ore needs to be added to each end. This can be calculated in the same manner as described in the first method. If the sections are not equidistant, the formula becomes:

$$V = \tfrac{1}{2}[d_1(A_1 + A_2) + d_2(A_2 + A_3) + d_3(A_3 + A_4) + \cdots d_{n-1}(A_{n-1} + A_n)]$$

where A_1, A_2, ... A_n are the cross-sectional areas and d_1 is the distance between A_1 and A_2, d_2 the distance between A_2 and A_3, and so on. The third method is known as Simpson's rule for volumes and can only be applied if all the cross-sections are equal distances apart and if there is an odd number of sections. The formula is:

$$V = \frac{D}{3}[A_1 + 4(A_2 + A_4 + \cdots + A_{n-1}) + 2(A_3 + A_5 + \cdots + A_{n-2}) + A_n]$$

As in the cases before, volumes need to be added to the last cross-sections to take into account ore extending beyond the end sections.

Let us consider an actual example to compare the various methods. Figure 9.7 shows a small stratiform copper deposit outlined by a number of vertical diamond drill-holes and we are required to calculate drill-indicated ore reserves. The simplest method is to use regular blocks. In Fig. 9.8, the deposit has been marked out in rectangular blocks by drawing lines midway between the holes. Calculations carried out as described for this method give 1 867 780 tonnes at 3·30% Cu using an average s.g. of 2·70. Figure 9.9 shows the deposit marked out in polygons in the manner shown in Fig. 9.3(b). Calculations carried out in the manner described give 2 137 900 tonnes at 3·48% Cu using an average s.g. of 2·70. Figure 9.10 shows the deposit marked out for the triangular method. Note that narrow rectangular blocks have been added around the edge to take in fringe ore. In this case the boundaries chosen are one-third the distance to the adjacent barren holes. The grades and thicknesses of these rectangular blocks are simply the respective values from the holes near the centres of the blocks. The percentage method was used to determine grades of the triangular areas and values were weighted according to the angles in the manner described previously. The total reserves come to 1 832 340 tonnes at 3·23% Cu using an average s.g. of 2·70. Figures 9.11 and 9.12 show the isopachyte map and metre-% map respectively. The various areas enclosed by each contour and the calculations are shown in the figures. Total ore reserves determined by the contour method are 1 919 210 tonnes at 3·58% Cu. Figure 9.13 shows a series of cross-sections through the ore body. The areas determined for each cross-section together with average grades are shown in the figure. As mentioned earlier there are several different

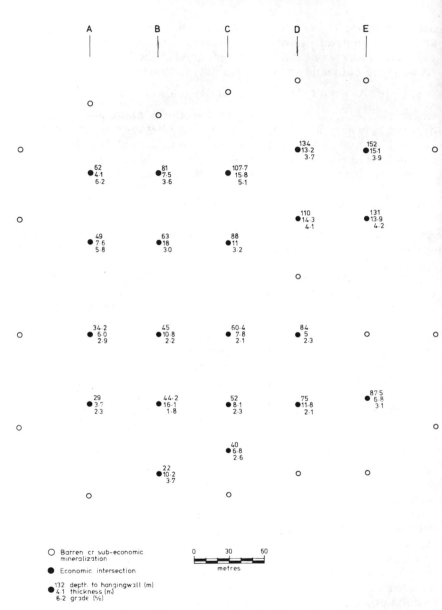

FIG. 9.7.　Small stratiform copper deposit defined by diamond drilling.

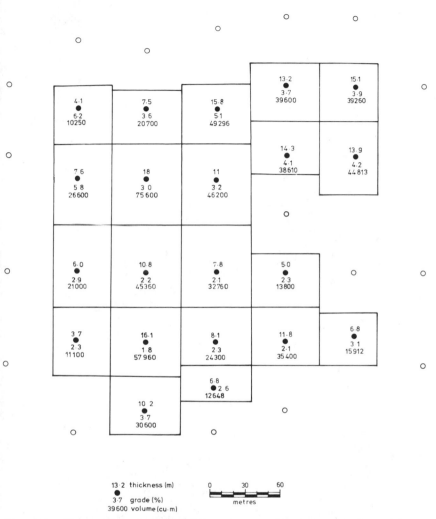

FIG. 9.8. Copper deposit in Fig. 9.7 marked out for ore reserve calculations by the method of regular blocks.

ways these can be used to determine the final tonnage and grade. One method is to multiply the cross-sectional areas by the distances midway between adjacent cross-sections. The calculations for the example in Fig. 9.13 are given on p. 446 using one-third of the distance to the adjacent non-ore section at each end.

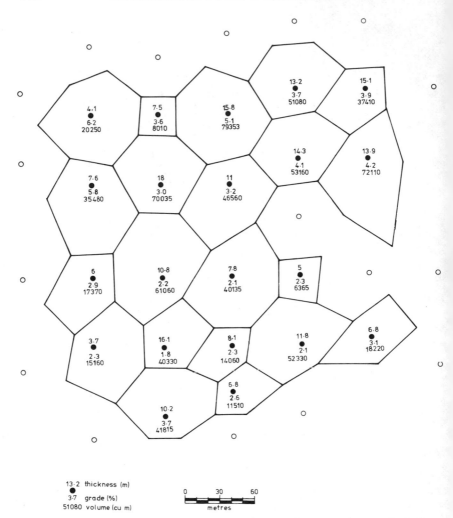

FIG. 9.9. Copper deposit in Fig. 9.7 marked out for ore reserve calculations by the polygonal method.

Section	Volume	Volume × grade
A	1292 × 50	1292 × 50 × 4·34
B	3907 × 60	3907 × 60 × 3·21
C	2939 × 60	2939 × 60 × 3·48
D	2059 × 60	2059 × 60 × 3·27
E	1751 × 50	1751 × 50 × 3·92
Totals	686 450	2 393 687

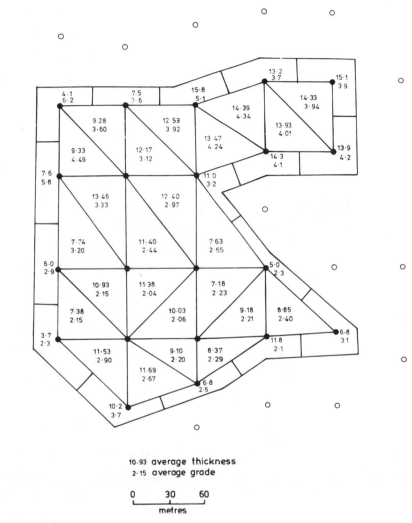

10·93 average thickness
2·15 average grade

0   30   60
metres

FIG. 9.10. Copper deposit in Fig. 9.7 marked out for ore reserve calculations by the triangular method.

$$\text{Average grade} = \frac{2\,393\,687}{686\,450} = 3\cdot49\%$$

$$\text{Tonnage} = 686\,450 \times 2\cdot7 = 1\,853\,415 \text{ tonnes}$$

Using a variation of the trapezium method for volumes the calculations are as follows:

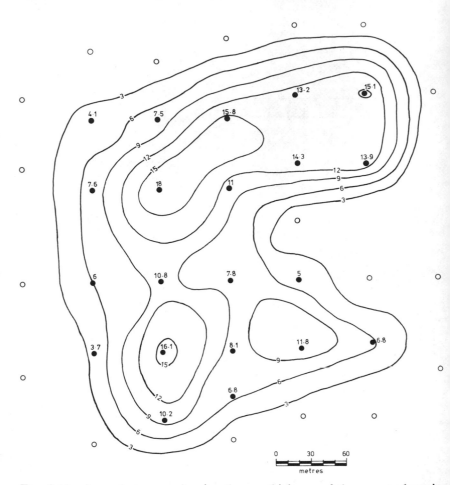

FIG. 9.11. Isopachyte map showing the ore thickness of the copper deposit in Fig. 9.7.

Section	Volume	Grade	Volume × Grade
Outside A	1292×20	4·34	112 146
A + B	$\dfrac{1292 + 3907}{2} \times 60$	$\dfrac{(1292 \times 4 \cdot 34) + (3907 \times 3 \cdot 21)}{1292 + 3907}$	544 463
B + C	$\dfrac{3907 + 2939}{2} \times 60$	$\dfrac{(3907 \times 3 \cdot 21) + (2939 \times 3 \cdot 48)}{3907 + 2939}$	683 076
C + D	$\dfrac{2939 + 2059}{2} \times 60$	$\dfrac{(2939 \times 3 \cdot 48) + (2059 \times 3 \cdot 27)}{2939 + 2059}$	508 820
D + E	$\dfrac{2059 + 1751}{2} \times 60$	$\dfrac{(2059 \times 3 \cdot 27) + (1751 \times 3 \cdot 92)}{2059 + 1751}$	407 906
Outside E	1751×20	3·92	137 278
Totals	686 450		2 393 689

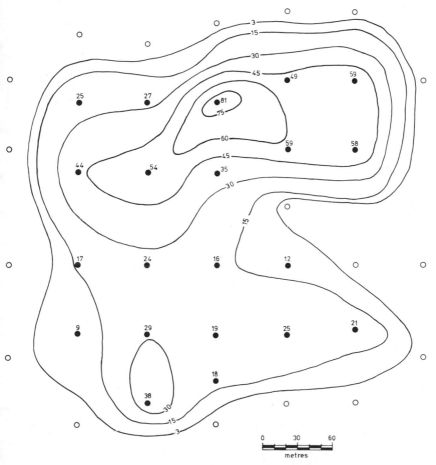

FIG. 9.12. Metre-% map of the copper deposit in Fig. 9.7.

$$\text{Average grade} = \frac{2\,393\,689}{686\,450} = 3\cdot49\%$$

$$\text{Tonnage} = 686\,450 \times 2\cdot7 = 1\,853\,415$$

This gives the same answer as the first method, though it is rather more cumbersome.

Using Simpson's rule for volumes we can calculate the tonnage as follows:

$$V = \frac{60}{3}[1292 + 4(3907 + 2059) + 2(2939) + 1751] = 655\,700$$

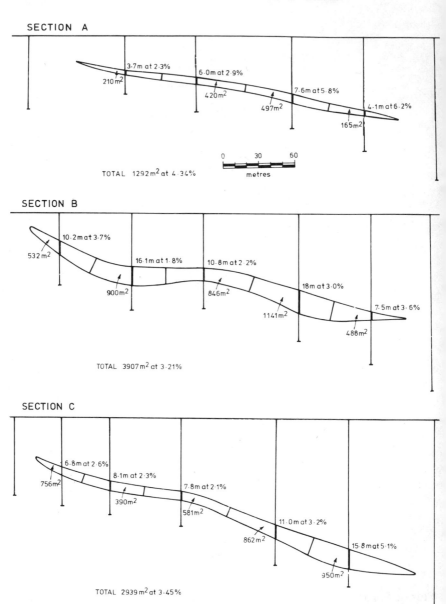

SECTION A

3·7m at 2·3%
210 m²
6·0m at 2·9%
420 m²
497 m²
7·6m at 5·8%
4·1m at 6·2%
165 m²

TOTAL 1292 m² at 4·34%

0   30   60
metres

SECTION B

10·2m at 3·7%
532 m²
16·1m at 1·8%
900 m²
10·8m at 2·2%
846 m²
18m at 3·0%
1141 m²
7·5m at 3·6%
488 m²

TOTAL 3907 m² at 3·21%

SECTION C

6·8m at 2·6%
756 m²
8·1m at 2·3%
390 m²
7·8m at 2·1%
581 m²
11·0m at 3·2%
862 m²
15·8m at 5·1%
950 m²

TOTAL 2939 m² at 3·45%

FIG. 9.13. Cross-sections through the copper deposit in Fig. 9.7 for ore reserve calculations by the cross-sectional method. (All sections drawn to same scale.)

SECTION D

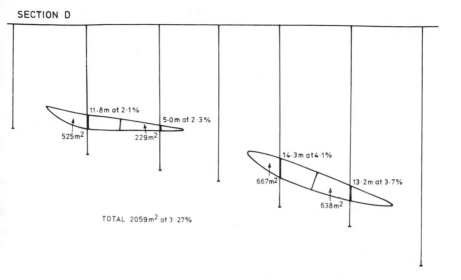

TOTAL 2059 m² at 3·27%

SECTION E

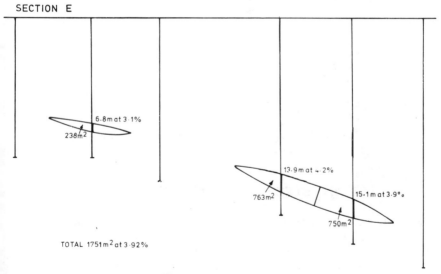

TOTAL 1751 m² at 3·92%

FIG. 9.13.—*contd.*

Then, adding volumes for ore outside the end cross-sections, the total volume becomes:

$$1292 \times 20 + 655\,700 + 1751 \times 20 = 716\,560$$
$$\text{Tonnage} = 716\,500 \times 2 \cdot 7 = 1\,934\,712 \text{ tonnes}$$

To calculate the average grade:

$$(1292 \times 20 \times 4\cdot34) + \frac{60}{3}\,[1292 \times 4\cdot34 + 4(3907 \times 3\cdot21 + 2059 \times 3\cdot27)$$
$$+ \, 2(2939 \times 3\cdot48)] + (1751 \times 3\cdot92) = 2\,449\,909$$
$$\text{Average grade} = \frac{2\,449\,909}{716\,560} = 3\cdot42\%$$

The results for the various methods are given in Table 9.1 where it can be seen that they compare very favourably with tonnages varying from 1·83 million to 2·14 million and average grades from 3·23% to 3·58%.

TABLE 9.1

COMPARISON OF ORE RESERVES OBTAINED FOR THE COPPER DEPOSIT IN FIG. 9.7 USING DIFFERENT METHODS

Method	Tonnage	Grade
Regular blocks	1 867 780	3·30%
Polygonal	2 137 900	3·48%
Triangular	1 832 340	3·23%
Contour	1 919 210	3·58%
Cross-sectional (1)	1 853 415	3·49%
Cross-sectional (2)	1 853 415	3·49%
Cross-sectional (3)	1 934 712	3·42%

9.4   STEEPLY DIPPING ORE BODIES

When dealing with ore bodies dipping at more than 20°, it is no longer satisfactory to use ordinary plan methods as they underestimate the total reserves. At a dip of 20° this underestimation is 6%, at 25° it is 9%, at 30° it is 13% and at 40° it is 23%. To avoid this it is best to use cross-sectional methods to make accurate reserve estimations of steeply dipping ore bodies. Plan methods can be used, however, if they are modified and *vertical thicknesses* used in place of *true thicknesses*. This relationship is illustrated for two dimensions in Fig. 9.14. Vertical boreholes X1 and X2 have intersected an ore body R dipping at $\theta°$. The vertical thicknesses in X1 and X2 are x_1 and x_2

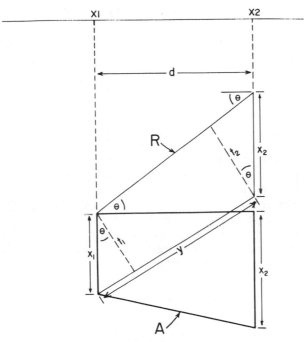

FIG. 9.14. Diagram to compare the effects of using true and vertical thicknesses of an ore body in reserve calculations by plan methods.

respectively with corresponding true thicknesses t_1 and t_2. The representation in plan of the ore in X1 and X2 will be represented by the figure A with the heavy outline.

Now:

$$\text{Area of } R = \left(\frac{t_1 + t_2}{2}\right)y$$

$$\text{Area of } A = \left(\frac{x_1 + x_2}{2}\right)d$$

$$t_1 = x_1 \cos \theta, \ t_2 = x_2 \cos \theta$$

$$d \doteq y \cos \theta$$

$$R = \left(\frac{x_1 + x_2}{2}\right)y \cos \theta$$

$$A = \left(\frac{x_1 + x_2}{2}\right)y \cos \theta$$

$$\therefore R = A$$

9.5   ORE BODIES OF VARIABLE DIP

If the dip of an ore body shows considerable variation, the reserves cannot be determined accurately by ordinary plan methods. Cross-sectional methods could be used quite effectively, but it is possible to use plan methods either by using vertical thicknesses as just described or, better still, by making use of *unrolled sections*. In effect this is a process of flattening or 'unrolling' the ore body so that it lies in one plane. Figure 9.15 (a) shows a cross-section of an ore body with a variable dip and the curved distance d has been measured and redrawn as a straight line to produce the unrolled cross-section of the ore body in Fig. 9.15 (b). A plan of the ore body is then drawn using the 'unrolled' measurements; this is the final unrolled section. Ordinary plan methods can then be used as if it were a flat, tabular ore body. It will be appreciated that unrolled sections can only be produced if the strikes of the ore body are parallel or very nearly parallel. If the strikes and dips are very variable, the ore body cannot be flattened or 'unrolled' without distortion.

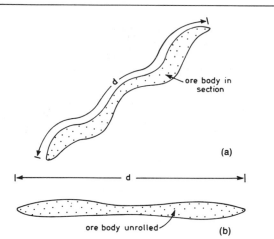

FIG. 9.15.   Diagram to illustrate the method of producing an unrolled section.

9.6   USE OF ORE BLOCKS

In addition to the methods described above, ore reserve calculations can be made by dividing the ore body into a large number of cubes or rectangular prisms. This is a useful method for very irregular or pipe-like bodies. There are no set rules for doing this and the division may be quite arbitrary. It is generally useful to break the ore body into a series of horizontal slices and sub-divide these into the ore blocks. Grade and tonnage factors can be assigned to each block by interpolating between data points. Ore reserve calculations are made by computers using this method and programs have been written for generating the blocks from drill-hole data for specific ore deposits.

9.7   CUT-OFF GRADES

Before determining ore intersections for reserve calculations, it is important to define the minimum grade which should be included. This is known as the cut-off grade. The value chosen is usually somewhat below the actual minimum economic grade as it is generally found that the higher grades in the deposit will raise the overall average to the economic grade required. For example, if the economic grade in a particular copper deposit is 3%, the cut-off grade used in the ore reserve calculations might be 1%. Small changes in the cut-off grade used might make big differences in the overall tonnage and this could affect the economic viability of the deposit. For example, the ore reserves of a nickel deposit might be 3 million tonnes at 2·4% using a 1% cut-off and 10 million tonnes at 1·3% using a 0·5% cut-off. The larger tonnage of lower grade ore might be a more attractive economic proposition than the smaller higher grade tonnage. For this reason, it is normal practice to carry out calculations using various cut-off grades for comparative purposes. Table 9.2 compares the total ore reserves and tonnes of contained copper metal for a large low-grade copper deposit obtained for various cut-off grades.

Let us consider some examples. Figure 9.16 shows two drill-hole intersections sampled in 50 cm lengths. In case A using a 0·5% cut-off gives an intersection of 4·5 m at 1·78% Cu, whereas a 1% cut-off gives

TABLE 9.2

ORE RESERVES FOR A LARGE LOW-GRADE COPPER DEPOSIT
AT DIFFERENT CUT-OFF GRADES

Ore reserves (millions of tonnes)	Grade(%)	Cut-off grade(%)	Tonnes of copper
957	0·70	0·1	6 699 000
749	0·84	0·3	6 291 600
435	1·14	0·6	4 959 000

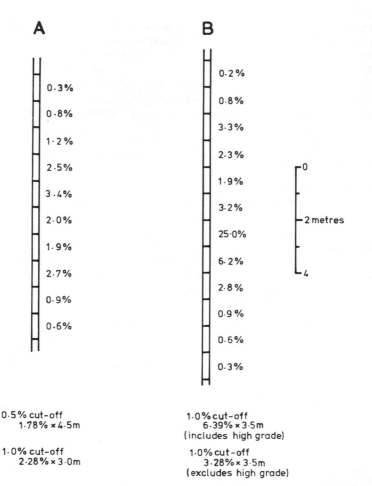

FIG. 9.16.   Examples of two drill-hole intersections of copper mineralization
with overall grades determined for different cut-offs.

an intersection of 3·0 m at 2·28%. In case B there is one sample with a
very high value, several times greater than any of the others. In a case
like this it is normal practice to discard the unusually high value as it
is unrepresentative of the intersection. Using a 1% cut-off and including the high value gives an intersection of 3·5 m at 6·39%, whereas
using the same cut-off and discarding the high value gives an intersection
of 3·5 m at 3·28%. The sample is not actually totally discarded, but rather
replaced by the average value obtained over the remainder of the
intersection (in this case 3·28% over 3·0 m).

In addition to specifying a cut-off grade it is also necessary to use a minimum grade x thickness to determine what can be considered as ore. This is done to take into account minimum mining widths. In the past when mining was labour intensive, widths as narrow as 1 m or less could be mined economically. This is still done in some parts of the world in small mines, but, as mining has become more mechanized and productivity per man-shift has had to be increased to cover high labour costs, economic mining widths are usually of the order of 3 m or more. In some cases it is still possible to mine widths less than this, but it is rare for them to be less than 1·5 m. Thus, in calculating the ore reserves of a copper deposit we might use cut-offs of 1% Cu and 3 metre-%, i.e. a grade of 1% is required over a minimum width of 3 m. Table 9.3 gives some intersections of copper mineralization which have been defined as 'ore' and 'non-ore' according to cut-offs of 1% and 3 metre-%.

TABLE 9.3

COPPER ORE INTERSECTIONS CLASSIFIED AS ORE AND NON-ORE
ACCORDING TO MINIMUM CUT-OFFS OF 1% Cu AND 3 METRE-%

True thickness (m)	Grade (%)	Metre-%	Ore	Non-ore
2·31	1·51	3·49	X	
4·53	3·20	14·50	X	
1·20	2·73	3·28	X	
8·27	0·81	6·70		X
5·20	1·73	9·00	X	
1·34	2·20	2·95		X

In determining ore intersections or ore zones it is important to remember not to include significant widths of barren or very low grade material to make up wide intersections of 'ore', even if the overall grade so obtained is greater than the minimum required. Two examples of drill-hole intersections in copper ore are shown in Fig. 9.17. In case A there is a very narrow barren section which it would be permissible to include in the ore intersection, but in case B we are dealing with two distinct and separate upper and lower ore zones. As they are quite rich, there is a temptation to include the middle section which gives an overall intersection of 2·68%, doubles the reserves and still gives an 'ore' intersection greater than 1% Cu and 3 metre-%. It is not permissible to do this, however, because it includes 5 m of virtually barren rock at 0·04% Cu. This could cause serious difficulties during mining operations by resulting in the delivery of considerable amounts of barren material to the mill during extraction of the

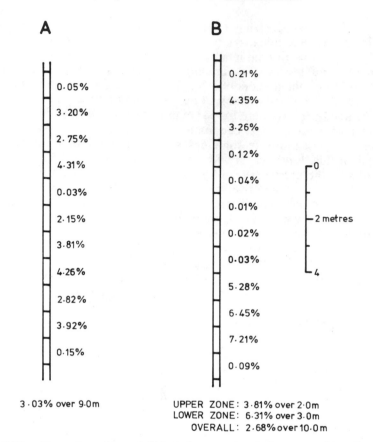

FIG. 9.17.   Examples of two drill-hole intersections of copper mineralization with barren materials in the middle of each intersection.

so-called ore zone. It should be remembered that extraction of ore does not necessarily take place over the full width of the ore body as a single operation.

9.8   TONNAGE FACTOR

In converting volumes to tonnages it is necessary to multiply by a tonnage factor. In Imperial measure this is usually quoted in cubic feet per ton (either short or long, though in North America the short ton of 2000 lb is generally used). In metric units the tonnage factor is simply the specific gravity or density which converts cubic metres to metric tons or tonnes (1000 kg). Tonnage factors need to be determined experimentally from density measurements in the field or lab. Density measurements on drill core or rock samples are easily carried out using the standard procedure of weighing the specimen in air and then weighing it immersed in water where the density is given by:

$$\frac{\text{Weight in air}}{\text{Weight in air} - \text{Weight in water}}$$

Rock densities are often quoted as dry or saturated densities. In the case of hard, compact rocks there will be little or no difference between dry and saturated densities, but in the case of porous rocks there may be a significant difference. Since the major portions of most ore bodies lie below the water table, saturated densities should be used, though for most mineralized rocks there is little difference between dry and saturated densities.

There can be big differences in tonnage factors for different types of ore and it is important to use the correct tonnage factor or evaluation of total reserves may be inaccurate. A number of typical tonnage factors are given in Table 9.4, but it must be remembered that

TABLE 9.4
SOME TYPICAL TONNAGE FACTORS FOR VARIOUS TYPES
OF BASE METAL ORES

Ore type	Typical tonnage factor	
	tonnes/m³	ft³/short ton
Porphyry copper (chalcopyrite + 5% pyrite)		
0·5% Cu in granite	2·78	11·52
1% Cu in granite	2·82	11·36
Copper in argillite (bornite)		
4% Cu	2·90	11·05
8% Cu	3·05	10·50
Pb–Zn in limestone		
5% Pb, 5% Zn	3·10	10·33
10% Zn, 10% Pb	3·50	9·15
Massive sulphide (pyrite, chalcopyrite, sphalerite, 15% silicate gangue)		
6% Zn, 1% Cu	4·55	7·04
Massive sulphide (pyrite, sphalerite, chalcopyrite)		
12% Zn, 2% Cu	4·75	6·74
6% Zn, 1% Cu	4·90	6·54

there can be a big variation due to variable amounts of heavy gangue minerals such as barite, pyrite or magnetite, in addition to the ore minerals.

Since the tonnage factor generally varies with grade, it is important to be able to select the correct tonnage factor for the average grade of the deposit. Therefore, during the course of normal assay work it is advisable to have specific gravity determinations carried out on a comprehensive selection of core samples. Table 9.5 lists a series of s.g. measurements on some copper ore samples together with corresponding assays. These values are shown plotted as a scatter diagram in Fig. 9.18 and it can be seen that they fall roughly along a straight line showing a good degree of correlation. The regression line can be drawn through them by eye or, better still, it can be calculated by the method of least squares which gives the coefficients of the regression line, $y = bx + a$, as follows:

$$b = \frac{n \sum x_i y_i - \sum x_i \sum y_i}{n \sum x_i^2 - \left[\sum x_i\right]^2}$$

$$a = \bar{y} - b\bar{x}$$

Using the data in Table 9.5 the regression line is

$$y = 0 \cdot 0406x + 2 \cdot 729$$

Thus, if the average grade of the deposit was calculated at 3·22% the correct tonnage factor to use would be

$$y = 0 \cdot 0406(3 \cdot 22) + 2 \cdot 729 = 2 \cdot 860$$

With complex ores of very mixed mineralogy, there may not be a particularly good correlation between grade and tonnage factor. In cases like this it is often best to determine separate tonnage factors for the various parts of the deposit, but this becomes very cumbersome. The problem is illustrated by Fig. 9.19 which relates the s.g. of a lead–zinc deposit (galena and sphalerite in limestone) to combined grade. If there is a big variability in the deposit between sphalerite and galena content a plot of grade against s.g. might show a wide scatter of points between the two 100% lines.

For eluvial or alluvial deposits tonnage factors are rather difficult to determine owing to the great variability of such deposits as regards degree of compaction, grain size and water content. For this reason reserve figures for eluvial or alluvial deposits are often expressed simply in terms of volume. For example, the grades and reserves of an alluvial tin deposit may be quoted in kilograms of cassiterite per cubic metre with total reserves of x m^3. The volumetric measurement

TABLE 9.5
S.G. MEASUREMENTS AND ASSAY
VALUES OF CORE SAMPLES
FROM A COPPER DEPOSIT

s.g.	% Copper
2·75	1·02
2·79	0·61
2·78	1·93
2·81	2·51
2·85	2·36
2·83	3·20
2·87	3·11
2·87	3·64
2·95	4·03
2·87	4·22
2·87	4·84
2·97	4·80
2·94	5·33
2·93	5·72
3·02	5·85
2·95	6·43
3·04	7·16
3·08	8.61
3·14	9·73

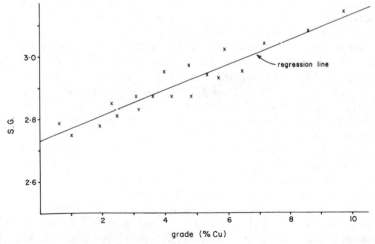

FIG. 9.18. Plot of copper assays against s.g. from a copper deposit.

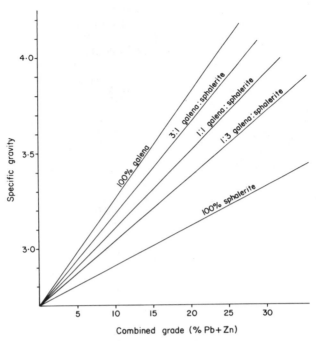

FIG. 9.19.   Diagram to illustrate how there can be poor correlation between grade and tonnage factor in complex ores.

is satisfactory for most eluvial or alluvial deposits as the grades are generally determined by a mineralogical separation rather than a chemical assay (kilograms of cassiterite, ilmenite, rutile, zircon, etc. per cubic metre). If assays are used, a tonnage factor is necessary, since assays are expressed in terms of weight percent. To determine a tonnage factor, measurements have to be made in the field. The simplest way to do this is to dig a circular pit, weigh all the excavated material and compute the volume of the pit accurately by careful measurement of its dimensions. Another method which is employed to determine the volume of the pit is to pour a quantity of dry measure such as maize or any other grain into the pit until it is full. Then the volume of the pit is simply the volume of dry measure used. It has been found that the volumetric measure with grain is quite accurate. If the moisture content is likely to show a seasonal variation or even variation within the deposit, it is best to determine a dry tonnage factor by drying the excavated material before weighing. Tonnage factors for eluvial and alluvial deposits generally vary from $1 \cdot 5 – 3 \cdot 0$ m^3/tonne (13–26 ft^3/short ton), depending on the type of deposit.

9.9 SAMPLING FOR GRADE DETERMINATION

Sampling a deposit would appear to be a relatively simple matter, but, in fact, it is a highly complex business which has absorbed the attention of a great number of workers and experts and has resulted in the development of new statistical theories and methods. Before considering some of the problems, the various types of samples will be described below.

Bulk sample. This is a large sample taken for the purposes of mineral dressing or metallurgical tests. Bulk samples generally weigh anything from a few tonnes to several hundred tonnes. The actual weight necessary to constitute a bulk sample is arbitrary, but a sample should not really be referred to as a bulk sample unless it weighs at least a tonne, although samples of several hundred kilograms are sometimes referred to as bulk samples by geologists.

Channel sample. This is the type of sample familiar to all mining geologists. As its name implies, it consists of cutting a channel across the strike of a vein, a mineralized bed, down the wall of a pit or along the bottom or side of a trench. The size of the sample depends upon the nature of the material being sampled, but to be taken correctly it should be of constant width and depth so that it is truly representative. The geologist should make a sketch in his field notebook of each channel sample taken.

Chip sample. This type of sample consists of taking a series of rock chips along a line or over a given area. It is useful in initial assessment work. For example, in evaluating limestones in a certain area as possible raw material for cement manufacture, chip sampling is a good method for locating the potentially interesting limestones for more detailed sampling later. Small chips are taken from all outcrops within 1 or 2 m on either side of a series of lines run across the strike. Instead of lines, areas can be marked out and chips taken from outcrops within the defined areas. A problem with chip sampling is to avoid biased sampling. There is always a tendency to take chips from the more 'eye-catching' or interesting outcrops which results in an unrepresentative sample.

Composite sample. This is simply a sample made up by combining two or more individual samples.

Drill-core sample. This is the sample of most importance to exploration geologists. The procedures of sampling are described in Chapter 7.

Grab sample. As the name implies, this is a sample taken more or less indiscriminately at any place. At best it may give a rough indication and at worst it is meaningless. It is the type of sample often taken at an operating mine where, for example, a sample may be 'grabbed' from an ore car or conveyor to obtain an idea of millhead grade. The grab sample is best avoided in exploration work and any evaluations based on grab samples should be regarded with the greatest degree of caution.

Groove sample. This is really a mini-channel sample and, as the name implies, consists of cutting a narrow groove across the strike of a vein, mineralized bed, etc.

Panel sample. This is the type of sample taken when there are no defined limits to the mineralization as, for example, in a winze or drift which is entirely in ore. Panel samples are also frequently taken from the sides of trenches. To take panel samples a series of rectangular or square 'panels' is marked out on the face to be sampled and each sample is cut to a uniform depth within each panel.

Sludge and other drill samples. Sludge, chip and dust from percussion, rotary and, in some instances, diamond drills are important types of samples. They are all described in Chapter 7.

The errors and problems involved in sampling a deposit may be discussed under the following headings:

   (i) Insufficient sample data points
   (ii) Biased and unrepresentative sampling
  (iii) Incorrect assumptions regarding distribution of sample values
  (iv) Inherent sample variability
   (v) Inherent variability of the deposit
  (vi) Inaccurate assays

Insufficient sample data points

The exploration geologist often has to determine grade and tonnage from a few widely spaced drill holes. For example, if we assume that mineralization intersected in a hole truly represents ore within a radius of 3 m of the hole, then on a 30-m grid the drill core samples represent 3·1% of the ore. On a 60-m grid the core samples represent 0·79% of the ore and on a 120-m grid they only represent 0·20% of the ore. In actual fact BX diamond drill holes on a 30-m grid only sample 0·0002% of the ore. If we consider the size of the sample that is actually assayed, there is an even bigger disparity. Assume that a drill hole with 46 mm core intersects 10 m of ore with an average s.g. of 2·72. If the hole is assumed to represent ore within a 30-m radius, then 45·2 kg of core samples 76 906 tonnes of ore. The core is split for assay so that the sample weight is only 22·6 kg or 0·000 029% of the ore. Now, if the core is assayed in 1-m lengths and 5 g used for each assay, which is carried out in duplicate, a total of 100 g is assayed. Thus, the grade of the block of ore in the example is based on an assay determined on 0·000 000 13% of the total material.

It is very difficult to determine when there is sufficient data to make a reasonably accurate estimate of grade and tonnage. It depends upon the inherent variability of the deposit, which itself has to be estimated from the available data. If the mineralization is uniformly distributed, a good estimate of reserves can be made from a few drill holes, but if there is great variability in the deposit, even a high density of drill holes may result in a poor estimate. Mistakes due to insufficient data can be extremely costly. There is a case on record of a lead–zinc deposit which was estimated to contain some 20 million tonnes of ore on the basis of just over 20 diamond drill holes. The mineralization occurs within a definite stratigraphic unit and it was thought to be stratiform and fairly uniform. On the basis of these assumptions and calculations shaft sinking and underground exploration were undertaken. This work revealed that the mineralization was not stratiform, but rather was controlled by a series of minor faults and the actual tonnage was only about a tenth of that estimated from the drilling. The high cost of underground exploration could have been saved if more drilling had been done to enable a better initial assessment of the economic viability to have been made. In addition to over-optimistic estimates, deposits are sometimes under-estimated and grades during mining are found to be better than those indicated by the exploration drilling.

In the determination of grades at the exploration drilling stage it is important to be able to make a reasonable estimate of the margin of error involved and thus decide whether there is sufficient data or if more drilling is required. In addition to various statistical methods for doing this, sound geological judgment is very important. This is often neglected in favour of a purely statistical approach, but a good geological interpretation based on the style of mineralization and its setting is invaluable.

Biased and unrepresentative sampling
There are set procedures in sampling to avoid, or at least minimize, biased sampling. Nevertheless, it is very hard to achieve unbiased sampling and, indeed, a bias may be introduced inadvertently simply by the choice of a certain drill grid pattern. The exploration geologist should always be aware of the dangers of biased sampling and try to ensure as far as possible that the sampling programme undertaken will be representative of the type of deposit being evaluated. Some simple examples of the type of thing to avoid, or at least be aware of, in sampling are shown in Fig. 9.20.

Incorrect assumptions regarding distribution of sample values
In the examples of ore reserve calculations given earlier in this chapter, arithmetic means were used to calculate the average grade. This is only strictly valid for deposits in which the distribution of

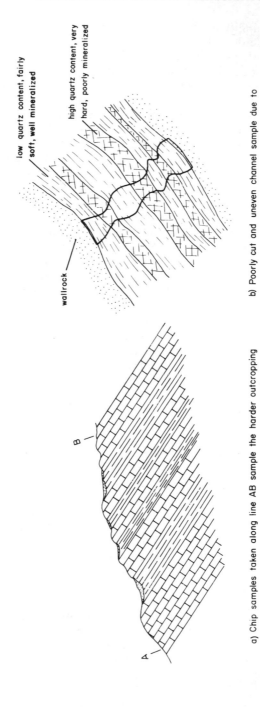

low quartz content, fairly soft, well mineralized

high quartz content, very hard, poorly mineralized

wallrock

a) Chip samples taken along line AB sample the harder outcropping limestone in preference to the softer poorly exposed shale

b) Poorly cut and uneven channel sample due to variable hardness of vein

FIG. 9.20. Examples of biased or unrepresentative sampling.

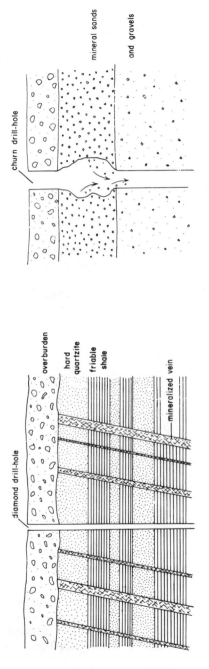

c) Poor core recovery in the soft friable shale results in inadequate sampling and the orientation of the drill-hole means that there is a poor probability of intersecting the veins

d) Lack of casing in hole results in contamination of deeper samples by near-surface gravels

FIG. 9.20.—contd.

assay values is normal. In deposits where this is not so, the use of the arithmetic mean will give an incorrect grade. It has been found that frequency distributions of most deposits approximate either to normal or log-normal patterns. For deposits in which the element of interest occurs as a major constituent, such as in iron deposits, the distributions of assays are almost always very nearly normal and even in other high-grade base metal deposits, there is still a tendency for distributions to be approximately normal. In deposits in which the element of interest occurs as a minor constituent, such as in gold or molybdenum deposits, the distributions of assay values show a strong tendency to be log-normal. In these cases the geometric mean or log mean should be used to give a true evaluation of the mean grade as the arithmetic mean will always over-evaluate a deposit in which the distribution of values is log-normal. This effect was minimized in the past by the practice of discarding unusually high values, but this is not necessary if the log mean is used.

Before calculating reserves and overall grade, it is good practice to plot a frequency distribution graph of assay values to determine whether the deposit approximates to a normal or log-normal pattern. Figure 9.21 shows a histogram of P_2O_5 values from an eluvial phosphate deposit over a carbonatite. As the distribution is very nearly

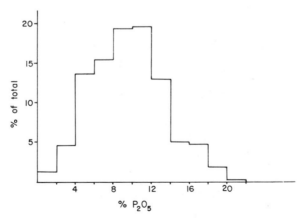

FIG. 9.21. Histogram of phosphate values in 414 1·5 m-channel samples taken from the West Valley, Sukulu carbonatite, southeast Uganda.

normal, arithmetic means can be used in determining the average grade. Figure 9.22 shows an example of a cumulative frequency plot on log–probability paper of SnO_2 values from a pegmatite tin deposit and it is clear that it approximates to a log-normal distribution. In this

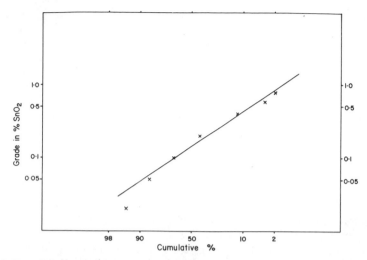

FIG. 9.22. Cumulative frequency plot of cassiterite content in 32 samples from a deposit in southwest Uganda.

case the log means should be used to determine average grade. The reserve calculations are made in the normal manner with the exception that the logarithms of the assay values are used. The average grade is then simply the antilog of the final value.

One problem is plotting frequency distributions is that the samples are often not the same size. It is common practice, for instance, to sample drill core in lengths related to variations in the mineralization. This is useful from a geological point of view, but, as it usually results in samples of unequal length, it is not satisfactory from a statistical point of view. Unfortunately, one has to make do with the data available and single assays have to be taken as single sample points even though they may represent different lengths.

Inherent sample variability
The nature of mineralization varies considerably with the type of deposit and from deposit to deposit. At one extreme there are the rich fine-grained ores uniformly distributed throughout the gangue or host rock and at the other extreme are the lean coarse-grained ores sparsely and very unevenly distributed throughout the gangue or host rock. All gradations exist between the two extremes and it is obvious that sample reliability is very much dependent on the style of mineralization. A small sample from an ore block with uniformly distributed mineralization may be a very good guide to the grade of the block, whereas a small sample from an ore block with very

irregularly distributed mineralization is likely to be a very poor guide to the grade of the block. Gold and tin deposits are notorious in this latter respect and geologists familiar with a particular deposit often put more store by the style of mineralization in a drill-hole intersection than by the actual grade indicated. A number of workers have given considerable thought to the problems and have devised various formulae for determining adequate sample sizes. The best known is probably that of Gy (1965), who has devised a special slide rule for performing the calculations. Gy's formula, which relates the sample variance to a number of factors is stated as follows:

$$S^2 = \left(1 - \frac{m}{M}\right) fclgd^3/m$$

where S^2 is the variance; m the sample weight in grams; M the weight of the bulk; f a shape factor, usually 0.5(0.2 for gold); c a mineralogical composition factor; l a liberation factor; g a grain size factor, usually 0.25; and d the size of the largest ore fragment in centimetres (equivalent to the screen opening which holds 5% of the ore). For bimineralic ores

$$c = (1 - a)\left[\frac{(1 - a)}{a}\sigma_1 + \sigma_2\right]$$

where a is the average amount of the ore mineral (100%$\equiv$1), σ_1 is the density of the ore mineral and σ_2 the density of the gangue.

The liberation factor l is related to d/L where d is the top particle size and L is the liberation size of the mineral of interest. Some values of l for various d/L are given below:

d/L	l
1	0.8
4	0.4
10	0.2
40	0.1
100	0.05

Usually m/M is very small and may be dropped so that the formula becomes:

$$S^2 = \frac{Cd^3}{m}$$

where $C = fclg$. For gold ores $C = 19/a$.

The slide rule has three scales, one for common ore, a second for alluvial gold and a third for coal.

Let us consider a couple of examples using Gy's formula. A 10 kg drill core sample of 1% copper ore consisting of chalcopyrite in granite is crushed to 0·25 cm top size. The chalcopyrite grains average about 0·05 cm across and we are required to determine what range of copper values to expect at the 95% confidence level in a 1/8 split.

$$1\% \; Cu \equiv 2\cdot9\% \; chalcopyrite$$

$$d/L = \frac{0\cdot25}{0\cdot05} = 5, \quad l = 0\cdot38$$

$$c = (1 - 0\cdot029)\left[\frac{(1 - 0\cdot029)}{0\cdot029}4\cdot2 + 2\cdot65\right] = 139$$

$$C = fclg = (0\cdot5)(139)(0\cdot38)(0\cdot25) = 6\cdot60$$

$$S^2 = \frac{6\cdot60(0\cdot25)^3}{1250} = 0\cdot0001$$

$$S = 0\cdot0091$$

At the 95% confidence level the required value lies within two standard deviations of the mean. Thus, the values should lie between 0·98 and 1·02%.

A tin ore consisting of cassiterite and quartz grading 0·5% SnO_2 with coarse cassiterite grains approximately 0·5 cm across is crushed to 2 cm top size. How large a sample has to be taken for further reduction to ensure that analyses are within ± 10% of the true grade at the 95% confidence level?

$$d/L = \frac{2}{0\cdot5} = 4, \quad l = 0\cdot4$$

$$c = (1 - 0\cdot005)\left[\left(\frac{1 - 0\cdot005}{0\cdot005}\right)6\cdot9 + 2\cdot65\right] = 1369$$

$$C = fclg = (0\cdot5)(1369)(0\cdot4)(0\cdot25) = 68$$

Now
$$2S = 0\cdot05$$

$$S = 0\cdot025$$

$$S^2 = 0\cdot0006$$

$$0\cdot0006 = \frac{68(2)^3}{m}$$

∴
$$m = 907 \; kg$$

Alternatively, if a 50 kg sample of the tin ore is taken what will be the expected range of values at the 95% confidence level?

$$S^2 = \frac{68(2)^3}{50\,000} = 0\cdot01$$

$$S = 0\cdot10$$

$$2S = 0\cdot20$$

The expected range of 95 samples out of 100 will be from $0 \cdot 3 - 0 \cdot 7\%$ SnO_2.

Inherent variability of the deposit

Classical statistical methods are based on the concept of random sampling and there are a number of reasons why they are often not applicable in determining average grades of mineral deposits. Mineral deposits are rarely sampled on a random basis. They are sampled in a systematic manner for obvious reasons. This does not matter, however, provided that the mineral particles are randomly distributed. In other words any method of sampling will result in random data if the various items of interest are themselves randomized. In the majority of mineral deposits this is not the case as the mineralization often displays definite trends and variations, known as the inherent variability of the deposit. For the sampling to be random the following conditions must be met:

(i) Equal sample lengths or, if sample lengths are unequal, there must be no correlation between assays and lengths
(ii) Negligible density contrast or variation between ore and gangue or between ore minerals
(iii) No trend or pattern to mineralization

Effects of unequal sample lengths

As mentioned earlier in the chapter, individual samples should have the same volume for a correct statistical evaluation to be carried out. This is not so important if it can be shown that there is no correlation between assay values and sample sizes. If there is a correlation between sample sizes and assay values, the effects can be minimized by using the percentage method in preference to the metre-% method when weighting the grades.

Effects of variations in density

In making reserve calculations grades are generally determined by using linear-%, area-% and volume-% relationships. Assays are expressed in terms of weight and unless the weight/volume ratio is constant or varies within a narrow range, average grades calculated by volume-% will be in error. The problem is illustrated by the example below which gives some drill-hole intersections for a hypothetical lead–zinc deposit (galena and sphalerite in limestone).

Length (m)	Pb + Zn Grade (%)	s.g.	Length × Grade	Length × Grade × s.g.	Length × s.g.
2	5·0	2·90	10·00	29·00	5·80
1	5·1	2·91	5·10	14·84	2·91
1·5	8·2	3·01	12·30	37·02	4·52
1	7·6	2·99	7·60	22·72	2·99
2	15·3	3·27	30·60	100·06	6·54
2	14·9	3·26	29·80	97·15	6·52
1·5	18·1	3·37	27·15	91·50	5·06
2	16·3	3·29	32·60	107·25	6·58
1	12·3	3·16	12·30	38·87	3·16
1	11·7	3·15	11·70	36·86	3·15
15			179·15	575·27	47·23

$$\text{Average grade by volume-}\%^* = \frac{179\cdot15}{15} = 11\cdot94\%$$

$$\text{Average grade by weight-}\% = \frac{575\cdot27}{47\cdot23} = 12\cdot18\%$$

In the example above, a proportional increase of s.g. with grade has been assumed, but, if the deposit shows more variation, the disparity between the two methods will be greater. Let us consider the same example but with a wider range in s.g. making the assumption that the higher grades have a higher proportion of galena.

Length (m)	Pb + Zn Grade (%)	s.g.	Length × Grade	Length × Grade × s.g.	Length × s.g.
2	5·0	2·81	10·00	28·10	5·62
1	5·1	2·83	5·10	14·43	2·83
1·5	8·2	2·93	12·30	36·04	4·40
1	7·6	2·89	7·60	21·96	2·89
2	15·3	3·46	30·60	105·88	6·92
2	14·9	3·37	29·80	100·43	6·74
1·5	18·1	3·55	27·15	96·38	5·33
2	16·3	3·38	32·60	110·19	6·76
1	12·3	3·02	12·30	37·15	3·02
1	11·7	2·94	11·70	34·40	2·94
15			179·15	584·96	47·45

$$\text{Average grade by volume-}\% = \frac{179\cdot15}{15} = 11\cdot94\%$$

$$\text{Average grade by weight-}\% = \frac{584\cdot96}{47\cdot45} = 12\cdot33\%$$

These examples illustrate how s.g. can be as important as the assay values in calculating the average grade. Many deposits exhibit only a

*Volume is proportional to the sample lengths.

fairly narrow range in s.g. and the volume-% methods for calculating average grades as shown earlier in this chapter are quite satisfactory. In deposits which exhibit a wide range in assay values and s.g.'s, there may be significant differences in average grade calculated by the two methods and the weight-% method should be used. Since s.g. increases with grade in almost all types of deposit, the use of the volume-% method will result in an under-evaluation of the average grade. As s.g. determinations are rarely carried out on each sample assayed, a graph such as that shown in Fig. 9.18 can be plotted from a selected number of samples chosen over a wide range of assay values. If there is a good linear trend, a regression line can be drawn and then used to determine the s.g. for any assay value.

Effects of trend
Mineral deposits, in which the minerals of interest are randomly distributed, are rare. Most deposits display definite trends and therefore cannot be evaluated properly by classical statistics. The problems have been studied by many workers in different parts of the world. Some of the better known studies include those by Krige (1951) and Sichel (1952) in South Africa, de Wijs (1952 & 1953) in Holland, Hazen (1958) and Becker and Hazen (1961) in the United States and Matheron (1962, 1963a & b) and Serra (1967) in France. To take account of trend in mineral deposits Matheron introduced the concept of the *regionalized variable* in the early 1960's and developed from it his theory of geostatistics. In classical statistics individual samples are selected at random from a population in which space has no meaning, whereas in geostatistics the location of a sample in space is important. In other words a regionalized variable is an actual function taking a definite value for each point in space and is quite distinct from an aleatory (based on chance) variable on which traditional statistics is based. The mathematical expression of the regionalized variable for most deposits would be incredibly complex if it could be determined, but, as it is always unknown, the problem is to estimate the properties of the regionalized variable from a sample as is done in traditional statistics when the unknown properties of a large population N are estimated from a small sample n drawn randomly from it. To do this Matheron makes use of the variogram which is a mathematical function derived from the sample data giving the degree of natural dispersion of assay values. The variogram can be defined for one, two or three dimensional cases, but only the unidimensional variogram will be described here. If we have a series of samples along a drift or from a series of drill holes across a deposit, the variogram function is given by:

$$\gamma(h) = \sqrt{\frac{(x_2 - x_1)^2 + (x_3 - x_2)^2 + \cdots + (x_n - x_{n-1})^2}{2(L - h)}}$$

where n is the number of samples, h is the distance between samples, L is the total sampled length and x_1 the first sample, x_2 the second sample and so on. The various types of variograms are shown in Fig. 9.23, but only those with good continuity (conforming to the de Wijsian model) will be considered here. The absolute coefficient of dispersion (α) can be obtained directly from the slope of the vario-

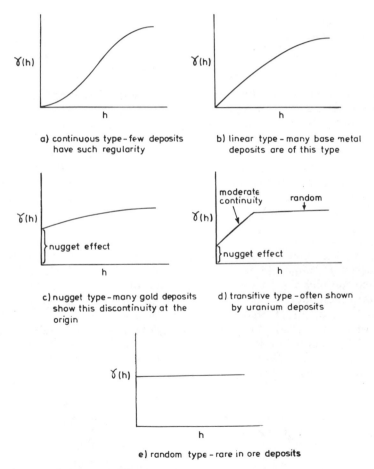

a) continuous type - few deposits have such regularity

b) linear type - many base metal deposits are of this type

c) nugget type - many gold deposits show this discontinuity at the origin

d) transitive type - often shown by uranium deposits

e) random type - rare in ore deposits

FIG. 9.23. Diagram to illustrate the various types of variograms (after Matheron, 1963b).

gram or from de Wijs's formula

$$\sigma^2 = \alpha \log_e \left(\frac{V}{v}\right)$$

where σ^2 is the log variance of assay values, V is the volume of the deposit and v the volume of the samples. This is only valid when the deposit and samples are geometrically similar. If this is not so, for example a tabular ore body sampled by diamond drill holes, the Matheron–Wijs formula

$$\sigma^2 = 3\alpha \log_e \left(\frac{D}{d}\right)$$

has to be used where D and d are the linear equivalents of the ore deposit and samples respectively.

$$D = A + B + 0\cdot7C$$
$$d = a + b + 0\cdot7c$$

where A, B and C are the dimensions of the ore block and a, b and c the dimensions of the sample with $A > B > C$, and $a > b > c$.

It should also be appreciated that the variogram may not be the same along different directions of the ore body. Thus, it is normal practice to determine variograms along at least two directions, one along strike and one across dip. Isotropic ore bodies will have similar variograms across the two directions, but some ore bodies may show a marked anisotropy with very different variograms for the two major directions.

Kriging

Kriging is a method of weighted moving averages proposed by Matheron (1963a) and named in honour of D. G. Krige who first applied the technique to the gold deposits of the Witwatersrand. It is a useful technique for determining average grades of deposits with a strong nugget effect characterized by erratic values. The effect of kriging is to upgrade low values and downgrade high values using smoothing factors or kriging coefficients obtained from semi-variograms. If z is the unknown grade of a block of ore, then an estimator z^* is determined by kriging in the form:

$$z^* = \Sigma a_i x_i$$

where a_i are the sample weights or kriging coefficients and x_i the sample assays, with the proviso that:

$$\Sigma a_i = 1$$

and the kriging variance (i.e. the variance of z by z^*) should take the smallest possible value.

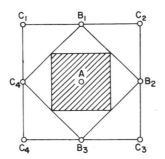

FIG. 9.24. Diagram to illustrate the principle of discontinuous kriging applied to a square grid.

In mineral exploration the problem is generally to estimate grades of blocks of ore from drill holes. In Fig. 9.24 the shaded block in the centre could be assigned a grade equal to the grade indicated by drill hole A, but this does not take into account the grades in surrounding drill holes which can be used to make a better estimate of the block. Matheron gives methods for determining kriging coefficients for a large number of different drilling patterns, but only the determination of kriging coefficients for a square grid will be shown here. In Fig. 9.24 the grades in drill holes B_1, B_2, B_3, B_4 and C_1, C_2, C_3, C_4 have an influence on the grade of the shaded block centred on drill hole A. Further rings of drill holes could be considered, but their effects are negligible compared with the first two rings of holes. For the example in Fig. 9.24 the estimated grade of the block is given by:

$$z^* = (1 - \lambda - \mu)u + \lambda v + \mu w$$

where λ and μ are kriging coefficients and u is the grade of A, v the weighted average grade of the B's and w the weighted average grade of the C's. The kriging coefficients are determined from the graph in Fig. 9.25 where the values along the abscissa are given by h/a, where h is the average thickness of the ore and a is the drill-grid size. The kriging variance is given by multiplying the value determined on the $2\sigma/3\alpha K$ curve by three times the absolute dispersion (α).

Kriging applied to widely separated sample data points such as a series of diamond drill holes is known as discontinuous kriging. When detailed sampling is undertaken by a series of closely spaced samples such as in underground workings, the method is known as continuous kriging. The principle is illustrated by Fig. 9.26 which shows a block of ore sampled around all four sides. In traditional methods of estimating the grade of the block each sample is given equal weight, but in kriging the samples are weighted according to their position.

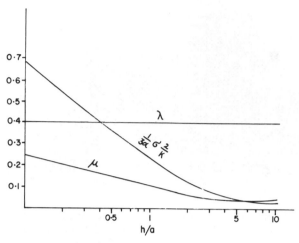

FIG. 9.25.   Graph for determining kriging coefficients for a square drill grid (from Matheron, 1963).

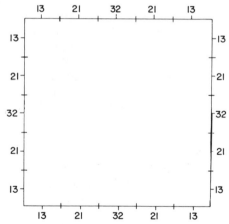

FIG. 9.26.   Diagram to illustrate the principle of continuous kriging.

The corner samples have less influence on the block than the ones at the mid-points of the sides and are therefore assigned lower kriging coefficients than the mid-point samples. The example shows possible kriging coefficients determined for a particular deposit.

Inaccurate assays
It is obvious that good estimates of the grade of a deposit are dependent on reliable sample assays. A wide range of assay methods

can be used depending upon the type of material, concentrations and degree of accuracy required. In assay work there are two main sources of error: (i) the sampling error in taking the aliquot required for the actual assay and (ii) the accuracy of the assay itself. The sampling error has already been described under inherent sample variability and it is important to minimize it as an error introduced at the sample splitting stage can never be removed no matter how precisely and accurately the assay is carried out. The accuracy of the assay is dependent upon the method and the skill and experience of the assayer so it is important that samples only be sent to reputable laboratories with competent analysts. Even so, it is advisable in any sampling programme to send some duplicate samples to at least one other laboratory as a check on the accuracy of the assays. A number of studies of the reliability of assays have been made by sending standard samples to numerous laboratories all over the world. Lister (1977) gives the results of such a survey which shows that there may be surprising aberrations. Nevertheless, standards of analysis have improved in recent years and variations between laboratories are generally within an acceptable margin of error.

Summary
The various problems encountered in sampling ore deposits have been described to illustrate the pitfalls in calculating ore reserves and average grades from exploration data. A brief outline of the latest geostatistical methods has been given to introduce the reader to the methods being used to overcome many of these problems. In most exploration work, however, advanced geostatistical methods are not necessary and the ordinary procedures described in this chapter can be used. At the feasibility study stage where a detailed evaluation is required, geostatistical methods may become important and the reader is referred to the literature for details which are beyond the scope of this book.

REFERENCES AND BIBLIOGRAPHY

Becker, R. M. and Hazen, S. W. Jr. (1961). Particle statistics on infinite populations as applied to mine sampling, *U.S. Bureau of Mines Rept. of Investigations* No. 5669.

Berning, J., Krisic, S. M. and Saunders, J. H. (1966). Assessment of Palabora ore reserves by computer techniques, *Trans. Instn. Min. Metall.*, Lond., 75, A189–202.

Blais, R. A. and Carlier, P. A. (1968). Application of geostatistics in ore evaluation, *Ore reserve estimation and grade control*, C.I.M. Spec. vol. 9, 41–68.

Gilmour, P. (1964). A method of calculating reserves in tabular orebodies, *Econ. Geol.*, **59**, 1386–1389.

Gy, P. (1965). *Calculateur d'Echantillonnage* (Gy's sampling slide rule), Société de l'Industrie Minérale, Saint Etienne, France, 16 pp.

Gy, P. (1966). *Sampling of Materials in Bulk—Theory and Practice*, vol. 1, Société de l'Industrie Minérale, Saint Etienne, France.

Gy, P. (1971). *Op. cit.* vol. 2.

Hazen, S. W. Jr. (1958). A comparative study of statistical analysis and other methods of computing ore reserves utilizing analytical data from Maggie Canyon manganese deposit, Artillery Mountain region, Mohave County, Arizona. *U.S. Bureau of Mines Rept. of Investigations* No. 5375.

Jones, M. J. (ed.) (1974). *Geological, Mining and Metallurgical Sampling*, Instn. Min. Metall., Lond. 268 pp.

Krige, D. G. (1951). A statistical approach to some mine valuation and allied problems on the Witwatersrand, M.Sc. Thesis, Univ. of Witwatersrand, South Africa.

Lister, B. (1977). Second inter-laboratory survey of the accuracy of ore analysis. *Trans. Instn. Min. Metall.*, Lond., **86**, B133–148.

Matheron, G. (1962). *Traité de Géostatistique Appliquée*, Tome 1: Théorie Générale, Editions Technip, Paris, 333 pp.

Matheron, G. (1963a). *Op. cit.*, Tome II: Le Krigeage, Editions Technip, Paris, 171 pp.

Matheron, G. (1963b). Principles of geostatistics, *Econ. Geol.*, **58**, 1246–1266.

Serra, J. (1967). Echantillonnage et estimation locale des phénomènes de transition miniers, D.Sc. Thesis, Univ. de Nancy, France, 671 pp.

Sichel, H. S. (1952). New methods in the statistical evaluation of mine sampling data, *Trans. Instn. Min. Metall.*, Lond. **61**, 261–288.

de Wijs, H. J. (1951). Statistics of ore distribution, part 1, *J. Roy. Netherlands Geol. Min. Soc.*, Nov., 365–375.

de Wijs, H. J. (1953). *Op. cit.*, part 2, *J. Roy. Netherlands Geol. Min. Soc.*, Jan. 12–24.

CHAPTER 10

Evaluation of Prospects

When exploratory work results in delineation of interesting grades and tonnage, it is necessary to evaluate the economics of the prospect. The principal factors to consider are:

(1) Treatment of the ore.
(2) Mining methods.
(3) Disposal of product.

10.1 MINERAL DRESSING

Ores are rarely mined in a saleable form and some degree of concentration is necessary, i.e. the minerals of interest have to be separated from the gangue, a process known as mineral dressing. There is a wide choice of methods depending on the nature of the mineralization and tests have to be carried out to determine the amenability of the ore to treatment. Grades and tonnage of a deposit may be very good, but, if the ore is difficult to treat, the deposit may prove to be uneconomic. Thus, it is normal practice to carry out metallurgical or mineral dressing tests on bulk samples from the prospect being evaluated. In many cases such samples will consist of composite samples of split drill core. The main stages in the mineral dressing process and the principal methods in use are given in Table 10.1.

Crushing and grinding
The first step in the mineral dressing of most ores is to break them up into small enough particles so that the individual ore grains are separated from the gangue. This is usually done in stages and the breaking up of the largest fragments is known as primary crushing. Jaw breakers and gyratory crushers are usually used for this; both work on the principle of squeezing the ore fragments in a wedge-shaped space between a fixed plate and movable one (Fig. 10.1). The next stage in crushing, known as

TABLE 10.1

OUTLINE OF THE MAIN STAGES AND PRINCIPAL METHODS USED IN
MINERAL DRESSING

I *Comminution*

A. Crushing

jaw crusher or breaker, gyratory crusher, hammer mill, rolls, cone crusher

B. Grinding

ball mill, rod mill, stamp mill (now obsolete)

II *Classifying*

A. Screening

grizzly, fixed screens, shaking screens, trommels

B. Hydrocyclones

III *Sorting and Separating*

A. Hand sorting

B. Gravity separation

sluice boxes, jigs, shaking tables, Humphrey's spiral, heavy media

C. Flotation

D. Magnetic separation

E. Electrostatic separation

F. Hydrometallurgy

IV *Filtering and drying*

secondary crushing, is generally carried out by cone crushers and rolls, principles of which are shown in Fig. 10.1. To liberate the ore minerals from the gangue it is often necessary to break the ore down into very fine fragments. This process is known as grinding and the two principal types of equipment for doing this are the ball and the rod mill (Fig. 10.1).

Classifying
The process of separating the crushed ore into various size fractions is known as classifying. For the sizing of the largest fragments a grizzly may be used. In its simplest form this consists of a series of heavy steel bars fixed at equal distances apart and set at an angle of 25–50°

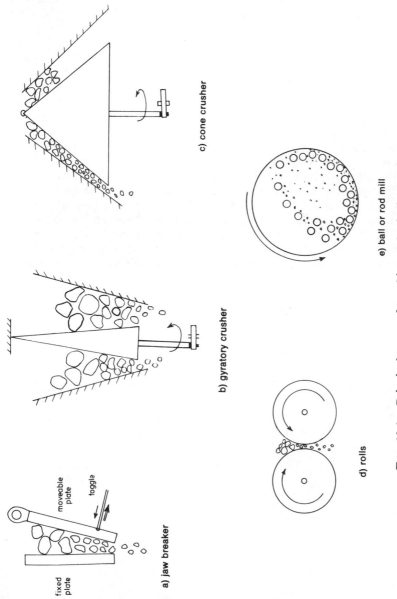

Fig. 10.1. Principal types of crushing and grinding equipment.

so that the oversize fragments can slide freely off it. Screens are used for handling intermediate-sized particles and there is a wide variety of types in use. They may be made from steel sheet with holes or slots punched in it (punched screens) or from steel or other metal wires woven into an open mesh (woven screens). The screens may be stationary (fixed screens), but they are usually provided with a mechanism for shaking or vibrating them (shaking screens) to prevent clogging. In addition, screens may be used with dry feed (dry screening) or wet feed (wet screening). In the case of wet screening jets of water are often sprayed onto the screen. Revolving screens in the form of a drum are often used and are known as trommels.

Screening is not practical below a particle size of 50 μm and other methods of classifying have to be used. The most widely used piece of equipment for handling these fine particles is the hydrocyclone or simply cyclone as it is often called (Fig. 10.2). A suspension of particles in water is pumped through a tangential feed tube at the side and a vortex is set up in the conical vessel. Through the action of centrifugal force the larger and heavier particles move out to the wall of the cone where they meet a zone of reduced pressure and settle towards the apex of the cone where they are discharged through an

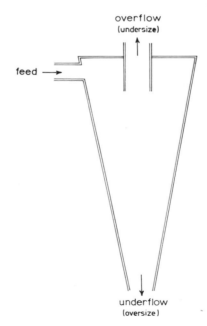

FIG. 10.2. Schematic diagram of a hydrocyclone.

opening. The finer and lighter particles remain near the centre of the vortex and are discharged through a central overflow orifice.

Sorting and separating

The simplest method of sorting is hand sorting, which is carried out on many small mines in parts of the world where labour is cheap. Even in other parts of the world the method can be economic in certain specialized cases when dealing with valuable ores though the method sounds very primitive.

The modern replacement of hand sorting is known as *photometric sorting*, which can be used to separate minerals with differences in reflectance levels down to a minimum particle size of about 10 mm. The equipment available has been developed to a high degree of sophistication and numerous types of ores can be treated. Ore fragments passing along a conveyor are scanned by a laser/rotating mirror/photomultiplier assembly and the reflectance level of each rock is signalled to an electronic processor which determines its position on the conveyor and decides whether to accept or reject it. Rejected rocks are blasted onto a separate conveyor as the scanned material passes over a blast manifold connected to a high-pressure air supply.

Many ore minerals have a higher density than the gangue and a number of gravity techniques can be used to effect a separation. These methods are very simple and work extremely well with ore minerals such as gold, cassiterite, wolfram and ilmenite. One of the simplest techniques that has been used since ancient times is the sluice box (Fig. 10.3). This consists of a long open box with battens, known as riffles, set across the bottom. The pulp to be treated flows in at one end and the larger and denser particles settle between the riffles. The method is crude, but it is very cheap and is still used in some parts of the world as a first step in upgrading low grade ores of heavy minerals such as cassiterite for later treatment by more efficient methods.

A widely used gravity technique for separating sand size particles is jigging. The basic principle of the jig is the pulsating of the particles on a screen. This is accomplished either by having a moving screen, or more commonly, by using a plunger mechanism to provide the pulsations (Fig. 10.3). The lighter particles are brought to the top and are carried out of the jig by the overflow, while the heavier particles collect on the screen and are drawn off into a launder under a gate which excludes the lighter particles resting on top of the heavies. Undersize particles which pass through the screen are known as the hutch product and collect in a compartment from which they are periodically drawn by a spigot. The hutch product consists of both dense and light fines and requires further treatment.

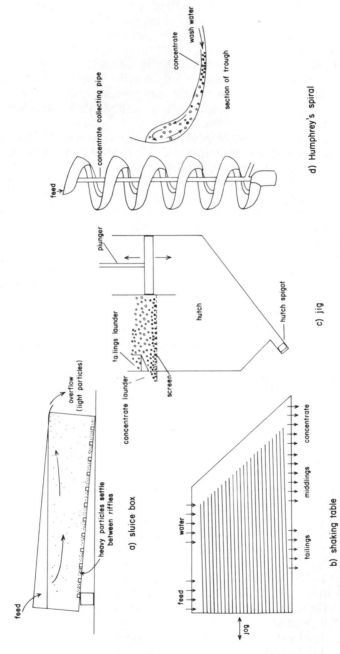

FIG. 10.3. Principal types of equipment for effecting gravity separations.

For separating particles too fine to be treated by jigs shaking tables may be used. These consist of a large flat surface or table inclined slightly forwards and sideways which is vibrated with a horizontal motion. Each cycle consists of relatively slow forward motion and rapid backward motion. The surface of the table is divided up by numerous, small, longitudinal riffles. Heavy particles move along the riffles and are washed off at and near the end of the table. Light particles move down with little horizontal movement and are washed off near the back of the table (Fig. 10.3).

Another method of gravity separation that is widely employed in treating mineral sands is the Humphrey's spiral. The appliance consists of five turns of troughing in the form of a spiral. The separating action is complex, but essentially it results in the lighter particles being washed and carried out towards the outer edge of the trough, while the heavier particles remain near the centre (Fig. 10.3).

In mineral separations carried out in the laboratory heavy liquids are frequently used, but all are too expensive to be used industrially. However, dense pseudo-liquids can be prepared by using a suspension of a finely divided solid in water. The use of such a process is known as heavy media or dense media separation (DMS). Solids that are used include galena, ferro-silicon and magnetite. The technique has been used for a number of different minerals, but it is most commonly employed in coal washing plants. High-grade coal floats in a liquid of specific gravity 1·12–1·35 and medium-grade coal in a liquid of specific gravity 1·35–1·6. A laboratory test to separate coal from carbonaceous shale is to use carbon tetrachloride (specific gravity 1·5) as the heavy liquid. On an industrial scale a magnetite–water suspension with a carefully controlled s.g. is used as the heavy 'liquid'. Magnetite has the advantage that it can be readily recovered by a magnetic separator and recycled.

It will be appreciated that the gravity methods with the exception of the DMS technique do not separate simply according to s.g. The separation of heavier particles from lighter ones is a function of both size and s.g. Thus, gravity methods act as classifiers in addition to separators. For this reason it is normal to limit the feed to jigs and tables to a narrow size range so that the gravity separation effected is mainly due to differences in s.g.

Flotation

This technique depends essentially on the property of surface tension which can be expressed as the 'wettability' of mineral particles. If air is bubbled through an aqueous suspension of mineral particles, bubbles will adhere to those particles which have a low wettability and thus cause them to float. Air bubbles will not adhere to easily wettable particles and they will sink. Since the surface properties of ore and gangue minerals vary within narrow limits, it is necessary to treat them selectively so that the desired minerals can be made to float. Organic reagents known as *collectors* are used for this and include oils (e.g. kerosene, fuel oils, creosotes), organic acids (e.g. oleic acid), salts of organic acids (e.g. xanthates—the most successful collectors) and organic bases (e.g. amines). Reagents known as *conditioners* are usually added to the pulp to control pH within narrow limits. Common conditioners are inorganic acids and bases. Two other classes of reagents are known as *activators* and *depressors*. As the names imply one is to render a mineral amenable to the action of a collector and the other is to render it inactive towards a collector. An example of an activator is the use of copper sulphate in the flotation of sphalerite. The cupric ions, which replace the zinc on the surface of the sphalerite grains, adsorb the xanthate collector so that the sphalerite becomes floatable. An example of a depressor is cyanide which is a powerful depressant for pyrite and sphalerite. The other important reagent for flotation is the frother or *frothing agent*. A common frothing agent is pine oil, but a number of new synthetic frothing agents are available and their use is increasing. The principle of the flotation cell is shown diagrammatically in Fig. 10.4.

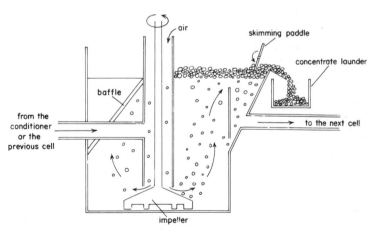

FIG. 10.4.   Schematic diagram of a flotation cell.

Flotation cells are usually set up in series for repeated treatment to ensure recovery of all the floatable product. Before being fed into a cell, the pulp is mixed with the various reagents in conditioning tanks. Banks of rougher cells are used to provide a rough concentrate from the earliest treated material and banks of scavenger cells are used to recover the last traces of floatable product before discarding the final tailings. Flotation is a cheap and versatile process which is used for the various sulphide minerals in addition to a wide range of other minerals such as apatite, kyanite, barite, graphite, mica, pyrochlore and zircon. The process has many variables and there are numerous complicating factors, and skilled and experienced operators are required to ensure correct and efficient flotation. For flotation to be carried out efficiently the mineral particles need to be less than 250 μm with a lower limiting size of approximately 10 μm. These extremely fine particles, known as *slimes*, are difficult to treat by any method and generally have to be discarded with the tailings even if they contain ore minerals.

Magnetic separators
A wide variety of equipment has been designed and built to exploit the magnetic properties of certain minerals in effecting separations. Both permanent magnets and electromagnets are used and both wet and dry material can be subjected to magnetic separations. In their simplest applications magnetic separators are used to remove ferromagnetics such as magnetite and tramp iron, but they can be used for separating a number of minerals with much smaller differences in magnetic susceptibility. In this latter regard, magnetic separators are commonly employed in treating zircon and monazite sands and in separating tantalite or columbite from cassiterite concentrates.

Electrostatic separators
Electrostatic separators, or high tension separators (HTS) as they are often known, are used in the separation of small dry particles by exploiting the differences in conductivity of mineral grains using an electrostatic charge. If mineral particles are charged by static electricity and then allowed to come into contact with a conductor at earth potential, weakly conducting grains will remain in contact with the earthed conductor by electrostatic attraction longer than good conductors which readily lose their electrostatic charges. Quite a wide range of minerals are treated by HTS, but probably their widest application is in the treatment of beach sands containing zircon, rutile, ilmenite and monazite. Figure 10.5 shows the principle of an electrostatic separator and Fig. 10.6 gives a rough outline of how magnetic and electrostatic separators are used in treating beach sands.

Hydrometallurgy
Instead of using physical properties in effecting mineral separations, chemical methods can be used and are conveniently classed as hydrometallurgical methods. Some minerals are not amenable to

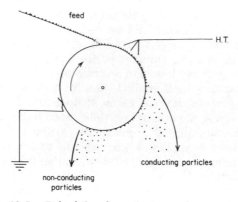

FIG. 10.5.   Principle of an electrostatic separator.

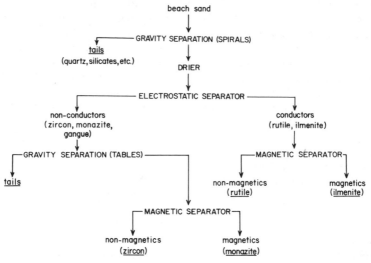

FIG. 10.6.   Outline of a flowsheet for treating beach sands.

physical separations, either because the grain size is too small, or because the properties of the minerals do not lend themselves to easy separation by physical methods. One of the oldest hydrometallurgical techniques is cyaniding of gold ores. A weak solution of potassium, sodium or calcium cyanide when aerated will readily dissolve gold by forming the complex anion $Au(CN)_2^-$. The gold is recovered from the solution by adding powdered zinc which replaces and precipitates the gold.

Acids can be used to dissolve certain minerals to separate them from the gangue and such techniques are widely used in the leaching of copper and uranium minerals. Sulphuric acid is generally used and is an important method for treating copper oxide ores which are difficult to treat by other methods. In many cases leaching is carried out in situ so that the method can be classified as a mining technique as well as a mineral dressing method. In addition to using sulphuric acid, bacterial action can be used to accelerate the leaching of copper ores. Certain bacteria such as *Ferrobacillus ferro-oxidans* feed on ore minerals in an acid environment and convert ferrous sulphate, produced by the oxidation of pyrite, to ferric sulphate which reacts with any copper sulphides present to form copper sulphate. The bacteria also promote the formation of sulphuric acid which becomes available for leaching copper oxide minerals. For successful bacterial leaching, pyrite has to be present in minerals being treated. In addition to the use of acids for leaching, carbonate leaches in alkaline solution are used in treating some uranium ores.

The metals may be recovered from solution by precipitation. For example, copper is precipitated from copper sulphate solution by adding scrap iron, often in the form of old tin cans from which the tin and lacquer coatings have been removed, to the solution. To concentrate the metals in the leach liquors, ion-exchange resins or solvent extraction techniques may be used. This is particularly common for recovering uranium, but may also be used for other metals. In the solvent extraction technique a suitable immiscible organic solvent is added to the leach liquor and transfer of the metal ions to the organic solvent results in a marked concentration of the metal in question. Hydrometallurgical techniques have become highly refined and there are a wide range of methods that can be applied depending on the ore being treated.

Amalgamation
This is a technique which is used to recover gold and silver from ores. If a finely ground ore containing native silver and gold is passed over copper plates coated with mercury, the mercury absorbs the gold or silver by forming an alloy known as amalgam. The precious metals can then be recovered by driving off the mercury in retorts. This technique is cheap and simple, but is is only effective in treating simple ores with fairly coarse-grained native gold or silver. Complex ores containing sulphides are not amenable to amalgamation techniques.

Filtering and drying
After a concentrate is produced, it is usually necessary to dry it since most concentrating processes are undertaken on wet material. In the

case of sand-sized particles drying is generally a simple process since a large part of the water content drains away freely and the concentrate can be force dried over a fire directly. In the case of fine-grained concentrates, such as those produced by flotation, it is not so simple to reduce the water content. The concentrate slurry is usually first passed into a thickener, which is a large tank in which the sludge settles out under the influence of gravity. Instead of a thickener, a cyclone can be used by setting it so that all solids pass through the underflow. Once the excess water has been removed, the concentrate sludge is passed through a filter and then dried.

Summary
Although the exploration geologist cannot expect to have a detailed knowledge of mineral dressing techniques, it is important for him to have an understanding of the basic principles so that he can understand reports on metallurgical testing of samples. Let us consider some examples to illustrate some of the methods used in treating samples. Two samples of magnetite-rich material from different parts of the Sukulu carbonatite in Uganda were submitted for testing as a possible iron ore. Both samples contained large amounts of phosphate which rendered them unsuitable in a raw state as potential iron ores and tests were carried out to see to what extent the phosphate content could be reduced in concentrates.

The raw samples gave the following assays:

	Sample 1	Sample 2
Fe	45·5%	57·4%
P_2O_5	3·6%	3·1%

Both samples were treated by grinding in a laboratory rod mill at 50% solids to −72 mesh. The +350 mesh fraction was removed by wet screening and the magnetic fraction removed with a hand magnet. The −350 mesh material was deslimed in a laboratory cyclone by removing the −15 μm fraction at a pressure of 1·40 kg/cm². The cyclone underflow was treated in a similar manner to the +350 mesh fraction to produce a magnetic and non-magnetic product. The results are given in Table 10.2.

The results in the table can be summarised as follows:

1. Total magnetics give the concentrate.

Sample 1

Mags 55·77 weight%

$$\%P_2O_5 = \frac{(5·0 \times 20·18) + (3·3 \times 35·59)}{55·77} = 3·9\%$$

TABLE 10.2

RESULTS OF TEST WORK ON IRON ORE SAMPLES FROM SUKULU, UGANDA

Sample 1

Product	Weight %	Assay P_2O_5	Fe	Distribution P_2O_5	Fe
Mags + 350 mesh	20·18	5·0	52·0	28·1	23·1
Non-mags + 350 mesh	12·19	2·0	41·7	6·8	11·2
Mags − 350 mesh	35·59	3·3	46·1	32·7	36·1
Non-mags − 350 mesh	3·52	2·2	43·4	2·2	3·3
Cyclone overflow	28·52	3·8	42·0	30·2	26·3
Head calc.	100·00	3·6	45·5	100·0	100·0

Sample 2

Product	Weight %	Assay P_2O_5	Fe	Distribution P_2O_5	Fe
Mags + 350 mesh	72·18	0·7	63·0	16·5	79·2
Non-mags + 350 mesh	3·65	15·0	29·0	17·9	1·8
Mags − 350 mesh	10·34	0·8	63·7	2·7	11·5
Non-mags − 350 mesh	4·49	23·4	13·6	34·4	1·1
Cyclone overflow	9·34	9·3	39·5	28·4	6·4
Head calc.	100·00	3·1	57·3	100·0	100·0

$$\%\mathrm{Fe} = \frac{(52 \times 20\cdot18) + (46\cdot1 \times 35\cdot59)}{55\cdot77} = 48\cdot2\%$$

Sample 2

Mags 82·52 weight%

$$\%P_2O_5 = \frac{(0\cdot7 \times 72\cdot18) + (10\cdot34 \times 0\cdot8)}{82\cdot52} = 0\cdot7\%$$

$$\%\mathrm{Fe} = \frac{(63\cdot0 \times 72\cdot18) + (10\cdot34 \times 63\cdot7)}{82\cdot52} = 63\cdot1\%$$

2. Total non-magnetics plus cyclone overflow give the discards or tailings. Phosphate and iron grades can be calculated in the same manner as above.

3. Total Fe distribution in the magnetics gives the recoveries.

Sample 1   59·2% recovery

Sample 2   90·7% recovery

Thus, a concentrate grading 48·2% Fe and 3·9% P_2O_5 with an iron recovery of 59·2% is produced from sample 1 and a concentrate grading 63·1% Fe and 0·7% P_2O_5 with an iron recovery of 90·7% is produced from sample 2. The results show that sample 1 is quite unsuitable as a potential iron ore; less than 60% of the iron is recovered, the iron grade is poor and no reduction in the phosphate content is achieved. Sample 2, on the other hand, is quite encouraging; a good iron recovery in excess of 90% is achieved, the iron grade is good and a big reduction has been made in the phosphate content. The phosphate content is still rather high for an iron ore, but it could be tolerated in certain circumstances.

As a further example let us consider two samples of copper ore from a deposit in Zambia. Sample A contains a high proportion of sulphide minerals and sample B is a mixed oxide–sulphide ore. The head assays were:

Sample A   Sample B
3·21% Cu   2·94% Cu

The 2-kg samples were ground in a ball mill until 50% passed 200 mesh and then both were subjected to a single (rougher) flotation test, but in the case of sample B it was subjected to conditioning with sodium sulphide to render the oxide minerals amenable to flotation. This is a process known as sulphidization. The results are given in Table 10.3.

Sample A shows that it is very amenable to treatment with an overall copper recovery of 97·6% and concentrate grade of 27·3% Cu. Sample B, on the other hand, is obviously difficult to treat. A good concentrate grade was obtained, but only 55·7% of the copper was recovered. Such mixed ores can be extremely difficult to treat and, although a number of techniques, such as leaching the oxides after a sulphide float, can be applied, a good copper recovery may never be achieved. Thus, good assay grades indicated at the exploration stage can be effectively reduced to levels that may be uneconomic. In terms of recoverable grades indicated by the tests given, sample A is reduced from 3·33 to 3·25% Cu and sample B from 2·92 to only 1·63% Cu.

TABLE 10.3
RESULTS OF SOME TEST WORK ON COPPER ORES FROM A DEPOSIT IN ZAMBIA

Sample A

Product	Weight %	Assay (% Cu)	Distribution
Rougher concentrate	11·9	27·3	97·6
Final tailings	88·1	0·1	2·4
Calculated head	100·0	3·33	100·0

Sample B

Product	Weight %	Assay (%Cu)	Distribution
Rougher concentrate	4·7	34·6	55·7
Final tailings	95·3	1·36	44·3
Calculated head	100·0	2·92	100·0

10.2 SMELTING AND REFINING OF ORES

Smelting is the process of melting and separating metals from ores and concentrates and it thus consumes large amounts of energy. There are many different types of smelting processes and variations on them, depending on the type of ores and final products. In order to recover the capital investment in building a smelter and ensure profitability, it is necessary to be able to deliver a steady supply of concentrates over a long period of time. For this reason only the largest mines are able to justify the additional investment required for constructing their own smelters, though mines located in remote areas with high transportation costs may have smelters built, since finished metal is able to stand up to longer transport than the lower valued concentrates. Most small and medium-sized mines, however, sell concentrates directly to smelters for processing and have no further interest in their products beyond the concentrate stage. After smelting, there are a number of processes for refining the metal product. Electrolytic methods of refining are used for a number of metals and result in a very pure product.

If concentrates are sent to a smelter, the producer is paid for the metal contained in the concentrate on the basis of the metal market price, less costs incurred by the smelter together with a profit margin. The price paid to the mine for the metal contained in the concentrate is known as the Net Smelter Return (NSR). For example, if the market price for copper was 75 ¢/lb, the NSR received by a mine producing copper concentrates might be 63 ¢/lb of contained copper metal. The actual pricing is quite variable and depends on the metal and district, but the payment by the smelter generally varies from 60 to 90% of the value of the contained metal. In addition bonuses may be paid for small amounts of precious metals which can be recovered at the refining stage and, vice versa, penalties can be incurred for the presence of elements such as As and Bi; the presence of Hg in the concentrate may even make it unsaleable.

10.3 MINING METHODS

At the evaluation stage it is necessary to decide on the basic mining method to determine costs. Initially this only needs to be known approximately and details of the method to be employed can be worked out by mining engineers if the economics of the deposit appear favourable. Near-surface deposits will be mined by surface methods and deeper deposits by underground methods. For obvious reasons underground methods are much more expensive than surface methods and economic trends today with rapidly rising costs favour large open-pit mines in which economies of scale can be employed. For example, it is not many years since dump trucks of 100-tonnes capacity were unusual, whereas today 200-tonne trucks are in use at many mines and trucks of 400-tonne capacity are planned. By moving large quantities of material at a time unit costs can be brought down, but the high capital costs involved require extremely large deposits to ensure the necessary high rate of production.

Surface methods
Alluvial or placer deposits are worked by dry or wet surface mining methods. Wet techniques include hydraulic mining and dredging. Hydraulic mining employs the use of powerful jets of water for breaking up soft material and washing it down to treatment plants. The water jets are directed by large nozzles known as *monitors*. Dredging is commonly employed in offshore mining or in naturally flooded areas, but it can also be used in dry areas by flooding and maintaining small lagoons for floating the dredges. Such techniques are widely used in alluvial tin mining in Malaysia, where the dredges not only mine the tin-bearing gravels, but are fully equipped with mineral dressing machinery for recovering the final cassiterite concentrate. Dry mining methods for unconsolidated or soft material employ a wide range of excavating machines which include bulldozers, front-

end loaders, bottom excavators, drag-line excavators and bucket wheel excavators.

Hard rock deposits are worked from open pits and have to be drilled and blasted to recover the ore and remove waste. Enormous holes may be excavated in this manner by working from a series of benches. The largest open-pit mines are hundreds of metres deep with lengths and widths of several kilometres. An important consideration in the economics of open-pit mining is the *stripping ratio*, which is defined as the ratio of total waste removed to total ore mined.

A surface mining technique commonly employed for flat-lying, shallow, bedded deposits such as coal seams is known as strip mining. With this method extremely large excavators (drag-lines or bucket wheel excavators) remove the overburden to expose the seam or ore bed for working. As the seam or ore is progressively extracted, the overburden can be replaced to restore the landscape. In strip mining an important consideration is the *overburden ratio*, which is defined as the ratio of the vertical thickness of overburden to the vertical thickness of ore.

Underground methods

There are a wide range of underground mining methods that can be employed depending upon the type of deposit. Access to the ore may be by shafts, inclined shafts, declines or adits if the topography is favourable. For depths up to 500 m access by declines with gradients of 1 in 7, which allows truck haulage, is the most economical method, but shafts are generally used for access to deeper deposits. Once a method has been selected and mining engineers have planned the detailed layout, site work can begin. Before any ore can be extracted, a considerable amount of development will be required to gain proper access to the ore body. This development work of sinking shafts, putting in declines, roadways and ore draw points is costly and, since no ore is produced, no revenue is obtained. Therefore, it is extremely important that this initial development is carried out efficiently and without delays to ensure smooth and steady production of ore as soon as possible.

Underground openings from which ore has been extracted are known as *stopes* and the process of removing ore is known as *stoping*. Ore may be worked from a lower level to a higher one (overhand stoping) or from a higher level to a lower one (underhand stoping). There is a wide range of stoping methods that can be used depending upon the width of the ore body, the attitude of the ore body, the strength of the ore body and the strength of the country rock.

Open stoping is used in steeply dipping vein deposits with good rock strengths. As the name implies no ground support is used and the stopes are left as permanent openings. In practice some support is necessary, but it generally consists of leaving some ore behind as pillars. If ground conditions are not very good, timber supports known as *stulls* may be put in and the stope is known as a *stulled stope*.

For mining thick ore bodies a *shrinkage stope* might be used if ground conditions are good. This is a type of overhand stoping which makes use of the fact that broken ore occupies up to 40% more space than ore *in situ*. Draw points are put into the ore from a lower level and the ore is broken from the backs and falls down to the bottom of the stope. Sufficient ore is drawn off to leave a space under the backs for the miners to carry on working. The remaining broken ore supports the miners and gives temporary support to the walls. When the stope is finished, the ore is drawn and the stope left empty. Instead of using a temporary filling of broken ore, waste material may be used to fill the stope permanently as the ore is worked. This is known as *cut and fill stoping*.

If ground conditions are poor, it will be necessary to provide timber supports in the stopes. The use of stulls has already been mentioned, but if the rock is very weak, a more elaborate system of timber frames which fit into each other and provide mutual support will have to be used. Such timbered stopes are known as *square-set stopes*. For obvious reasons this type of stoping is expensive and should be avoided if possible.

For mining flat-lying ore bodies a method of stoping known as *room and pillar* can be used. With this method the ore is extracted from a series of 'rooms' with a regular pattern of pillars for support. When the ore has been mined out, a certain amount of ore can be recovered from the pillars during the retreat. Room and pillar mining usually results in 60–80% extraction of the ore. The room and pillar method was commonly used in coal mining, but it has been replaced in many mines by *longwall stoping* which is more productive. In the longwall method a long face is developed and worked along its full length. Ground support is provided by a series of special hydraulic props and as the face advances the props are moved forward allowing the roof to settle behind. Both advance longwall stoping, in which the working face progresses away from the access shaft, and retreat longwall stoping, where development haulages are put in to the extremities of the ore to be worked and the working face retreated towards the access shaft, are practised.

For working large ore bodies underground caving methods can be used if the rock is weak enough so that it can be induced to cave and run under gravity. In the *block caving* method development levels are put in under the ore body to be mined with ore draw points at strategic places. The ore body is then undercut and allowed to cave.

Drawing off the caved ore should induce further caving until all the ore has been drawn in that particular section of the mine. The advantages of block caving are that mining costs are low and a high rate of production can be achieved. The disadvantages are that there can be no selective mining of the ore, dilution is a problem and the extensive development work required prior to extraction of the ore is expensive. Another caving method that can be employed in mining wide ore bodies in moderately good ground overlain by ground which will cave readily, but still support itself temporarily over small openings, is known as *sub-level caving*. Essentially it is a method of underhand stoping in which a number of sub-levels are developed in the ore at vertical intervals of 4–8 m and ore is blasted down successively in descending order between the sub-level drives. The walls and cap rock break up and follow the operation down. The advantages of sub-level caving are that it permits selective mining of the ore body, less development is required than in the case of block caving, the method is productive and no pillars of ore are lost. The disadvantages are that high dilution has to be tolerated if good recovery is to be obtained, surface subsidence is caused and ventilation of stopes is difficult.

Mining costs

It is not possible to give precise costs for the various mining methods as they vary so widely from one part of the world to another. In a district where mining is well established accurate mining costs are usually known, but in an area where mining is a new venture it may be difficult to estimate mining costs accurately. This can be a big problem in evaluating a prospect since reliable mining costs are an important factor in determining the economic viability of a deposit. The best that can often be achieved is an estimate determined by mining engineers using all available data from a roughly comparable district. Underground mining costs in 1978 varied anywhere from $5·00 to $25·00/tonne depending upon the size of operations and part of the world. Open pit mining costs in 1978, on the other hand, varied from $0·50 to $2·50/tonne of material. On top of the mining costs there are other operating costs which include administrative and servicing overheads, milling and concentrating, and transport.

If approximate mining costs are known for a particular area, it is useful to prepare graphs relating *in situ* grade to total costs per tonne of metal for various types and scales of mining operations. This is a very useful exercise as it defines the type of exploration target on which it is worth spending time and money in detailed evaluation. Figure 10.7 shows an example of such a graph prepared by the writer for small and large underground copper mines on the Zambian Cop-

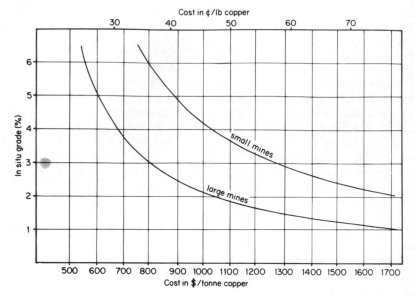

FIG. 10.7. Total copper production costs in 1976 related to *in situ* grade for large (> 20 million tonnes) and small (5–10 million tonnes) underground mines on the Zambian Copperbelt. A 90% copper recovery and 10% dilution is assumed.

perbelt. It can be readily seen from this graph that small deposits with grades of less than 4% copper are unlikely to be economic with the high operating costs that pertain in Zambia, but, in the case of large deposits with lower unit costs, somewhat lower grades should be profitable.

10.4   ECONOMIC FEASIBILITY STUDIES

To illustrate the methods used in a preliminary feasibility study an actual example of a study of the Mwambashi stratiform copper deposit in Zambia (Fig. 10.8) will be given. Original ore reserve calculations showed 8 million tonnes grading 2·80% Cu and depths of the mineralization indicated that it would have to be an underground operation. Although the grade and tonnage appeared to make it an attractive proposition, it was obvious that the mixed 'oxide'–sulphide mineralization would be difficult to treat and weak rock strengths might make mining difficult. Accordingly metallurgical tests were undertaken and a geotechnical study was made. The

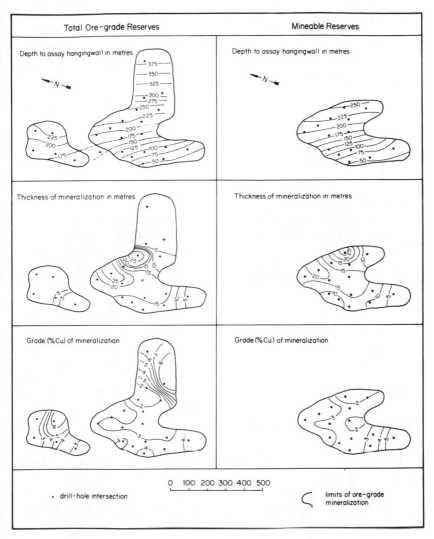

FIG. 10.8. The Mwambashi stratiform copper deposit on the Zambian Copperbelt. The mineralization occurs in sandstones and conglomerates of the Lower Roan formations of the Katangan Super-group. The principal ore minerals are chalcocite, chalcopyrite, chrysocolla and malachite with lesser bornite and native copper.

metallurgical tests showed that, although the sulphide ore was amenable to treatment, copper recovery from the high 'oxide'–sulphide material was poor, ranging from 56 to 75%. The geotechnical study showed that the mineralized zone is composed mainly of competent rocks, but areas of the hanging wall are very weak. This led to elimination of the northern mineralized zone and the necessity of leaving ore in the backs as a support to parts of the hanging wall. This mining hanging wall varies from 0 to 4 m below the assay hanging wall and resulted in final mineable reserves of 5·66 million tonnes at 2·72% Cu.

Owing to the extremely weak nature of the hanging wall formations, it was decided that caving methods should not be employed and one of the standard stoping methods with or without postfilling would be used. It was also considered that on average 20% non-recoverable pillars would have to be left as ground support, though in parts of the deposit as much as 50% pillar support might be necessary.

The main mineralized zone lies between 50 and 250 m depth and on the basis of composite split drill core samples taken for metallurgical testing the deposit can be divided broadly into the following zones:

zone	copper recovery (%)	reserves (millions of tonnes)	grade (%)	depths (m)
1. good sulphide	97	2·23	3·01	175–250
2. mixed 'oxide'–sulphide	75	1·43	2·06	150–175 & 225–250
3. high 'oxide'	56	2·00	2·88	50–150

The method for determining the final mineable tonnages and millhead grades assuming a 10% dilution with 0·5% Cu material is as follows:

	zone 1	zone 2	zone 3
total reserves	2·23 at 3·01%	1·43 at 2·06%	2·00 at 2·88%
less 20% pillar loss	0·45	0·29	0·40
	1·78	1·14	1·60
plus 10% dilution	0·18 at 0·5%	0·11 at 0·5%	0·16 at 0·5%
	1·96 at 2·78%	1·25 at 1·92%	1·76 at 2·66%

A mining rate of 1200 tpd (tonnes per day) or about 420 000 tpa (tonnes per annum) was assumed as a feasible target considering the size of the deposit and ground conditions, which gives the mine a life of about 12 years. A sequence of extraction given in Table 10.4 was planned and the following costs used:

Total working costs: $22/tonne ore
NSR: 65¢/lb or $1433/tonne copper

TABLE 10.4

PROPOSED SEQUENCE OF ORE EXTRACTION AT THE MWAMBASHI COPPER DEPOSIT

| Year | Tonnes mined | | | Levels worked |
	High oxide ore	Mixed ore	Sulphide ore	
1	420 000	—	—	50–150 m
2	420 000	—	—	50–150 m
3	210 000	—	210 000	50–150 m 175–250 m
4	210 000	—	210 000	50–150 m 175–250 m
5	120 000	—	300 000	50–150 m 175–250 m
6	120 000	—	300 000	50–150 m 175–250 m
7	—	210 000	210 000	150–175 m
8	—	210 000	210 000	150–175 m
9	—	120 000	300 000	200–250 m
10	—	300 000	120 000	200–250 m
11	—	320 000	100 000	200–250 m
12	260 000	90 000	—	150–175 m
Totals	1 760 000	1 250 000	1 960 000	—

Using these costs and other data given, the yearly copper production and cash-flow figures shown in Table 10.5 were determined. Note that no account has been taken of increased production costs due to inflation as it is assumed that the metal prices will also be affected and therefore rise accordingly. It can be seen that the mine starts operations at a loss and then moves into a profitable position until year 12 when a small loss is made. So far no account has been taken of capital development costs which need to be considered for the complete evaluation. It was estimated that $25 million would be needed to bring the mine into production and would be spent over a 3-year period with $5 million in year 1 and $10 million in years 2 and 3. This investment capital is finally recovered in year 9 and the total excess over the initial capital investment is only $3·44 million.

It is now necessary to determine the rate of return on the investment to complete the economic feasibility study. The capital required for bringing the mine into production may come from a variety of sources, but essentially there are two main sources: loan capital and equity capital. The loan capital may be put up by banks, governments or large mining houses, whereas equity capital is obtained through

TABLE 10.5

YEARLY COPPER PRODUCTION, WORKING COSTS AND REVENUE FOR A PROPOSED MINING OPERATION AT MWAMBASHI. COSTS IN $MILLIONS

Year	Tonnes mined	Production cost	Average grade(%)	Average copper recovery(%)	Tonnes copper	Revenue	Annual profit/(loss)	Cumulative profit/(loss)
1	420 000	9·24	2·66	56	6 256	8·96	(0·28)	(0·28)
2	420 000	9·24	2·66	56	6 256	8·96	(0·28)	(0·56)
3	420 000	9·24	2·72	77	8 796	12·60	3·36	2·80
4	420 000	9·24	2·72	77	8 796	12·60	3·36	6·16
5	420 000	9·24	2·75	85	9 818	14·07	4·83	10·99
6	420 000	9·24	2·75	85	9 818	14·07	4·83	15·82
7	420 000	9·24	2·35	86	8 488	12·16	2·92	18·74
8	420 000	9·24	2·35	86	8 488	12·16	2·92	21·66
9	420 000	9·24	2·53	91	9 667	13·85	4·61	26·27
10	420 000	9·24	2·17	81	7 382	10·58	1·34	27·61
11	420 000	9·24	2·12	80	7 123	10·21	0·97	28·58
12	350 000	7·70	2·47	61	5 273	7·56	(0·14)	28·44
Totals	4 970 000	109·34			96 161	137·78	28·44	

subscription of shares by private individuals or institutions or through the stock market by sale of shares to the public. In the case of loan capital the principal will have to be paid back in addition to an interest charged on lending the money and in the case of equity capital the shareholders will expect a return on their investment in the form of dividends paid out of the future mine's profits. In both cases the return on the investment can be measured by the Discounted Cash Flow or DCF. Essentially this is an annual interest charged on the unrecovered investment capital. Groundwater (1967) gives a good account of DCF methods in assessing capital projects. Today most investors would expect a DCF return of at least 15%. Let us return to the example we have been considering to see how far it falls short of producing a DCF of 15%. The cash flow is tabulated in Table 10.6

TABLE 10.6

CASH FLOW FOR THE MWAMBASHI COPPER DEPOSIT FOR A DCF RETURN OF 15%. ALL COSTS IN $MILLIONS

Year	Investment capital	Interest	Debt	Working revenue	Total debt
−3	5·00	0·75	5·75	—	5·75
−2	10·00	2·36	18·11	—	18·11
−1	10·00	4·22	32·33	—	32·33
1	—	4·85	37·18	(0·28)	37·46
2	—	5·62	43·08	(0·28)	43·36
3	—	6·50	49·86	3·36	46·50
4	—	6·98	53·48	3·36	50·12
5	—	7·52	57·64	4·83	52·81
6	—	7·92	60·73	4·83	55·90
7	—	8·39	64·29	2·92	61·37
8	—	9·20	70·57	2·92	67·65
9	—	10·15	77·80	4·61	73·19
10	—	10·98	84·17	1·34	82·83
11	—	12·42	95·25	0·97	94·28
12	—	14·14	108·42	(0·14)	108·56

where it can be seen that the mine runs up a total debt or loss of $108·56 million for its owners on the basis of an expected DCF of 15%. In fact the deposit produces a DCF return of 1·5%, very far short of a reasonable return.

The economics can be improved by taking a higher metal price or lower mining costs, but one should be extremely cautious to avoid being over-optimistic. If mining costs are known reasonably well in a particular district, it is highly unlikely that they will be reduced in a new mine unless new and more productive methods are being introduced. These may increase the capital investment required, however, and effectively offset any gains. In the case of metal prices a reasonable market average should be taken for the evaluation. Peak prices should not be used as these are almost certainly bound to be unrealistic. Likewise, bottom prices should not be used either as this is unduly pessimistic and may 'kill' a prospective viable mine.

Evaluation of open pit deposits
In the evaluation of open-pit mines an extremely important factor is the stripping ratio, which will affect the economics of the deposit. In many evaluations the overall stripping ratio, calculated to the final planned depth, is used, but consideration should be given to the progressive stripping ratio which may be of greater importance. The principle is illustrated by an example of two hypothetical open-pit copper deposits worked at the same rate to the same depth over a period of five years (Fig. 10.9). Both deposits have the same overall stripping ratio to the end of year 5, but in deposit A the ore is at some depth below the surface and no ore is recovered in year 1, whereas in deposit B the ore is at the surface and ore is mined from the start.

An evaluation of the deposits is based on the following parameters:
Mining costs (a): $1·25/tonne of material
Milling and concentrating (b): $2·25/tonne of ore
Services and administration (c): $1·85/tonne of ore
Average grade (g): 2% Cu
Recovery (r): 90%
Dilution (d): 10%.
Total working costs per tonne copper:

$$\frac{a(1+s)+b+c}{g(1-d)r}$$

$(s$ = stripping ratio)
Capital cost: $18 million
Concentrate grade: 30% Cu
Transport: $60/tonne concentrate
NSR: $1200/tonne contained copper
DCF of 15% is required.

The economics are summarised in Table 10.7 where it can be seen that deposit B meets the requirements for a DCF of 15%, whereas deposit A does not. In fact, the DCF on deposit A is approximately 6%. Both deposits have identical total working costs, transport costs, copper production and revenue. The example serves to show how it is important that production commence as early as possible so that revenue from the sales of product can be used to recover the capital investment before accumulation of interest pushes the total debt too high.

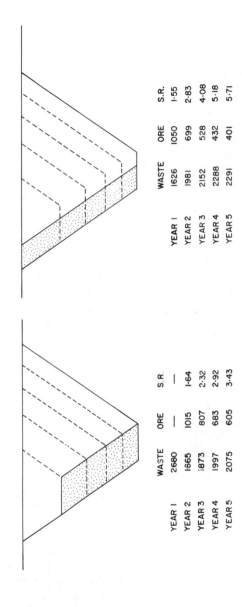

	WASTE	ORE	S.R.
YEAR 1	2680	—	—
YEAR 2	1665	1015	1·64
YEAR 3	1873	807	2·32
YEAR 4	1997	683	2·92
YEAR 5	2075	605	3·43

DEPOSIT A

	WASTE	ORE	S.R.
YEAR 1	1626	1050	1·55
YEAR 2	1981	699	2·83
YEAR 3	2152	528	4·08
YEAR 4	2288	432	5·18
YEAR 5	2291	401	5·71

DEPOSIT B

Waste and ore in 000's tonnes. Overall stripping ratios both 3·31.

FIG. 10.9. Two hypothetical open-pit copper deposits with identical reserves and overall stripping ratios, but different progressive stripping ratios.

TABLE 10.7
ECONOMICS OF THE TWO OPEN-PIT COPPER DEPOSITS GIVEN IN FIG. 10.9. COSTS IN MILLIONS OF DOLLARS DCF OF 15% EXPECTED

DEPOSIT A

Year	Working costs	Transport costs	Tonnes copper	Revenue	Loan	Profit/(loss)
1	3·35	—	—	—	18·00	(21·35)
2	7·51	3·29	16 443	19·73	24·55	(15·62)
3	6·66	2·61	13 073	15·69	17·96	(11·54)
4	6·15	2·21	11 065	13·28	13·27	(8·35)
5	5·83	1·96	9 801	11·76	9·61	(5·64)
Totals	29·50	10·07	50 382	60·46		

DEPOSIT B

Year	Working costs	Transport costs	Tonnes copper	Revenue	Loan	Profit/(loss)
1	7·65	3·40	17 010	20·41	18·00	(8·64)
2	6·21	2·26	11 324	13·59	9·94	(4·82)
3	5·52	1·71	8 554	10·26	5·54	(2·51)
4	5·11	1·40	6 998	8·40	2·88	(0·99)
5	5·01	1·30	6 496	7·80	1·14	0·35
Totals	29·50	10·07	50 382	60·46		

10.5   EXAMINATION OF PROPERTIES

As part of the exploration geologist's job it is frequently necessary to examine prospects on which a certain amount of mine development has been undertaken. This may consist of limited surface or underground workings which expose the mineralization or of old abandoned mine workings which may have some potential for further production. In both cases it will be necessary for the geologist to undertake some mapping and possibly some sampling. In the case of old mine workings accurate plans may be available on which to plot geological observations, but, more often than not, it will be necessary to survey the prospect. Methods of surface surveying have already been dealt with in Chapter 8 and it is necessary to consider some of the aspects of underground surveying.

Underground surveying
Both theodolite traverse and compass and tape surveys are used in underground surveying. Plane tabling, whilst perfectly adequate for surface surveying, is impractical underground, though small plane tables have been used in special cases. A theodolite is obviously necessary for really precise work, but compass and tape surveys can be sufficiently accurate for many evaluation purposes in small mines where there is no magnetic disturbance.

When mapping a prospect where a good plan of the underground workings is available, it is quite easy for the geologist to work on his own. The end of the tape can be hooked onto survey marks or other identifiable positions and the geological features can be plotted directly on the base map. If it is necessary to produce a topographic survey at the same time as the geological one, however, it is essential to have at least one assistant, who will be required to hold a lamp for taking readings to the forward survey station and to hold the tape while the geologist takes offset measurements and makes geological observations.

Since underground workings give one a three dimensional view, it is important to be consistent when plotting features on a plan. When mapping along drifts and crosscuts, it is the normal convention to plot everything at approximately waist height. This is often quite difficult to do accurately and underground surveying is a specialized skill; the ability to sketch accurately and quickly is a useful asset. The difficulties can be appreciated by considering how the positions of important features may vary; sometimes they may only be visible in the walls and at other times only in the backs or floor.

For a compass and tape survey it is useful to use two stout sticks, each about 1 m or slightly more in length (pick-axe handles are useful), in addition to the compass, tape and clinometer (Abney level). The sticks are held vertically at each of the survey stations between which measurements are being made. (Rocks can be piled around the sticks to help support them.) Both a compass bearing and vertical angle reading are taken from the top of one stick to another. Then, while the tape is held tight between the tops of the sticks, offsets are taken to the sidewalls so that the configuration of the workings can be plotted. It is important to do this and not simply draw the sides of a drift as straight lines. For the measurements of offsets a thin wand or rod about 1·5–2 m long marked in 10 cm intervals is very useful. To avoid any magnetic effects on the compass it is important not to hold a lamp, whether it be electric or a carbide lamp, too close to the compass while taking a reading. For this reason candles, which can be used in most metal mines where there is no danger of gas, are useful for reading the compass. In abandoned workings the survey stations can be marked by a peg and/or small pile of rocks on the floor, but in operating mines the conventional survey markers should be used. These consist of copper or aluminium tags, on which numbers are stamped, fastened to the backs or timber sets by a masonry nail. It is normal practice to fasten the tag to a suitable place in the backs and then plumb in the survey station precisely on the floor.

When taking bearings along steeply inclined sights up raises or in stopes, the old-fashioned *hanging compass* is a very useful instrument. This consists of a large compass mounted and pivoted between two supports which can be hooked over a string. When this is done, the compass swings into a level position and the reading gives the bearing of the supporting string. To take a measurement a string is secured taut between the two survey stations and the compass hung on the line to take the reading. Both forward and back bearings can be taken by turning the compass around at either survey station.

It will be appreciated that underground surveys consist largely of open traverses and it is important when surveying along different levels to close the survey through raises whenever possible. It is also important to plot the survey as work progresses so that it can be checked for errors and amendments made if necessary. When accurate base plans are available, the geological survey should be plotted directly on the base map and not plotted later from notes. When the geologist has to conduct his own topographic survey, it is not possible to do this, but each part of the survey should be plotted as soon as possible so that the survey can be checked on site. For reference, standard symbols are given in Fig. 10.10 and an example of underground survey is given in Fig. 10.11.

As a word of caution it should be remembered that old mine workings can be very dangerous. There may be accumulations of bad

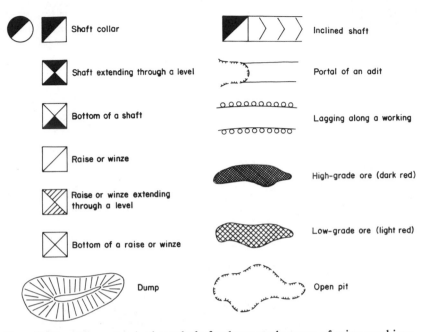

FIG. 10.10. Some standard symbols for large scale maps of mine workings.

air, timbering may be rotten and loose rock may fall or workings collapse. If there is any doubt, the geologist should only enter workings that have been checked by a competent mining engineer or experienced miner.

Definitions of some mining terms not already given in the text are listed below:

Back(s). The upper side of a drift or stope or any underground passage. If the underground working is compared to a room the backs would correspond to the ceiling. The term 'roof' is used in coal mining instead of backs.

Crosscut. A level or underground passage driven across the strike of a vein or ore body.

Drift or *drive.* A level or underground passage driven along the strike of a vein or ore body.

Face. In any adit, drive, crosscut or stope the end at which work is in progress or was last done.

Heading. In any adit, drive or crosscut the end which is being advanced.

Lagging. Timbering along the sides of a drift or crosscut to prevent loose material falling into the workings.

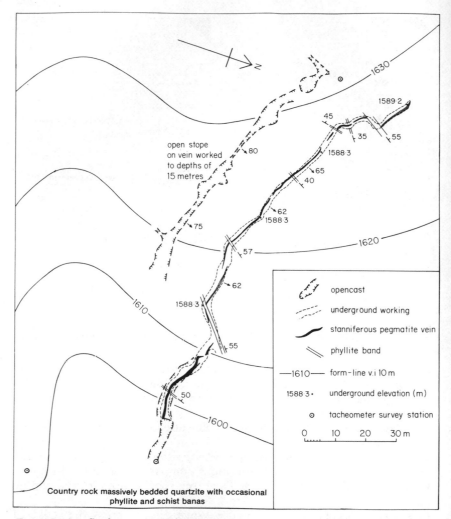

FIG. 10.11.   Surface and underground workings on the No. 1 vein, Kitembi tin mine, Ankole, Uganda. Surface surveying by tacheometry, underground survey by compass and tape; surveyed by J. H. Reedman 1967.

Level. Group of workings all at approximately the same elevation. Levels are usually put in at depth intervals of 15–60 m. Workings put in for a short distance between levels are known as *sublevels.*
Ore pass. An opening in a mine through which ore is passed from a higher level to a lower one.

Pillar. A piece of ground or ore left to support the roof or hanging wall. *Rib pillars* are vertical pillars between stopes. *Crown pillars* are horizontal pillars left along the upper part of a stope to protect drives above. *Sill pillars* are horizontal pillars left in the bottom of a stope to protect drives below and to give sufficient height to put in draw points.

Raise. An opening like a shaft to connect a lower level to a higher one.

Sill. The bottom or floor of an underground opening.

Spiling. A type of timbering put in while advancing in very bad ground. The spiles are driven into loose ground ahead to permit removal of material without caving.

Winze. An opening like a shaft to connect a higher level with a lower one.

Other aspects of a property examination
In addition to geological considerations which determine the extent or probable extent of the mineralization, it is important to include a report on a number of non-geological features which will affect the economics, to complete the property examination. These include:
1. Location. Availability and cost of transport and type of transport. Distance to ports or important centres where product can be sold or processed.
2. Availability of adequate water supplies.
3. Availability of power—cheap electric power is a big asset.
4. Availability of labour, both unskilled and skilled.
5. Availability of suitable support timber (important for underground mines).
6. In the case of abandoned mines the amount and condition of equipment, plant and buildings that might be on site. It is also important to mention the state and condition of the mine workings.

REFERENCES AND BIBLIOGRAPHY

Gilchrist, J. D. (1967). *Extraction Metallurgy*, Pergamon Press, Oxford, 291 pp.

Griffith, S. V. (1960). *Alluvial Prospecting and Mining*, Pergamon Press, London, 245 pp.

Groundwater, T. R. (1967). Role of discounted cash flow methods in the appraisal of capital projects, *Trans. Instn. Min. Metall.*, Lond., 76, A67–82.

Lewis, R. S. and Clark, G. B. (1964). *Elements of Mining*, John Wiley and Sons Inc., New York, 768 pp.

McKinstry, H. E. (1948). *Mining Geology*, Prentice-Hall Inc., New York, 680 pp.

Peele, R. (ed.) (1945). *Mining Engineers' Handbook*, 2 vols., John Wiley and Sons Inc., New York.

Pfleider, E. P. (ed.) (1968). *Surface Mining*, A. M. M. & P. E., New York, 1061 pp.

Pryor, E. J. (1958). *Economics for the Mineral Engineer*, Pergamon Press, London, 254 pp.

Pryor, E. J. (1965). *Mineral Processing*, Elsevier Publishing Company, Amsterdam, 844 pp.

Richards, R. H. and Locke, S. B. (1940). *Textbook of Ore Dressing*, McGraw-Hill Book Co., New York, 608 pp.

Taggart, A. F. (1951). *Elements of Ore Dressing*, John Wiley and Sons Inc., New York, 595 pp.

Thomas, L. J. (1973). *An Introduction to Mining*, Hicks Smith & Sons, Sydney, 436 pp.

Winiberg, F. (1966). *Metalliferous Mine Surveying*, Mining Publications, London, 456 pp.

Index